MW00634731

The Ecology of Agroecosystems

Related Titles by Jones and Bartlett Publishers

Bioethics: An Introduction to the History, Methods, and Practice, **Second Edition**
Nancy S. Jecker, Albert R. Jonsen, & Robert A. Pearlman

Botany: An Introduction to Plant Biology, **Fourth Edition**
James D. Mauseth

Environmental Oceanography: Topics and Analysis
Daniel C. Abel & Robert L. McConnell

Environmental Science, **Eighth Edition**
Daniel D. Chiras

Environmental Science: Systems and Solutions, **Fourth Edition**
Michael L. McKinney, Robert M. Schoch, & Logan Yonavjak

Invitation to Organic Chemistry
A. William Johnson

Organic Chemistry, **Third Edition**
Marye Anne Fox & James K. Whitesell

Outlooks: Readings for Environmental Literacy, **Second Edition**
Michael L. McKinney & Parri S. Shariff, eds.

Plant Cell Biology
Brian E. S. Gunning & Martin W. Steer

Plants, Genes, and Crop Biotechnology, **Second Edition**
Maarten J. Chrispeels & David E. Sadava

Plant Structure: A Color Guide, **Second Edition**
Bryan G. Bowes & James D. Mauseth

Principles of Environmental Chemistry, **Second Edition**
James E. Girard

Tropical Forests
Bernard A. Marcus

The Ecology of Agroecosystems

John H. Vandermeer

Asa Gray Distinguished University Professor & Arthur Thurnau Professor
University of Michigan, Ann Arbor

JONES AND BARTLETT PUBLISHERS
Sudbury, Massachusetts
BOSTON   TORONTO   LONDON   SINGAPORE

World Headquarters
Jones and Bartlett Publishers
40 Tall Pine Drive
Sudbury, MA 01776
978-443-5000
info@jbpub.com
www.jbpub.com

Jones and Bartlett Publishers Canada
6339 Ormindale Way
Mississauga, Ontario L5V 1J2
Canada

Jones and Bartlett Publishers International
Barb House, Barb Mews
London W6 7PA
United Kingdom

Jones and Bartlett's books and products are available through most bookstores and online booksellers. To contact Jones and Bartlett Publishers directly, call 800-832-0034, fax 978-443-8000, or visit our website, www.jbpub.com.

Substantial discounts on bulk quantities of Jones and Bartlett's publications are available to corporations, professional associations, and other qualified organizations. For details and specific discount information, contact the special sales department at Jones and Bartlett via the above contact information or send an email to specialsales@jbpub.com.

Copyright © 2011 by Jones and Bartlett Publishers, LLC

All rights reserved. No part of the material protected by this copyright may be reproduced or utilized in any form, electronic or mechanical, including photocopying, recording, or by any information storage and retrieval system, without written permission from the copyright owner.

Production Credits
Publisher, Higher Education: Cathleen Sether
Acquisitions Editor: Molly Steinbach
Senior Editorial Assistant: Jessica S. Acox
Editorial Assistant: Caroline Perry
Production Assistant: Lisa Lamenzo
Senior Marketing Manager: Andrea DeFronzo
V.P., Manufacturing and Inventory Control: Therese Connell
Composition: Glyph International
Cover Design: Kristin E. Parker
Assistant Photo Researcher: Carolyn Arcabascio
Cover Image: Top: © Dozet/Dreamstime.com, Bottom: Courtesy of John H. Vandermeer
Printing and Binding: Malloy, Inc.
Cover Printing: Malloy, Inc.

Library of Congress Cataloging-in-Publication Data
Vandermeer, John H.
    The ecology of agroecosystems/John H. Vandermeer.
        p. cm.
    ISBN 978-0-7637-7153-9
    1. Agricultural ecology. 2. Food—Quality—Moral and ethical aspects. 3. Soil chemistry. I. Title.
    S589.7.V36 2010
    630—dc22

                                                            2009049561

6048

Printed in the United States of America
14 13 12 11   10 9 8 7 6 5 4 3 2

Dedicated to the small-scale farmers of the world and those who support them.

BRIEF CONTENTS

CONTENTS

Chapter 3   Competition and Facilitation Among Plants: Intercropping, Weeds, Fire, and the Plow   63

Chapter 4   Soils and the Emergence of the Industrial Approach   117

Chapter 8 Toward a Sustainable Future   309

Preface

In 2009, the First Lady of the United States, Michelle Obama, inaugurated what some hope will be an annual tradition. She established a vegetable garden on the White House lawn. Recognizing the importance of how food is produced in general and the importance of agriculture explicitly, the gesture was appreciated by a broad range of people, and she was roundly applauded for the effort. In the eyes of some, however, she committed a sin. She said the "O-word"—organic. Almost immediately, she received an e-mail from the Mid America CropLife Association, representing the industrial agricultural system, encouraging her to use "crop protection products," which, of course, is a nice way of saying pesticides. The symbolism of a popular First Lady "going organic" was viewed as a challenge to the industrial agricultural system.

Although the White House vegetable garden attracts deserved attention, the issue is not a new one. The industrial agricultural system, maturing only since World War II, has spawned a host of problems—ecological, medical, social, political, and economic. In response, there have been calls to reevaluate the way in which food and fiber are produced and consumed in the world. Many of these calls are political. Many are also economic, cultural, health related, or moral/ethical. A key element in all of these calls, however, is the need for an approach that considers ecology, acknowledging that the agroecosystem is foremost an ecosystem that needs to be managed as such. The movement toward a more ecological and sustainable agriculture may have many names (e.g., organic, agroecological, sustainable, or low input), but it aims at a central core—the fundamental natural laws of ecosystems are involved and need to be taken into account in the design and operation of the agroecosystem. This book is about those laws.

Unlike other natural systems (e.g., a coral reef or tropical rain forest), the agroecosystem always contains a particular "keystone" species. That keystone species is *Homo sapiens*. It is a species that engineers its own environment far more than any other species in history, a fact that would make ignoring its ecology naïve; nevertheless, its "ecology" involves a structure that no other species has ever had. It has the ability (or is it a need?) to communicate ideas from individual to individual, perpetuated to heights that are magnitudes larger than any other organism in the history of life. That is, our species has language with which it creates structures of culture and society, of economics and politics. As the keystone species in the ecosystem, our special nature becomes part of that ecosystem. The agroecosystem is thus endowed with not only traditional subjects of ecology, but also with the immensely complicating aspects of this particular feature of the keystone species.

This book aims to present the ecology of agroecosystems, acknowledging not only the sometimes-perplexing complexity of ecosystems in general, but also the even higher levels of complexity stemming from the presence of that keystone species. Because in reality it is an ecosystem that touches on all of the human sciences as well as most of the natural sciences, it inherently requires an interdisciplinary approach. Thus, this book includes material from anthropology, economics, sociology, chemistry, and geography, as well as the core ideas from ecology. I have tried to weave the subject mater into a tapestry that reflects that interdisciplinary reality. In doing so, however, the inherent contradictions emerging from our current definitions of disciplinary boundaries sometimes become evident.

In teaching for many years an upper-level course in agroecosystem ecology, I have seen such contradictions emerge from an unusually broad range of expertise among matriculates. However, they generally fall into two categories: "I hate math" versus "I think the social sciences are a load of fluff"—math phobics versus those prejudiced against the social sciences. Although exaggerated for heuristic purposes, that is generally the pattern. Some students identify with the more sociopolitical aspects of the subjects and others with the more "technical" aspects. The agroecosystem is an integrated whole that defies the artificial and arbitrary distinctions among the contemporary disciplines and thus seeks to teach the mathematical subjects to the math-phobic student and the sociopolitical subjects to the more technically oriented student. As I contend in Chapter 1, understanding the agroecosystem is more about understanding the interactions among many of these disciplines than about any one of them. Although my emphasis is obviously on the more ecological topics, those topics are always immersed in the superstructure of sociopolitical forces that condition and are conditioned by the ecological forces.

Conversations with many people have contributed to my thinking on this subject. It would be impossible to mention all of them. A few stand out, however, as especially influential. First, my long-term colleague Ivette Perfecto constantly points me in the right direction in those many cases in which I get off on the wrong foot. Richard Levins introduced me to the topic initially many years ago and continues to inspire. Peter Rosset is always ready to challenge me on the more political issues involved and has clearly contributed a lot to the general philosophy of the book, as has Miguel Altieri. Steve Gliessman and Helda Morales, in two different ways introduced me to the idea of asking farmers why they do what they do. Luis Garcia-Barrios continues to generate insights as to how theoretical ecology relates to the practical issues of agroecosystems. Catherine Badgley brings up things both practical and theoretical that force me to rethink my intellectual ruts.

Other colleagues from the New World Agriculture and Ecology Group have contributed in innumerable ways, many of which I probably have come to think are of my own origination—Doug Boucher, Richard Lewontin, Dave Andow, Inge Armbrecht, Iñigo Granzow de la Cerda, Julie Jedlica, Deborah Letourneau, Tom Deitch, Julie Grossman, Gerry Smith, Katherine Yih, Kristen Nelson, and Chris Picone. My recent graduate students in their own work on agroecosystems are constant sources of information, in particular Stacy Philpott, Heidi Liere, Doug Jackson, Jahi Chappell, Bruce Ferguson, and Shalene Jha. I would also like to express my gratitude to those who reviewed drafts of the manuscript, whose feedback helped to shape the text in many ways:

Tor Arvid Breland, Norwegian University of Life Sciences
Shannon Cowan, University of British Columbia
Alan DeRamus, University of Louisiana, Lafayette
Ken Diesburg, Southern Illinois University
William Eisinger, Santa Clara University
John Erickson, University of Florida
Charles Francis, University of Nebraska, Lincoln
Richard Hazlett, Pomona College
Heather Karsten, Pennsylvania State University
Jennifer Ramstetter, Marlboro College
George Robinson, State University of New York at Albany
Mike Saunders, Pennsylvania State University
Karin Warren, Randolph College

Finally, approximately 15 years of groups of advanced undergraduates and graduate students have listened to me talk about this material and have criticized and confirmed in ways that have in the end put the structure to the text. I thank them all.

For Instructors

A PowerPoint® Image Bank is available online for instructors adopting this textbook. The Image Bank provides the illustrations, photographs, and tables (to which Jones and Bartlett Publishers holds the copyright or has permission to reprint digitally) inserted into PowerPoint slides. With the Microsoft® PowerPoint program, you can quickly and easily copy individual images into your existing lecture slides. For more information, please contact your Jones and Bartlett publisher's representative.

Chapter 1
Three Vignettes: Setting the Stage

Overview

In this opening chapter, agroecosystem ecology is introduced in a broad, interdisciplinary framework. Three vignettes are presented that embrace the extensive range of topics within which the subject naturally fits. After the three vignettes, a general view of agricultural ecosystems and their study is presented as two poles in a continuum of possible framings. Finally, an attempt is made to provoke thinking about how agroecosystems and the food and fiber that they produce are distributed in the sociopolitical world.

Introduction

One of my first doctoral students was told that his thesis proposal would not be approved because he was planning to do field work in an agricultural ecosystem. His idea was, according to the conventional wisdom in my department, not appropriate for a study

1

in "ecology," and if he wanted to pursue it, he was advised to transfer to some sort of agriculture department, by which I assume was meant agronomy or some other academic tradition that focused on agricultural production. The myopia was evident to me even then. However, it takes on special meaning today as the world is slowly coming to realize that we face huge problems in the Earth's agroecosystems and that many of those problems are the result of our failure to consider agricultural ecosystems as falling within the same scientific principles as other ecosystems. Granted, the language has been different— "herbivores" were sometimes "pests," predators and parasites could be "natural enemies," some plants were "weeds," others "crops," nutrients could be "fertilizers." Certainly, the culture of the practical in the traditional agricultural sciences is different from the more ethereal academic culture of ecology. Nevertheless, the past 35 years have seen a major transformation in this attitude, and with ever-increasing frequency, agroecosystems and the special problems that they present have become subjects for legitimate scientific study within the traditions of academic ecology.

This text is intended as an introduction to this new/old discipline—new in that it is still not common to see agroecosystems studied from an ecological point of view and old in that ecological principles apply as much to agroecosystems as to the rest of the world's ecosystems, irrespective of the underlying assumptions of some more traditional agricultural scientists. It has always been an important subject, even if it has not been enthusiastically embraced in the recent past. However, it takes on special meaning today. As we approach the end of the decade, the world has entered into yet another phase of food shortage. But this time the complexities that bring about that crisis are evident and openly discussed. No absolute shortage of food or potential production exists, but the people who need the food simply cannot get it. Furthermore, now overwhelming evidence shows that the agroecosystem, writ large, is involved in a host of other problems—from the contribution of industrial agriculture to greenhouse gases to the creation of hypoxic zones in many of the world's oceans, from the obesity crisis resulting from an unusual food delivery system to the cancer epidemic at least partially the result of pesticide abuse, from bacterial resistance conditioned by the unrestrained use of antibiotics in animal feed to the dramatic elimination of tropical biodiversity by agricultural intensification. The industrial agricultural system, which effectively was born at the end of the world wars and is thus not very old, has generated environmental problems that certainly will be hard to deal with. But at least they need to be recognized as symptoms of a system in need of change.

As might be expected, the extreme environmental insults resulting from this particular model have long been recognized by a small cadre of people. Even before Rachael Carson popularized the problems of pesticide abuse, alternatives to industrial agriculture were proposed and model systems put in place. With the rapid economic growth and massive social transformations of the past half century, the problems with the system have become more evident to more people and the alternatives more than ever before recognized as necessary. The consequent dramatic growth in research and development of alternative agricultural technologies since 2000 is thus not surprising. Codifying the intellectual background for these alternative technologies is the purpose of this text.

It is a strong assumption of this text that fundamental ecological principles apply to the agroecosystem as much as any other ecosystem. Consequently, standard topics of the science of ecology form the backbone of the text's structure. However, the agroecosystem is one of those ecosystems that exists only in the presence of one dominant species, *Homo sapiens*, and that invariably complicates the picture. One of my ecology professors, Larry Slobodkin, to make a point about the transcendence of natural processes, provided a trenchant anecdote. Indian farmers, at the time, produced an abundance of tomatoes. But at local markets, two types of tomatoes were available—on the one hand fresh and on the other old and rotting. Vendors receiving fresh tomatoes from farmers set their price high in hopes that a rich Indian consumer would be tempted to purchase one. The rate at which such purchases were made was low, but the price of the fresh tomato was so high that it made sense to wait for the possibility that one of those rich consumers would be attracted to your stall, no matter how low the probability. Naturally, as time passed, the fresh tomatoes began to rot, at which time the vendors admitted that they were not going to attract any of the rich consumers, and thus lowered the price concomitant with the degree of rottenness of the tomato. Thus, the farmers produced an abundance of fresh tomatoes, but the bulk of the population could eat only rotten ones.

This simple anecdote exemplifies the interpenetrating roles of society and nature. To understand the shortage of fresh tomatoes, it was not enough to understand either the physical fact that things rot nor the political fact that India had an exaggerated inequality of purchasing power. It was the organic connection between the social and natural that revealed the truth, and an argument over whether the problem was that tomatoes rot or that riches were unequal would be pointless and misplaced.

The species *Homo sapiens* has a suite of characteristics that compel it to generate social relations that in turn interact with the ecological forces operative in the agroecosystem—that is, there are inevitable dialectical interactions between the social and ecological that are almost the definition of "the agroecosystem." Although this text obviously emphasizes the ecological, it is not possible or advisable to ignore the evident socioecological interactions. Consequently, much of the text is couched in the superstructure of sociopolitical forces that condition and are conditioned by the ecological forces.

Furthermore, the subtext of production, which is to say the assumption that technological advance is equivalent to increases in production, is explicitly rejected. The purpose here is to explore the ecology of the agroecosystem in the framework of its interdisciplinary complexity. Focusing on the amount of biomass produced in a unit area is of limited utility. Just as we would find the question of how much biomass is produced by a coral reef only one of many interesting questions to ask about that ecosystem, the amount of biomass produced in a hectare of wheat is only one in a large set of interesting questions to ask about that ecosystem. My focus is on how the agroecosystem works and does not necessarily imply anything about how it should work, unless such a normative goal is explicitly stated as such (which is indeed frequently the case). As explained later in this chapter, my focus is explicitly not productionist. Indeed, a more general rejection of the productionist mentality is something I seek to promote.

To give a flavor of the interdisciplinary focus that is appropriate for understanding the dynamics of agroecosystems, I begin this text with a series of vignettes. First, instead of examining the process of providing food for a population, I examine one of the cases in which the system failed to do so, the famous Irish potato famine. Second, rather than lament the loss of biodiversity resulting from the expansion of tropical agricultural frontiers, I look at the complicated case of biodiversity transformations in the coffee agroecosystem. Third, in contrast to the post-WWII industrial agricultural system, I examine the experience of Cuba, forced to undertake a massive reorganization of its system to a more ecologically friendly model after the collapse of the Soviet Union. These three vignettes offer a collective introduction to the complex interdisciplinary nature of this most complicated of all terrestrial ecosystems.

The Three Vignettes

The Irish Potato Famine[1]

Ireland presents a spectacular case of failure in an agroecosystem. The Irish population had risen to approximately 8 million people by the mid 19th century. Within a period of 5 years, it was reduced to less than 6 million, and today still stands below the population density before the great famine. The socioecological conditions that led to the famine, as well as its enduring effects, are spectacular historical examples of the sorts of complex interactions among people and their environment that characterize agroecosystems. To fully appreciate the significance of this event, it is necessary to understand first, the agricultural system of Ireland before the famine, and second, the politically restrained response of the British to the famine, and third, the worldwide ramifications.

Western Ireland was the outback, so to speak. Most of the best agricultural lands were located in the East, and the sparse population of the 17th and early 18th century West was unevenly distributed over the landscape. That landscape was composed mainly of peat bogs, rocky hills, and valley bottoms. The agroecosystem was based on the Baile system, in which a small collection of farmhouses was congregated in a valley bottom, surrounded by a fenced "infield" and an "outfield" outside of the fence. The infield was generally restricted to the flat land, containing relatively more fertile soils. It was devoted to intensive farming of grains in the "rundale" system, in which strips of land were cultivated by individual families, sometimes alternate ridges in the same field, each owned by a different family in the Baile. The Baile was largely kept together based on extended kinship ties. The rundale was periodically redistributed to make sure that each family was able to take advantage of any microclimatic or microedaphic idiosyncrasy that may have existed in the infield. Individual home gardens also dotted the infield landscape. The infield was regularly fertilized with livestock manure, peat, and seaweed in quite an intensive operation. The outfield was organized based on the "collop" or share, in which each family had the right to graze a certain quantity of livestock on the common property of the outfield. Frequently, especially in the summer months, the outfield was extended high into the surrounding hills, and summer lodgings were built for the stock herders, who were generally young boys or girls. The importance of livestock for the Baile was central to the operation of the system. The basis for establishing the collop was the amount of livestock owned, which in turn was the basis for establishing cultivation rights in the infield, as maintenance of the rundale was dependent to a large degree on the amount of manure available.

Into this background the potato arrived on the scene in the early 16th century. Originally it was a luxury crop, cultivated in the gardens of the well-to-do, but it rapidly became a mainstay, especially in western Ireland, and encouraged massive expansion of agricultural lands and population growth. The potato is easy to produce, as anyone who has ever gardened can attest. Compared with oats, for example, which must be grown in a field that is plowed and ridged and furrowed, a seed potato can be planted in a piece of land opened by a spade, with little or no fertilization. Furthermore, harvested potatoes need not be further processed. The need for grinding grain and the social structure associated with this need disappear as the potato comes to dominate.

This new crop was, in a sense, a miracle crop. Because it could be so easily grown even in marginal lands and because it provided a reasonable diet even for extremely poor people, its cultivation spread like wildfire throughout western Ireland. It changed both the society and the ecology of that part of the island. The Baile system gradually became replaced by the cottier system in which the bottomlands were devoted to grain cultivation, mainly for export to England, and the marginal areas devoted to potato production to maintain an increasingly marginalized peasantry. No longer did the exquisitely managed interfurrow system of the rundale provide reliable and varied food, and no longer did the complex Baile provide the social cohesion from which social stability resulted. Indeed, even before the failure of the potato crop, famines existed, lesser in extent to be sure, but famines nonetheless, produced largely by the combination of the social organization of agriculture and unfavorable climatic conditions. That is, with the elimination of the Baile and rundale system, the population of Ireland, especially the western part of the island, became vulnerable.

The potato conditioned certain basic economic facts. Given its ease of cultivation and preparation as food, it provided a way in which the theoretical lower wage limit could be very closely approximated. If the goal is to drive the wage of the underclass as low as possible, which is always the case in a profit-driven system, the potato was far more efficient in this respect than the former mixed system of grains and livestock. What became known as the "potato wage" enabled the extraction of the most labor for the least cost, driving the cost of labor close to its theoretical minimum.

Then disaster struck. The microorganism *Phytopthera infestans* arrived, presumably from the Americas, and found the effective monocultures of potatoes an ideal background environment for rapid spread. Potatoes rotted in the fields and the storehouses. In 1845, the potato harvest was reduced by about a third because of this pathogen, and following years were even worse. The poor, so completely dependent on this one species, were devastated. Many tried to eat what appeared to be the less damaged parts of the potatoes and got sick. Many tried to separate the rotten parts and make flour or porridge out of the other parts and got sick. Many resorted to wild herbs and seaweed as alternative food sources and starved. All in all, historians now agree that about 1 million people starved to death in the 5-year period spanning from 1845 to 1850 and that about 2 million migrated to other parts of the world. At the time, Ireland had a population of about 8 million. Proportionally it would be as if 35 million people starved in a 5-year period in the United States, and another 70 million migrated to other countries. It was a disaster of unprecedented proportions. Even today, Ireland's population has not recovered from the devastation. An evocative sculptural representation of the famine today sits on Dublin's Custom House quay, where many famine refugees set sail for Liverpool (Figure 1.1).

The response of the English and their class representatives in Ireland is also important. It is clear that the problem was recognized at the highest levels of political power (as was the case in later famines in China and India, also conditioned by British Imperialism).[2] Debates about what to do about the crisis reached all levels of society—from politicians who effectively put forth a Malthusian line to biologists who sought cheap recipes for soup kitchens. For example, the minister in charge of famine relief noted, at the time of the famine, "except through a purgatory of misery and starvation, I cannot see how Ireland is to emerge into a state of anything approaching to quiet or prosperity," meaning that Ireland should be left to starve to death "for its own good." Nutritionists were brought in to invent recipes that would produce food out of local materials (excluding, of course, oats and other grains, which were being exported to England). The most important starvation food available at that time was so-called Indian corn, which was coarse-ground maize, mainly imported from America. This food source, however, had to be further processed before it became an acceptably healthy food—if eaten raw, as many of

Figure 1-1 An evocative sculpture by Rowan Gillespie in Dublin honoring the victims of the Irish Potato famine.

the poor were forced to do since they lacked the resources to purchase fuel for cooking, it was worse than nothing because the coarsely ground grain fragments were like small knives that cut the wall of the intestine.

The final political structure that must be kept in mind is that, like all famines for which the British Empire ultimately bears responsibility (e.g., China and India), grain was being produced in large quantities and shipped to England during the entire crisis. In the 1790s, about 16% of grain available to the English was imported, but by the 1830s, that figure had risen to 80%. There seems to be little historical evidence that this irony ever entered public debate, so fixed was the conventional wisdom on the idea of cheap food policy as a condition for industrial growth. Grains had to be imported in massive quantities from all of Britain's colonies to maintain the huge supplies that were required to keep the price of food artificially low so that the new working class in Britain had enough surplus from their wages to be able to purchase the products streaming out of the factories, even though their wages were driven to the lowest possible levels. Even if the Irish and Indians and Chinese had to starve by the millions, few ever questioned the fact that grain had to be exported from their countries even while massive starvation was occurring. That general structure was assumed to be a necessary background condition, and debating its legitimacy was outside of respectable limits. In the middle of the crisis, British politician Gladstone perceptively noted, "It is the greatest horror of modern times that in the richest ages of the world and in the richest country of that age, the people should be dying of Famine by hundreds." This viewpoint is precisely mirrored in the contemporary world, along with the remarkable censorship of all discussion of its true significance. The notion that we live in the "best of all possible worlds," even as poverty, ignorance, and human suffering remain, is perhaps one of the most enduring ideologies shared by all classes at all times. In the face of obvious structural deficiencies and political arrangements that encouraged the development of the problem in the first place, the British ruling class did what all ruling classes do—they sought refuge in blaming the victim. Gladstone's comment referred not to the failure of the world capitalist system but rather to the "slovenly" Irish and their Celtic ways.

The reason for beginning with this vignette is that such an incredible human tragedy was the consequence of a poorly planned agroecosystem. The monoculture of potato was a bad idea from the start, as is universally recognized today. But more important, the social and political structures that surrounded this biological fact were part and parcel of the agroecosystem, from the destruction of the Baile and rundale system to the export of grain, to the Malthusian assumptions about food aid, to the insistence of a cheap food policy elsewhere in the world. It was an integrated package that had to do with ecology, politics, economics, and culture, all interacting in complex ways. Agroecosystems are like that.

The Sun Coffee Crisis

Northern Latin America consists of the land mass embraced by Colombia to the south and Mexico to the north and includes Panama, Costa Rica, Nicaragua, Honduras, El Salvador, Belize, and Guatemala. Collectively, this area long produced the second largest amount of coffee in the world, second only to Brazil. In contrast to the Brazilian system, however, which evolved independently, when the coffee agroecosystem was established in northern Latin America, it was originally a mimic of the natural African forests in which coffee occurred before its domestication—an agroforest with shade trees creating an environment that was reminiscent of the forest understory in which the species had evolved originally. Sometimes the setup was only to clear the understory of a natural forest and plant coffee beneath the original forest canopy. More commonly, shade trees of various species were planted at the same time that coffee was planted. The result was that coffee production was traditionally an agroforestry system (Figure 1.2). As seen by air over the central valley of Costa Rica, for example, in the late 1960s, the vista below appeared to be forest, but was in fact many contiguous hectares of shade coffee production.

Agronomists who work on coffee are no different from agronomists who work on maize. They have a certain worldview that is only recently changing, and only very slowly. That worldview stems from the dawn of the Industrial Revolution when Malthusians and others saw the need for increased production efficiency in agriculture. Increasing "efficiency" of production, sometimes in terms of amount of produce per unit area and sometimes in terms of amount of profit per unit investment (not the same thing), remains the *raison d'être* of the agronomist mind, even though the modern sustainability revolution is slowly making inroads. For example, I attended a coffee conference in Costa Rica in 1989 in which a bevy of social scientists spent 2 days discussing the crisis in coffee that had emerged and was expected to intensify. At that conference, the attendees, from conservative

Figure 1-2 Overview of a traditional polycultural shaded coffee system in southern Mexico. The actual coffee plants are only barely visible under the complex shade cover of overstory trees.

economists to radical sociologists, completely agreed that the main, and pretty much only, problem that coffee producers faced was low coffee prices paid to the producers. Furthermore, all participants agreed as to the cause of that problem—a massive overproduction of coffee, saturating world markets. No other problem was even discussed as having even a small impact. The problem, the only problem, was overproduction. After the conference, I visited the Costa Rican center for technical research in coffee and talked to the agronomists doing research on coffee. I asked, "What is the purpose of your research?" The answer was invariably, and with complete lack of irony, "To increase production of coffee."

With such a mindset and with a background that deeply appreciates the details of photosynthesis and nutrient absorption but fails to comprehend the complexity of socioecological systems, technical experts would unsurprisingly develop an obsession with the fact that coffee was produced in shaded conditions. Because production of any plant increases with increased radiant energy (well, not exactly, but close enough) and because the only goal of the agronomist is to increase production, one could easily predict that they would see the shade above the coffee as the main enemy. Supported by international development agencies, vanquishing that enemy became the major goal in the 1970s for coffee research—find a way to produce coffee in the full sun.

With the development of new varieties of coffee, along with a technological package that included, at least indirectly, the application of chemical fertilizers and herbicides, the sun coffee system was in full promotion by the late 1980s. Especially in Costa Rica and Colombia, the transformation from shade to sun coffee was dramatic, and the forested vista of the central valley of Costa Rica changed dramatically. The "forests" that almost completely covered the landscape in the 1960s appeared from the air to be a typical row crop in 1990.

An important alert emerged in the 1990s. Researchers and bird watchers had already become alarmed at the decline in populations of songbirds in eastern North America. It was only a short step from this alarm to the realization that the winter feeding grounds of many of these birds had been the coffee agroforests of northern Latin America and that the massive conversions of these forests to sun coffee could be part of the problem of declining bird populations. Research by the Smithsonian Migratory Bird Center in Washington, DC seemed to confirm this suspicion,[3] finding a dramatic difference between sun and shade coffee with regard to bird biodiversity. More important, this research raised the general question of biodiversity and its relationship to such a dramatic ecosystem transformation.

Generally, when thinking of biodiversity in agriculture, two issues need to be taken into consideration: the planned/associated biodiversity contrast and the intensification gradient. The planned biodiversity is the diversity of the things that are purposefully placed under the care of the farmer, the diversity of crop types, the diversity of varieties of crops, the diversity of farm animals, and so forth. The associated biodiversity is that which arrives independently of the farmer's planning—the birds, the butterflies, the bacteria in the soil. Clearly, the birds in the coffee agroforests of the 1960s were part of the associated biodiversity.

The intensification gradient is a conceptual tool that is largely borrowed from anthropology. As early agriculture evolved into more intensive forms, it is easy to see a gradient from hunting and gathering to casual cultivation to settled agriculture to terraces and irrigation—a gradient of intensification (as described in detail in Chapters 2 and 7). This idea has been used as a rough conceptual tool when thinking about associated biodiversity in agriculture. In the case of coffee, the gradient is fairly obvious—the "rustic" coffee system, in which coffee is planted under the canopy of a natural forest, gives way to the traditional polyculture in which a diverse assemblage of timber species, fruit trees, and other trees dominates a relatively high shade canopy to the commercial polyculture in which the shade is less diversified in its species composition to a monocultural shade in which a single type of shade tree dominates the system to full sun coffee in which shade trees are completely eliminated, the classic formulation of which is reproduced in Figure 1.3.[4]

Given both of these conceptual framings, the question obviously arises: "What is the pattern of associated biodiversity with respect to the intensification gradient?" An implicit answer to this question had already developed in the conservation community: as soon as a natural system became agricultural, a massive decline in biodiversity could be expected, and later intensifications mattered little. An alternative implicit answer, however, had evolved among workers associated with the alternative agriculture movement that only the most intensive

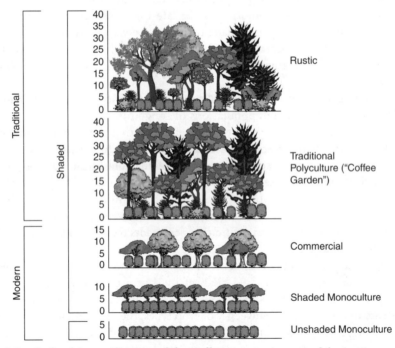

Figure 1-3 Stylized intensification of the coffee agroecosystem in Mexico.

Reproduced from: Moguel, P. and Toledo, V. M., "Biodiversity Conservation in Traditional Coffee Systems of Mexico," *Conservation Biology, 13*(2001):11–21, Fig. 1.

industrial systems resulted in high rates of biodiversity loss. That is, one could see two points of view regarding the response of associated biodiversity to the intensification of agriculture: the conservationist position that the transformation from natural habitat to agriculture was the critical point for biodiversity loss and the alternative position that the transformation from a nonchemical, more organic production to the modern technified system was the critical point for biodiversity loss. Although this dichotomy was dramatic, surprisingly little serious scientific work was actually done to provide evidence for one or the other position.

The pioneering work on this subject was undertaken more or less simultaneously and independently by Russ Greenberg of the Smithsonian Migratory Bird Center and Ivette Perfecto of the University of Michigan.[5] Greenberg documented the suspected fact that more traditional forms of coffee indeed did contain more bird diversity than the new sun coffee systems, and Perfecto did the same for ants and other insects. Especially in the case of insects, the pattern, with exceptions, appears to be only a small reduction in biodiversity as we change from a natural forest to rustic coffee but extremely dramatic reduction as we change from shade to sun coffee. Subsequent study leaves little doubt that in terms of biodiversity loss the main culprit is not the forest to agricultural transformation, but rather the shade to sun transformation.

These studies alerted the world to the importance of the agroecosystem for more traditional goals of biodiversity conservation. However, their emphasis on the agroecosystem as an ecological entity isolated from other entities was prejudiced in much the same way that the more traditional conservation community was. As we now understand, the coffee agroecosystem and more generally the managed ecosystems of the world are not only repositories of biodiversity themselves—they inevitably constitute the matrix within which fragments of natural habitat occur. Consequently, their condition is essential to the biodiversity conditions of the larger landscape. If organisms live in the fragments but the fragments are too small to sustain them, local extinctions (of populations from particular fragments) must be balanced by migrations from other fragments, and those migrations must occur through the matrix of managed ecosystems, which include the world's agroecosystems. This more subtle effect on biodiversity is only now beginning to be appreciated in conservation movements. Yet it represents probably an even more important reason to be concerned about the quality of the agroecosystem than the simple fact of the associated biodiversity that lives there permanently.[6]

As the world gradually came to realize that the shaded coffee system was "biodiversity friendly," a new political movement was born—biodiversity-friendly certification. Because conservation of biodiversity and coffee technification are clearly at odds, a new niche in the specialty coffee market has emerged. Bird-friendly coffee, or shade coffee, is the latest "special premium" effort to be added to organic and fair trade coffee. In mid 1990s, the Smithsonian Migratory Bird Center developed certification criteria for certifying bird-friendly coffee. Other brands tout a similar message (e.g., Rainforest coffee from Chiapas Mexico, Eco-OK coffee certified as shade grown by the Rainforest Alliance, E-coffee from Conservation International), and more will likely be seen in the future. The point is to develop an "intelligent" market in which consumers are willing to pay a price for biodiversity preservation.

Less well known but extremely important is a body of research that has recently been undertaken to demonstrate the effect that biodiversity has on ecosystem function in the shaded coffee system. Biodiversity of birds and arthropods has been demonstrated to have a significant impact on the control of coffee pests. This result brings up an interesting practical point. The role of biodiversity in controlling pests suggests that a positive relationship likely exists between shade and "sustainability." If the old agronomists are right about the negative relationship between productivity and shade (they are not completely right, but for the sake of argument, suppose they are), this means that there is a contradiction between production and sustainability—one goes down and the other up with increased shade. When such a contradiction exists in two quantitative variables, the joint maximization of the two variables is at some intermediate value. Thus, a combination of production and sustainability goals implies an intermediate level of shade for coffee, independent of the direct goal of biodiversity preservation. Optimization at intermediate levels of intensification, even with reduced production, may turn out to be a principle of major significance, not only for the coffee agroecosystem.[7]

The Cuban Experience

The post-WWII period has seen the evolution of a worldwide system of agriculture based on a model that is historically unique. In later chapters, we examine the details, but for now, simply note that the current industrial system is based on the use of industrial inputs, from the machines and the petroleum they consume, to the near elemental forms of soil fertilization assumed to be necessary, to the crude idea that poisoning the environment paves the way for increased productivity, to the even cruder idea that anarchistic planning will lead to efficiency in provisioning food and fiber. That conventional system is young, and despite triumphalist trumpeting, its long-term survival is under serious debate. It includes extensive monocultures designed for political purposes, as with potato monocultures in Ireland and sun coffee in Latin America, which have been combined with direct and evident poisoning of the environment, resulting in a public health crisis and destruction of the resource base on which food provisioning depends. Understandably, many people have concluded that this industrial system is too damaging to both the natural and social environment and needs to be transformed dramatically. Indeed, one purpose of this book is to help provide a small nudge in the direction of that transformation.

The debate on the transformation has a history as long as the industrial system itself, as examined in more detail later. It generally takes the form of the industrial versus alternative schools of thought, with both industrial and alternative being eclectic collections of ideas and analyses, but in the end, despite disagreements among proponents on either side, a clear dichotomy emerges—the industrial system versus an alternative. The industrial uses biocides to control pests, inorganic fertilizers to enhance soil fertility, and heavy machinery to cultivate and harvest, and has as its ultimate goal production value per unit value of investment. The alternative eschews the use of biocides, promotes organic fertilizer for soil enhancement, adapts machinery to the smallest scale possible, and has as its ultimate goal the provisioning of food and fiber to people. At the level of political propaganda, it is common to cite a figure of 20% decline in productivity when moving from the industrial to an alternative system, a figure that was invented for political purposes and has no known rational derivation. Regardless, it is often assumed, even by proponents of alternative agriculture, that there will be a price to pay when we move to the alternative system, and a lingering fear exists that the price will be so high that the transformation will ultimately never happen. Indeed, a major argument of industrial system promoters is that the cost of the alternative system would be so great that the only feasible means of food and fiber production is the industrial system—the set of possible worlds lies outside of the dreams of the alternativists.

The experience of Cuba since its 1959 revolution is an important case study in light of this contemporary debate. At the time of this writing, it is still not clear what will happen to Cuba's agricultural system as time progresses; however, whatever the future trend, the past 19 years provide us with a unique experiment in the alternative system. Although it has only a small population, some 10 million people, Cuba's foray into alternative agriculture represents a kind of experiment, the results of which will be important as the rest of the world moves toward the alternative system. As of this writing, the results are hopeful but certainly not yet cemented in place.

Since the revolution of 1959, Cuba's agricultural system evolved along the same lines as industrialized agriculture in the rest of the world. Technological progress was based on heavy use of mechanization and synthetic chemicals, much as in the Soviet Union and Eastern Europe. Large state farms were managed as factories. Proletarianized farm workers enjoyed considerable advantages over their counterparts in the capitalist world but were nevertheless tied to neither pieces of land nor particular production technologies and worked in order to receive an hourly pay, not to produce food or fiber from the land.

With the breakup of the European socialist economies, Cuba suddenly encountered conditions of extreme scarcity of basic inputs such as fuel, pesticides, and fertilizers, virtually the definition of the technical side of the industrial model. It is difficult for outsiders to imagine the extent of this crisis. Imagine any enterprise dependent on outside inputs losing about 80% of those inputs overnight. That is what happened to Cuba. A very special period thus presented itself to the country, a period so special that Cuban dissidents in Miami and elsewhere were convinced that the Cuban system would finally collapse and usher in their long awaited return to run this island nation. Entering this "special period," Cubans were rapidly forced to rethink their agricultural model. What they have done thus far holds many lessons for the rest of the world. Quite purposefully, a new model has been advanced in which sustainability has replaced profitability as the key guiding force in planning. For pest control, pesticides have been replaced with biological and cultural control forms. For soil management, biofertilizers have replaced most chemical fertilizers. For land preparation and cultivation, animal traction and small-scale mechanization has replaced the megamodel inherited from the Soviet Union.[8] The basic model, in an early form, as contrasted with the model that had been in effect since the Cuban Revolution, is shown in Table 1.1.

TABLE 1-1 Comparison of Classic Model with the New Cuban Model (Translated from a Cuban Ministry of Agriculture Chart Circulated to Planning Staff, Near the Start of the Alternative Movement in the Early 1990s)[9]

Classical Model	vs	Alternative Model

External dependence:
  of the country on other countries
  of provinces on the country
  of localities on the province and the country
		Maximum advantage taken of:
		  the land
		  human resources of the zone or locality
		  broad community participation

Cutting edge technology:
  imported raw materials for animal feed
  widespread utilization of chemical pesticides and fertilizers
  utilization of modern irrigation systems
  consumption of fuel and lubricants
		Cutting edge technology, but appropriate to the zone in which used:
		  organic fertilizers and crop rotations
		  biological control of pests
		  biological cycles and seasonality of crops and animals
		  natural energy sources: hydro (rivers, dams, etc. . . .), wind, solar, slopes, biomass, etc.

Tight relationship between bank credit and production: high interest rates		

Priority given to mechanization as a production technology		Animal traction:
		  rational use of pastures and forage for both grazing and feedlots
		  search for locally supplied animal nutrition

Introduction of new crops at the expense of:
  autochthonous crops
  autochthonous production systems
		Diversification of crops and autochthonous production systems based on accumulated knowledge

Search for efficiency through intensification and mechanization		Introduction of scientific practices that correspond to the particulars of each zone: new varieties of crops and animals, planting densities, seed treatments, post harvest storate, etc. . . .

To satisfy ever increasing needs has ever more ecological or environmental consequences, such as soil erosion, salinization, waterlogging, etc. . . .		Preservation of the environment and ecosystem:
		  need for systematic training (management, nutritional, technical)
		  systematic technical assistance

Classical Model	vs	Alternative Model
Real possibility of investing in production and commercialization		Promote cooperation among producers, with and between communities
Accelerated rural exodus		Obstacles to overcome: difficulties in the commercialization of agricultural products because of the number of intermediaries poverty among the peasantry the distances to markets and urban centers (lack of sufficient roads and means of transport, etc. . . .). illiteracy

The Cubans have taken advantage of many of the well-known alternative techniques developed throughout the world over the past century but have also been remarkably pioneering in other areas, combining sophisticated modern scientific research with grassroots innovativeness. Their continuously evolving system can be summarized by looking specifically at three issues: pest control, soil fertility management, and new or enhanced forms of integration such as organoponics and urban agriculture (Figure 1.4).

Figure 1-4 Typical "organoponico" city garden within the city limits of Havana. Raised beds are maintained with a target of 70% organic matter for the production of vegetables. Production is semiorganic, and reports are that gardens like these supply close to 100% of the fruit and vegetable needs of city residents.

Because the use of biocides has probably been the single most devastating aspect of the industrial system the world over, pest control is viewed as one of the most important components of the alternative model. Even before the crisis, Cuban entomologists and crop scientists had been concerned with this issue, and many techniques had been developed for reducing the use of pesticides. For example, control of the sweet-potato weevil is frequently accomplished through the physical relocation of nests of the predatory ant *Pheidole megacephala*. Banana stems are laced with sugar and placed in areas designated as reserves (because they naturally harbor high densities of *Pheidole megacephala* nests). The ants relocate their nests inside of the banana stems, attracted to the high levels of sugar. Subsequently, the banana stems, with ant nests inside, are placed along the rows of newly planted sweet potatoes. The sun heats up the stems, causing the ants to move their nests underground, where they encounter the sweet-potato weevils and prey on them. Such combinations of cultural and biological control are continually being developed in Cuba, having seen the light of day well before the special period.

What is most impressive about the pest management system is its organization. Throughout the country, there are "centers for the production of insect predators and pathogens" known by the Spanish acronym CREE. In these centers, a variety of products is produced, from pathogenic fungi that attack caterpillar pests to small parasitic wasps that attack the eggs of many insect pests. Each CREE typically produces five or six products specifically designed to manage the pests that are common in the area in which the CREE is located. Local farmers and farming cooperatives purchase these products directly. The CREEs are promoted and subsidized by the government but are expected to be self-sustaining to whatever extent possible. The general mix of government subsidy and price paid by farmers seems to be a negotiated issue involving all parties involved, not a simple matter of several parties each trying to maximize personal benefits.

In a similar manner, soil management has taken on many of the characteristics well known in the alternative movement in other parts of the world. For example, vermiculture, well known in organic farming, is now widespread and reportedly provides for almost all nitrogen needs in food production. But in addition to traditional organic techniques, Cubans have also pioneered new techniques, using several free-living, nitrogen-fixing bacteria, mainly *Azobacter* and *Azotospirillum*. In extremely precise laboratory work, Cuban technicians have developed "cocktails" of bacterial mixtures, each of which is fine tuned to a particular crop and set of production circumstances. These cocktails are now known as "biofertilizers" and are widely available to farmers and farming cooperatives. Indeed, several enterprises, with a varied mix of state and private funding, have emerged that are exporting these biofertilizers to other countries of Latin America.

In addition to strictly technological innovation, Cuba has developed a variety of new production systems, some of which are simply extensions or elaborations of already existing systems, albeit marginal in the rest of the world, others of which are entirely new. A most impressive example of the former has been the remarkable growth of the urban agriculture sector. It is estimated that close to 100% of fruits and vegetables consumed by urban residents now come from urban gardens. These gardens are not necessarily the small city plots normally associated with the urban gardening movement in the United States, but rather they are huge and highly organized operations, mainly at the peripheries of the cities. Another system, purely Cuban, is the organoponic agriculture movement (see Figure 1.4). On many marginal soils, especially in city areas where city rubble underlies the soil, beds are built up above the soil. These beds contain soil made up of almost pure organic material (thus giving rise to the term "organoponic"). Organoponics are especially common in urban areas, where stands advertising "organoponic produce for sale" dot the landscape.

Cuba thus represents something of a success story for alternative agriculture, at least at the level of developing and implementing alternative technical procedures. In this sense, Cuba can be seen as a potential model. On the other hand, the problems that Cuba faces in implementing its program are perhaps the most interesting aspects of the entire transformation. For example, workers in one of the CREEs report that local farmers are not enthusiastic about the new procedures because it is inevitably more work than the old techniques of simply spraying pesticides. This is actually not all that surprising. Anyone who has ever tried to grow tomatoes knows perfectly well that the devastating tomato hornworm is easily controlled by physically locating the caterpillar and picking it off by hand. This is a quick and efficient method of pest control—quick and efficient if you have a small patch of tomatoes in the backyard. If you have 100 hectares of tomatoes using this method would require a significant number of workers. Using a pesticide means that you would not have to hire labor. In this sense, the use of a pesticide is simply a labor-saving device, no different in principle from mechanization. Although there may be some exceptions, this principle may turn out to be disturbingly common in the future. Alternative techniques will usually be more demanding of labor than use of industrial techniques. Because the logic of the development of industrial techniques was based in large part on the need to reduce the cost of labor, it is most obvious to expect that the development of alternatives will require more labor.

Here the Cubans have faced up to what the rest of the world will likely be forced to face when the transition to a sustainable agriculture becomes a reality. The new agriculture will require more labor than industrial agriculture, and Cuba, like the rest of the world, has experienced a massive migration from rural to urban during the past half century. Rural society, the source of agricultural labor (with exceptions, of course), has become an endangered species in most of the developed world and is on its way toward the same fate in most of the Global South today. This debilitation of rural society is one of the major problems facing Cuba today. Although it is reported by academics of all stripes that the "peasants" of the Global South continue to refuse to disappear as the economists predict they must, there is no doubt that a combination of rural/urban migration and penetration of industrial inputs to agriculture has changed the existence of rural society in the past 50 years the world over. The concomitant loss of detailed local knowledge about agroecosysems that has been noted and lamented by Cuban planners will have to be faced by the rest of the world soon.

In fact, Cuban planners long anticipated the serious problems urbanization would create in a society that for the past 30 years had been mechanizing agriculture and inadvertently encouraging rural to urban migration. Moreover, social welfare programs in place since 1980 that aimed to equalize wages between city and countryside did little to stem the tide of rural–urban migration. Consequently, with the shift to the alternative model, the requisite labor force no longer existed in rural areas. Both short- and long-term plans are under consideration for dealing with this problem. But what is unique about Cuba is in the simple recognition that the problem exists and requires a rational solution. "Reruralization" has become a local buzzword in planning circles in Cuba.

The Cuban experiment with alternative agriculture must be judged a success, albeit the jury is still out since it is only 19 years old.[10] When the U.S. economic blockade is lifted, the propaganda machinery of pesticide sales staff will once again penetrate the island, and it is unknown whether it will be successfully resisted. If a new industrialization program is introduced that once again requires a cheap food policy and if cheap petroleum encourages environmentally destructive subsidy structures, the temptation to take the shortcuts of the industrial system, however damaging they may be in the long run, may be irresistible. Whatever the future outcome of this experiment, it has over its 19-year period shown the world that an alternative is possible.

The Two Views of Agriculture in Today's World

The overall picture painted by these three vignettes is one of enormous complexity and consequently, in the words of Richard Levins, the inevitability of surprise—the potato blight, the possible extinction of songbirds, the loss of agricultural inputs from the Soviet Union. The inevitable surprise is conditioned by the complexity operating in the ecological and the sociopolitical frameworks. It is perhaps in the best tradition of early science that one seeks to reduce complexity in framing questions about nature, and it is certainly undeniable that frictionless worlds have led to deep understanding—from the laws of Newton to those of Mendel. This attitude, laudable when first conceptualizing a problem, has led agricultural scientists of various sorts to simplify, to construct their own frictionless worlds, and to believe that they are true. Thus, the intellectual framework in which the agroecosystem has been couched has come to be dominated by a single idea, maximize production. Rarely is it acknowledged that such an idea is really quite new, probably with a history of less than a couple of hundred years, and only about 60 years old in its most modern form. One of the challenges to the industrial agricultural system is precisely the questioning of this basic assumption. Why should it be a goal at all to maximize production? We return to this question in a later chapter. For now it is sufficient to note that many analysts have concluded that goals such as environmental sustainability, social cement, cultural survival, and others are at least equally as important as maximizing production.

This fundamental contradiction in attitudes is reflected most strongly in the alternative agricultural movement, especially among those who study agriculture from an ecological perspective. A distinct feature of ecologically based agriculture needs to be recorded here. Ecological scientists in general tend to have a certain mindset, perhaps derived from the complexity of their subject matter. This mindset is most easily appreciated by comparing the mindset of a typical agronomist with a typical ecologist, especially an agroecologist. Both seek to understand the ecosystem that they are concerned with, the farm or the agroecosystem. Both have the improvement of farmers' lives as a general practical goal. However, the way they approach that goal tends to be quite different.

In the 1990s, a Guatemalan entomologist, Helda Morales, began research for her doctoral dissertation among traditional Mayan corn producers in the Guatemala mountains.[11] In seeking to understand and study traditional methods of pest control, she began by asking the question: "What are your pest problems?" Surprisingly, she found almost unanimity in the attitude of most of the farmers she interviewed: "We have no pest problems." Taken aback, she reformulated her questionnaire and asked, "What kind of insects do you have?" She received a large number of answers, including all of the main characteristic pests of maize and beans in the region. She then asked why these insects, known to be pests by professional entomologists, were not pests according to the Mayan farmers. Again, she received all sorts of answers, always in the form of how the agroecosystem was managed. The farmers were certainly aware that these insects could be problems, but they also had ways of managing the agroecosystem such that the insects remained below levels that would be categorized as pests. Morales' initial approach probably was influenced by her original training in agronomy and classic entomology, but her interactions with the Mayan farmers caused her to change her approach. Rather than study how Mayan farmers solve their problems, she focused on why the Mayan farmers do not have problems.

Here we see a characterization of what might be called the two cultures of agricultural science, that of the agronomist (and other classical agricultural disciplines such as horticulture and entomology) and that of the ecologist (or agroecologist). The agronomist asks, "What are the problems the farmer faces, and how can I help solve them?" The agroecologist asks, "Why are things that could be problems not?" This is not a subtle difference in perspective but rather a fundamental difference in philosophy. The admirable goal of helping farmers out of their problems certainly cannot be faulted on either philosophical or practical grounds. Yet with this focus, we see only the sick farm, the farm with problems, and never fully appreciate the farm running well, in "balance" with the various ecological factors and forces that inevitably are operative. It is a difference reflected in similar other human endeavors—preventative medicine versus curative medicine, regular automotive upkeep versus emergency repair, and so forth. The ecological focus of asking how the farm works is akin to the physiologist's focus of asking how the body works. The agronomist intervenes only when problems arise. The agroecologist seeks to understand how the farm works and thus how it is maintained (and, by implication, could be continuously maintained) free of problems. The agronomist sees himself or herself as a "fixer" of farmers' problems, whereas the agroecologist sees himself or herself as trying to prevent those problems in the first place.

The Moral Dead Zone—Reflections on a Nightmare

Although the bulk of this book aims to reveal the rich texture of ecological research that has brought understanding to the functioning of agroecosystems, such ecosystems by their nature are steeped in normative content. The uptake of nitrate or the oscillations of parasites will occur according to ecological principles, to be sure, and great value exists in understanding these ecological principles, the main purpose of this book. However, the agroecosystem involves human beings, and those human beings dramatically alter the underlying framework within which those ecological principles operate. Ignoring the normative, a universal goal of the ivory tower scientist, is ignoring one of the main drivers of the ecosystem. Thus, as stated earlier, there will be occasions for which delving into the normative becomes unavoidable. From the start, the agroecosystem has produced or reinforced human inequality the world over, one of those normative issues.

The astonishing inequalities that persist in today's world create a dead zone. It is a moral dead zone that is so stubbornly persistent as to make nothing more than frustration out of pedagogy. Twenty percent (or is it 30%?) of the world's population goes to bed hungry, and millions of children will die of some disease that can be tied to malnutrition, to say nothing of the approximately 90% of the world's population that lives with the reality of insecurity of where next week's food will come from. How do we teach about that? How can we challenge the dead zone? Does the evident complacency of the Global North mean that we live in a dream—a nightmare?

Speaking of nightmares, I had one recently that startled me with a vision of my own complacency. I was a kid again, and my family was sitting around the dinner table. My dad brought home the groceries and prepared the evening meal for me and my siblings and he gave me a steak, a generous helping of French fries, a side of delicious green beans, a big glass of fresh milk, plus a small bowl of cream of tomato soup and topped the whole thing off with a chocolate mousse for dessert, all of which came from organic, fair-trade biodiversity-friendly production from various parts of the world. He gave my sisters Kathy and Jenny two slices of bread, my brother Eddie a single slice of bread, and my other brother Jeb a quarter of a slice of bread from the heel of yesterday's loaf that had gone pretty stale. He pronounced, proudly, that he had provided for all of us in the best way he could, under the circumstances of the state of the world. Brother Jeb complained (as he usually did) that his portion was not even enough to keep him alive, let alone work in the mine all day. Kathy and Jenny told him to shut up (as they usually did) and be glad he got even as much as dad was willing to give him, because dad did not make the rules and was following the script that had been handed down and could not be changed or things would be even worse. Eddie was quiet (as he usually was). Maybe he thought that if he spoke up he would be remanded to a quarter of a slice. I remained in my dead zone, staring at my plate, gobbling up the delicious food and thinking of the fun I would have that day driving my race cars around. Jeb was a bother of a brother. My dreamy memory recalled that I used to sneak him pieces of my food, but all he ever did was ask for more, so I got sick of listening to him whine. I even loaned him money a couple times but he never paid it back so I thought, "Forget him. He's not my problem."

Dad explained the situation very clearly (as he apparently had done many times in the past, judging from his tone). The way the system works is that I get a lot of food and the rest get very little. If he cut down on my portion that would automatically cut down on everyone else's portion—it was a law of nature. Thus, how could I be so selfish to want my portion to be cut if that would mean that Jeb would get even less? Jeb got so angry that he lost his sense of moral balance and told dad that that law of nature needed to be changed. I thought to myself, "Wow, Jeb's really losing it." I kept thinking that the solution to Jeb's problem was obvious—dad needed to bring more food into the house. Pure and simple.

But then, as nightmares frequently do, the world in which my family was embedded began to shift, and Jeb's skin got darker and he was selling something in the street in some other country. Dad drifted away from me, but I could still see him in an office in a big sky-scraper in New York, or was it Beijing? Then I faced a new nightmare. As Jeb receded into the streets of that obscure country and dad receded into that skyscraper, I recognized my connection with the both of them less and less, and my complacency grew and grew. That was the real peak of the nightmare.

In the world of the past 500 years, we have seen the development of a model that now seems to be experiencing some troubles. Actually, the troubles are not new. Hunger and starvation have been recurring problems for the past 500 years, rarely in the developed world, but chronically in the underdeveloped world. The model is a form of capitalism, although I use that word in an expanded sense not to signify "free market economics" as it is sometimes used, but rather to indicate the political system that has evolved over the past 500 years as a worldwide system, a perspective owed to Wallerstein.[12] The trouble this model faces currently is in one sense the same as the one it has repeatedly faced in the past but in another sense different from the past because of a situational background that has not been faced before. At the risk of being misinterpreted, I will acknowledge up front that that situational background is, broadly speaking, the environment—the problems with overexploitation of natural resources and myriad forms of pollution, not the least of which is atmospheric greenhouse gases. The risk of misinterpretation I fear is that I will be seen as just another misinformed Malthusian, lamenting the fact that we will soon be faced with perhaps as many as 10 billion souls on the planet, way more than could be sustained over the long haul. That is not what I mean. What I mean is that the structures we were able to put into place in the past to resolve the problems created by the internal workings of the capitalist system may no longer be available to us because of new environmental constraints.

Consider the historical record since WWII. The Great Depression that was a contributor to the development of the war was itself resolved by a new economic model (although the war itself played a big part also). John Maynard Keynes perceptively acknowledged that the raw fibers of the idealized free-market capitalist system would normally lead to problems because of the internal contradictions in the system long ago noted by other economists, most notably Ricardo and Marx. Keynes' solution was to acknowledge what Adam Smith had already built into his generalizations—that social controls over such things as excessive exploitation of one sector of humanity by another were necessary backdrops to the efficient functioning of the idealized market system. The resulting Keynesian solution was to acknowledge the inevitable role of social structures, in this case through government intervention, to obviate market distortions. The Keynesian solution created unprecedented prosperity in the United States during the 1950s and early 1960s and extended in the form of government protection and subsidies to favored industries in Japan and Western Europe, which created the prosperity we now call "developed." Along with this evident prosperity came a new kind of analysis that generally became known as "underdevelopment."

Various interpretations of underdevelopment emerged during this period. On one hand, we see what might be referred to as the "time hypothesis," in which countries and regions are viewed as climbing the development mountain.[13] On top of the mountain were the United States, Australia/New Zealand, Canada, Western Europe, and Japan, and somewhere along the slope, actually quite far down the slope, were the countries of Asia, Latin America, and Africa. Some (like many African countries) were very far down the slope and others somewhat further up. The time hypothesis simply assumed that the countries lower down on the slope were in the process of climbing up and eventually, given enough time, would naturally rise to the summit in the same way that the United States did, through the normal functioning of the capitalist system. The alternative to the time hypothesis was the newly triumphant Keynesian system. With Japan as a most striking example, it became evident that governments could create structures that would grease the wheels of capitalist development. In particular, what became known as "export-led growth" functioned to propel Japan to the heights of industrial development such that it went from a devastated country immediately after WWII to one of the most developed countries in the world, positioned to challenge even the hyperdeveloped United States. Using tariff structures to protect its nascent industries, especially those industries that were uniquely positioned to take advantage of the explosively growing market in the United States, Japan's postwar growth was phenomenal. Toyota, to take the most evident example, has now become the world's largest automobile company, a direct result of the post-WWII arrangement that had the full support of planners in the United States. Climbing the development ladder could thus be seen as part of the capitalist system, as long as Keynesian constraints were in place.

Thus, the model of protecting local industries to generate massive exports became enshrined as the magic bullet that would solve the internal contradictions of capitalism. Other countries would follow, most notably South Korea and Taiwan, both of which continue their climb on the slopes of the development mountain, with considerably less success than Japan. Other Asian countries would follow—first South Korea, Taiwan, Hong Kong, and Singapore and somewhat later Malaysia, Indonesia, Thailand, and the Philippines. The collection became known as the Asian Tigers and during the late 1960s and early 1970s became beacons of hope for those countries still at the bottom of the mountain. Already at this point, the energy crisis of the 1970s signaled that the world's environment might be objecting to this hypergrowth. Eventually, however, an evident crisis of this export-led development caused a stagnation in the growth of the Asian Tigers, eventually resulting in the crisis of the 1990s for those very countries (mainly Malaysia and Indonesia, although the others were not immune from the problems) that had seemed to benefit most from the basic model.

An alternative to Keynesianism had been in development at about the same time as John Maynard Keynes was developing his major thesis. The theoretical apparatus of the alternative was provided by the Austrian political philosopher Friedrich von Hayek and was introduced throughout the U.S. academic community by Milton Friedman of the University of Chicago, which is why this new approach became known as the "Chicago school." The new system was based on the proposition that attempts to control markets would lead to inefficiency in the long run, even if it solved local problems in the short run. Unswerving in its faith in elementary principles of price theory, this new ideology was liberal in the formal sense of the word and certainly new in its claims and thus became known as "neoliberalism." It was not a philosophy that could gain much traction during the heyday of Keynesianism, where the post-WWII growth models seemed to be providing unprecedented wealth to all countries willing or able to apply them. However, with the stagnation of the Asian Tigers (beginning in the late 1970s, even though the ultimate crisis would not come for another decade or so), neoliberalism began to attract more adherents, especially in the United States and the United Kingdom. With the elections of Ronald Reagan and Margaret Thatcher, the stage was set for the complete application of the neoliberal model, the final stages of which were seen with the reelection of George W. Bush and the transformation of both the Democratic Party of the United States and the Liberal Party of the United Kingdom into somewhat more compassionate purveyors of the model.

What the complete application of the model had done was on one hand very successful but on the other hand devastating—very successful in transferring wealth from poorer sectors (social classes and countries) to richer sectors, but devastating to the aspirations of countries on the lower slopes of the development mountain. The rapid rise of Japan, still incompletely realized by South Korea and Taiwan, was eliminated from the set of possibilities by the establishment of the neoliberal model across the world. Protecting nascent industries became a "market distortion," and "comparative advantage" became the byword (as pointed out ironically by more than one economist, if the United States had adopted neoliberalism at the time of its independence, we would today be exporting furs and fish). The distortions of 2008–2009 may indicate even more intractable problems with the economic aspects of the model, to say nothing of the persistent and growing environmental limitations.

By the 21st century, the evident failure of the neoliberal model was one of the factors that led to what has become known as the BRIC block (Brazil, Russia, India, and China). Especially China and India, but with Russia and Brazil not far behind, a new hyperdevelopment phase was energized by a return to basic Keynesian principles within each country. Despite the dramatic economic performance of both India and China, it is not clear that they will be able to ascend the development mountain as rapidly as Japan did, nor even that they will ever be able to reach that level of development given the internal and international problems they currently must face. Furthermore, those problems are not just the normal problems of capitalist development (e.g., rising social inequalities) but involve something totally new—extreme environmental constraints. Eventually the climate crisis must be taken seriously, and the other evident environmental crises, from the cancer epidemic to the hypoxic zones in the world's oceans, are poised to stifle further human progress, even as the richer classes continue to benefit from this new combination of "neoliberal Keynesianism" (which is to say, agreement to play by the neoliberal rules internationally, but organizing production internally according to Keynesian principles).

In summary, climbing the development model seemed to be a possibility in the post-WWII Keynesian era, only to be stifled by the rise of neoliberalism, leading to the general idea that a return to Keynesianism would be the answer to the development problem (generally the left-wing point of view in both the United States and the United Kingdom). Today, however, a return to Keynesianism is not likely to work because of new constraints on capitalist development, namely environmental ones. This fundamental problem has led some analysts to the conclusion that we must look back even further in questioning how to (or whether to) climb that development mountain.

Endnotes

[1]Unless otherwise noted, all notes about the Ireland tragedy are from the classic account of Woodham-Smith, 1962.

[2]Davis, 2001.

[3]Greenberg et al.,1997.

[4]Moguel and Toledo, 1999.

[5]Greenberg et al., 1997; Perfecto et al., 1996.

[6]Vandermeer and Perfecto, 2007a.

[7]Perfecto et al., 2005.

[8]Funes et al., 2002.

[9]Vandermeer et al., 1993.

[10]A recent analysis by Wright provides a somewhat pessimistic assessment of the future, based on interviews with rural residents of Cuba. Wright, 2009.

[11]Morales and Perfecto, 2000.

[12]For an introduction to the world systems perspective, see Wallerstein, 2004. Also, see the appendix to Chapter 7.

[13]I owe this metaphor to Michael Pickard of CIEPAC (Centro de Investigaciones Económicas y Políticas de Acción Comunitaria) in San Cristobal de las Casas, Chiapas, Mexico.

Chapter 2

Constructivism and the Evolution of Agriculture

Overview

Beginning with the basic ecological concept of constructivism, the idea is transformed to deal with human social institutions, especially with the origin of agriculture. Then the two examples of agricultural origins in Mesoamerica and the Middle East are presented,

followed by a discussion of basic ecological processes involved in extensive food production systems. Finally, the chapter closes with an extensive discussion of the principles of
intensification of agriculture in both its historical and ecological contexts.

Introduction

In the beginning, there was no agriculture. Then it came to dominate all aspects of human
society. How that domination happened is partly a story about ecology, but it is mainly a
story about how humans interact with one another, a story located somewhere in the vast
world of the social sciences. Thus, the narrative of this text begins with a heavy emphasis
on the evolution of human societies, involving a variety of subjects normally thought of
as members of one or another academic discipline pigeonholed into the social sciences.

The purpose of all social sciences is to understand the human condition. If we were to
judge the social sciences based on their ability to predict the behavior of their subject
matter, they would fall far short of the natural sciences. If the test of "understanding" is
based on the ability to predict behavior, the social sciences are remarkably weak. However,
such a weakness should not be interpreted as weakness of the intellectual endeavor, but
rather should be acknowledged as the outcome of a remarkably complex subject. The
movement of planets and structural integrity of molecules are regular and ultimately
simple phenomena. The behavior of humans in their social space is, by comparison, infinitely more complicated. The simple fact that there are so many interacting components
is one factor impeding our understanding, but even more important is the fact that
changing our level of understanding changes the nature of those components. It is a sort
of super Heisenberg uncertainty principle. Whatever social circumstance I participate in,
my knowledge of how that social circumstance works changes the way it works. For
example, the continual struggle to understand the movement of financial markets invites
ridicule in its cycle of claim and retraction, at least partly because an understanding of the
market causes a change in the behavior of the individuals that make up the market, which
trumps the understanding.

The study of agroecosystems is unusual in that it sits on the border of the social and
natural sciences. Agriculture is not planting a seed and harvesting a crop. Agriculture is
making a contract among people to provide for one another, using seeds and harvests to
do so. Studying agroecosystems is not simply studying the way a crop uses nitrogen. It is
studying the way, for example, an economic blockade by the United States plus the failure
of the Soviet Union plus the geological background that led to oxisol formation plus the
culture of eating sweet potatoes together resulted in the development of new strains of
Azotobacter, which provides nitrogen to the sweet potatoes growing on Cuba's inherently
poor soils today.

While acknowledging that agroecosystems are inherently more complex than so-called
natural ecosystems, my intent is not to hide behind complexity as an excuse for ignorance.
Indeed, the subject of agroecosystem ecology provides an unusual opportunity for some
creative thought. Perched as it is on the border of the natural and social sciences, might

the study of agroecosystems benefit from a cross-fertilization? In the attempt to provide a general framework for agriculture's main dynamical features, some simple ideas borrowed from evolutionary biology might provide some clues, although I hope to avoid the naïve positivism that has sometimes emerged from such a focus in the past.

Today the foundation of all of biology is found in Charles Darwin's opus "The Origin of Species." The science of ecology itself had much of its agenda set in Chapter 4 of that work. Even though Darwin never uses the word, it is clear that his "force of selection" is equivalent to what we think of today as ecology. More recently, another aspect of Darwin's program has been reevaluated by population geneticist Richard Lewontin, who noted that Darwin's distinction between organism and environment, while brilliant and necessary in the mid 19th century, may be a stumbling block to understanding today.[1] He concludes that organisms, to a great extent, construct their own environments. His insights gave rise to a new trend in evolutionary biology known as constructivism, best illustrated in the recent work of Olding-Smee et al., *Niche Construction: The Neglected Process in Evolution.*[2] Constructivism[3] essentially views the organism and environment as a dialectically interacting pair, with environment generating selection pressure that causes genetic changes in the population of the organisms, and the organisms acting on the environment to alter it for future generations. In this way, both organism and environment are actually "constructed" from one another.

Although the recent interest in constructivism has focused on its application to evolution, there is a clear and important lesson for population dynamics at an ecological scale also. That is, if environment and organism are changing one another so as to effect changes at an evolutionary time scale, the same sorts of constructions must be happening at an ecological time scale also. This point of view is useful in all of ecology, but especially in understanding the human/environmental dynamics inherent in the agricultural ecosystem.

Basics of Constructivism

Constructivism in Ecology

It has long been appreciated that organisms affect their environment, and that the affected environment can thus have a reciprocal effect on other organisms, creating an environment different from what it would have been before having been changed. For example, Goldberg[4] provided an important framework for understanding competition between plant species by noting that competition could be divided into response and effect competition—a species "effects" changes in the environment to which another species must "respond." To put this insight into the framework of constructivism, species A constructs the environment in which species B must live. Similarly, the literature on "ecosystem engineers"[5] implicitly incorporates the same idea—that some species alter the physical environment to such an extent that others are profoundly affected.

Suppose that a given ecological niche form is dictated by some critical factor—for example, the depth of the water in a beaver pond, which of course depends on the size of the dam. On one hand, a given size dam requires a certain number of beavers for its maintenance. This number of individuals is the *necessary* population. On the other hand, the niche affects the organism and effectively dictates how many individuals can be sustained at a given level of constructed niche. More beavers can be sustained in a larger beaver pond than in a smaller one. This number is the *sustainable* population.

The relationship between the necessary and sustainable populations defines a clear dynamic for the population and its niche. This dynamic is illustrated in Figure 2.1a. Consider the point E* and N*; that is, consider a population of size N* with a niche of E*.

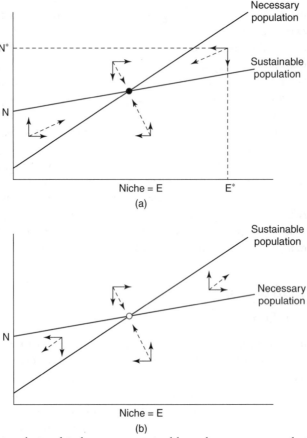

Figure 2-1 Basic relationship between sustainable and necessary population as a function of the ecological niche.[6] Some measure of niche (e.g., the depth of the beaver pond) on the x axis and the number of individuals in the population on the y axis. The two lines give population size as a function of the niche, one function for the necessary population, the other for the sustainable population. The dynamics (see text) are such that the intersection of the two functions is an equilibrium point, a stable equilibrium in *a*, and an unstable equilibrium in *b*.

That population will be above its sustainable level (not enough food in the pond for all the beavers that live there) and thus must decrease. That population, however, will be below the necessary population—the population density necessary to maintain the niche level at E*—and thus, the niche level must decline (there are not enough beavers to maintain the dam at its current height). Following this reasoning for the rest of the possible states of N and E, we see the overall dynamical behavior as illustrated in Figure 2.1a, wherein any population/niche combination will eventually come to lie at the intersection of the two lines representing the necessary and sustainable populations, indicated with a solid dot in Figure 2.1a.

If the relationship between the sustainable and necessary functions is reversed (Figure 2.1b), the dynamic reverses itself, and the intersection of the two functions becomes an unstable point. Thus, any combination of N and E will result in either the extinction of the population or a continuously increasing population and niche. The latter is obviously impossible in the real world and thus, some constraining force must be involved, discussed later. However, the basic principle of either a stable or an unstable relationship between N and E is clear at this level of abstraction.

The world is very rarely linear, and this abstraction is not likely to be an exception. Two sorts of nonlinearities could be involved in this formulation. First, it might be that increases in the necessary population at high levels of niche will increase more dramatically than at low levels; thus, it is impossible to increase niches without limit. Second, it could be that increases in sustainable levels will saturate at high levels of the constructed niche; thus, the population that can be supported by a constructed niche increases at a slower rate the "larger" the constructed niche (if you increase the beaver dam from 1 meter high to 2 meters high, you can support 20 instead of 10 beavers in the pond, but if you increase a larger dam from 2 meters to 3 meters, you can support only 21 beavers). Given these two nonlinearities, which seem most likely for all organisms except humans (as discussed later in this chapter), we obtain the picture illustrated in Figure 2.2. The basic pattern then from a constructivist approach is that a population may either form a globally stable situation (Figure 2.1a) or it may form a bistable case with a higher stable state for both N and E and a zero stable state for either N or E (Figure 2.2).

Looking at a population from this point of view leads to many insights in population ecology. Even so, the idea that a population constructs its own niche is of special importance to anyone concerned with the particular species *Homo sapiens*. The beaver may make a dam and the wasp an elaborate nest, but only the human generally lives in environments so completely constructed by it. It is in the origin of agriculture that we can see so clearly how this process operates as a generalization that incorporates both social and ecological forces as a dialectical whole, as we see presently.

Work and Agricultural Origins

Why people should have invented agriculture is not as clear as it seems on first glance. Even the seemingly obvious notion that agriculture evolved to have a larger or more secure supply of food is not true. Indeed, one agricultural origin was not for producing food at all.

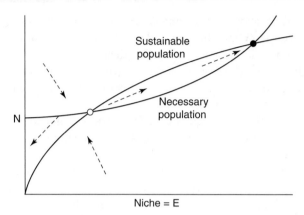

Figure 2-2   Incorporating nonlinearities in the basic construction dynamics. Assuming that the population cannot increase niches without limit, we reach a point in which adding ever more extra individuals to the population is necessary for even small changes in the niche, thus leading to the nonlinear form of the necessary population function. On the other hand, assuming that the sustainable population cannot be pushed higher and higher by more of the niche element but must reach some sort of asymptote leads to the non-linear form of the sustainable population function. Three equilibria result, one unstable one (the lower crossing of the two curves) and two stable ones (one at the upper crossing of the curves the other at zero).

Some early South American farmers initiated domestication with cotton and gourds, the first to produce nets and the second to produce floats for their nets—agriculture here was effectively an industrial activity only indirectly related to food production (fishing).

Even stranger is the fact that early agriculture did not offer an obvious advantage over food gathering. Cultural anthropologists seem to have come to agreement that the hunting–gathering way of life was actually more efficient than agriculture in subsistence activities, at least sometimes. A case in point is provided by the now classic work of Richard Lee on the !Kung of South Africa.[7] Living in small groups, !Kung women were responsible for most of the gathering of plant material. Lee estimates that on average it takes a single woman about 6 hours to gather enough food for her family for 3 days, a much shorter time than is normally required in primitive agriculture to produce the same amount of food. The available plant matter includes the energy-rich mongongo nut (Figure 2.3), but the diversity of food sources is quite high, including a great variety of fruits, tubers, and insects. The men hunt, but hunting seems more related to the social desire for meat, not to satisfy either a protein or a calorie requirement. Hunting is thus an occasional activity. Until a very recent shift to sedentary existence, the !Kung are reported to have lived in this way for the past 10,000 years, suggesting it is a relatively stable form of existence.

This group of hunters and gatherers is thought to be similar to most others, leading to the general conclusion that under the circumstances that probably existed when agriculture

Figure 2-3   Young !kung woman collects mongongo nuts.

was adopted, it did not make sense to adopt it. It is also clear from other groups of hunters and gatherers that even when the practice of agriculture is well known, a fully sedentary agricultural lifestyle does not necessarily follow. For example, the Huaorani people of the upper Amazon are mainly hunters, with monkeys being their principal source of protein and a diversity of other animals and fruits gathered from the forest, providing a relatively rich diet.[8] They also casually plant cassava, usually in naturally caused gaps in the forest, and they maintain at least a mental map of the position of their peach palm trees, almost all of which were planted by previous generations of Huaorani. Living in temporary small clearings, the peach palms are planted around the dwellings, along with the cassava. The cassava is harvested, but the peach palms may not even grow to fruiting size before the group moves on to another clearing, usually looking for more game. Thus, the forest contains scattered peach palm trees where the Huaorani formerly made a small clearing, the forest having taken over that clearing. So part of their activities is "gathering" the peach palms from trees that they had planted earlier. Is this agriculture?

Other groups of hunters and gatherers are known to provide a low level of husbandry for the plants they gather, encouraging some species and discouraging others. Is this primitive weeding? Does this husbandry count as a component of agriculture?

With these observations, it makes sense to think of agriculture not as suddenly emerging from a pure hunter and gatherer existence, but rather as a slowly evolving adoption of agricultural activities. The !Kung are an extreme, the Huaorani more the rule.

As agriculture becomes more a part of the life of a people, new challenges emerge. More time must be spent in subsistence activities, which likely translates into a desire for a larger family size to help with the work. The vagaries of the weather become far more important, and it certainly must be difficult to make the decision to abandon the home site for better hunting grounds, thus giving up the garden in which so much time and energy had been invested. After a commitment to sedentary life is made, the flexibility of moving to better territory is lost.

The Constructivist Model for Humans

The dynamic processes that must have operated at the time of this transition can be visualized with the same constructivist framework as in other populations, as described in the previous section. However, the peculiarities of *Homo sapiens* require significant modification. The metaphorical beaver dam is "the agricultural infrastructure." That agricultural infrastructure may be terraces built collectively by a whole society, it may be an irrigation system constructed by a single farm family, or it may be the collective equity of a modern farm. At its most elementary level, that agricultural infrastructure may be nothing more than the time devoted to agriculture, with the !Kung devoting zero hours a week and a fully sedentary agricultural family perhaps 40 hours per week. Alternatively, the agricultural infrastructure could be the fraction of available land actually devoted to agriculture. The entire idea is somewhat slippery and will be slightly different for different applications, but the idea of agricultural infrastructure is nevertheless a useful variable. It is zero in hunters and gatherers and increases with increasing agricultural intensification. As in some other applications (e.g., the relationship between biodiversity and agricultural intensification), the idea of "agricultural intensification" is virtually identical to "agricultural infrastructure."

Given this somewhat soft variable, the idea that a given number of people are necessary to maintain a given level of agricultural infrastructure is as clear as the idea that a given number of beavers are necessary to maintain a given sized dam. If we were to view humans as any other niche-constructing organism, we could simply say that when the actual population is below the necessary population, the constructed niche (the agricultural infrastructure) will decrease, and when the population is below the sustainable level, it will increase, resulting in the dynamics illustrated in Figure 2.1, where niche = E on the x axis is "agricultural infrastructure." The species *Homo sapiens*, however, is unique, and its particular social nature must be taken into account. This uniqueness results in considerable complication, indeed even a reversal of the basic niche construction dynamics (the reverse of what is shown in Figure 2.1). How can this be so when if there is not enough population to sustain a particular agricultural technology that technology must decrease to accommodate the population that in fact is there to support it? Worse, it is not necessarily the case that a population larger than the sustainable level will result in a decrease in the population, something seemingly at odds with basic thermodynamic principles. Indeed, in the case of humans, when the population is above the sustainable level, there is a tendency to increase it further!

To see the special case of *Homo sapiens*, consider a subsistence farming family in which the population is below the necessary population but above the sustainable population—in other words, there are not enough people to work the land that is set aside for agriculture and, worse, not enough is produced on that land to support the existing population. According to Figure 2.1 and likely true for all nonhuman animals, we would expect both the niche and the population to decline. Humans, however, have a social capacity unmatched in the nonhuman world. The family will view the situation from the point of view of how much food is available. Because the actual population is above the sustainable one, the family will see a food shortage. Unless we are dealing with a case of extreme starvation, it is unlikely that having a population above the sustainable one will generate a decline in population, at least over a short period. However, the experience of a shortage of food most probably will result in a decision to intensify agriculture in an attempt to produce more food. Thus, seemingly ironically from the point of view of Figure 2.1, a population above the sustainable one will tend to increase the amount of land in production. Similarly, if the actual population is below the necessary population, as we assume here, what will be the response of these sentient humans? If a certain number of people are needed to maintain the agricultural technology and that number is not available, there will be a tendency to try and increase the population, either by having more children or by convincing relatives to help with the work or extending the work day (which increases the person/hours and thus effectively the population size). This dynamic, like the response to a food shortage, is the opposite of what we expect in all other organisms.

Let us consider a numerical example. Suppose we have an extended family of !Kung gatherers, with a population of 100 people. To maintain their gathering lifestyle, let us suppose that they would need a minimum of 90 people and that with their gathering activities they could support 110 people. Thus, with a population of 100, they are just about right (more than the needed 90 and less than the maximum supportable of 110). As with almost all cases of hunters and gatherers, they know about agriculture. Suppose they know that if they were to plant a small patch of sorghum they could produce enough calories to sustain a population of 160 people (rather than the 110 that they can with their hunting and gathering habit). In order to prepare the land, plant the sorghum, and weed the patch, they would need a population of 170 people. Now let us suppose that relatives arrive on the scene so that the overall population becomes 165, and to feed all of those people, they all decide they indeed do need to plant the patch of sorghum. So, they produce enough to feed 160 people, meaning that they will experience a food shortage (as there are five more than the amount of food will support). To deal with this shortage next year, they are likely to try to increase the amount of land that they plant (to produce more food). They also know, however, that they already had to work too hard to produce that patch of sorghum (because it requires 170 people and they only had 165), which means their experience is of a labor shortage. To deal with this shortage, they are likely to look for more relatives to join the group (maybe have more children if they see it as a longer term prospect). If the sustainable population increases in the same way as it did before (an increase of 50 people, from 110 to 160) and if the necessary population increases the same way it did before (an increase of 80 people, from 90 to 170), then by trying to plant more sorghum, they would only exacerbate the problem. But, by exacerbating the problem they would continue to create conditions in which there would be a shortage of food as well as a shortage of labor. In other words, they would be on a sort of treadmill, constantly trying to produce more food by putting more land in production but constantly needing more labor to work that land. This situation with the particular numbers used in this example is illustrated in Figure 2.4.

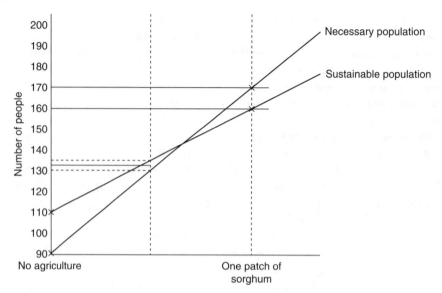

Figure 2-4 Simple numerical example for the niche constructive dynamics of *Homo sapiens* (dramatically different from other organisms), as explained in the text.

The situation would be completely different if, for example, only 32 relatives arrive (making the total population 132) and they decided to plant half of a patch of sorghum (see the other vertical line on the graph in Figure 2.4). With the same rules of how the sustainable and necessary populations increase with agricultural technology (i.e., the lines are straight in Figure 2.4), that would give about 135 people that could be supported and only about 130 necessary to maintain the production (see the dotted horizontal lines in Figure 2.4). So, here the experience of the population is one of food surplus (they produce enough for 135 people but only have 132) and of more than enough people to work the land. It makes no sense to continue producing a surplus, and thus, the amount of land in production the next year would likely be reduced. Because there was not very much work required to produce what was produced (i.e., 130 people would be needed to do the work and they actually had 135—more than enough), maybe the relatives could go somewhere else. Consequently, the general pressure is to decrease both the population and the amount of land in production. Because the necessary and sustainable population rules remain constant (the lines in Figure 2.4 are straight), this tendency will continue until there is no agriculture at all practiced, the population reverting to its hunter/gatherer existence.

Generalized Constructivism and Agricultural Origins

We can generalize these overall dynamics on a graph of population density versus agricultural infrastructure (Figure 2.5). The perspective first outlined by Boserup[9] is useful here and corresponds to the thought experiment outlined in the previous paragraph.

A population not involved in agriculture at all will have the sustainable upper limit of its population density set by the availability of naturally occurring resources. This is not to say that the actual population will ever reach that point, as most human populations are regulated well below what might be considered the "thermodynamic carrying capacity" (i.e., usually the actual carrying capacity of a population is determined by human social interactions, not ecology). Certainly, some theoretical limit exists, however, whether or not a population actually tends toward it. If the population ever goes above that theoretical limit, shortage is experienced. Some agricultural activity clearly increases the sustainable

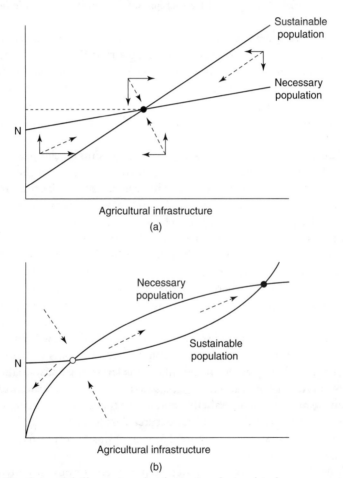

Figure 2-5 Simple model illustrating the dynamic relationship between agricultural infrastructure and population density, effectively the reverse dynamics of nonhuman organisms (see Figures 2.1 and 2.2). The curvilinear aspects of the two functions are also likely to be reversed from the condition with other organisms, although this is not necessarily the case (see text).

population level, even if only by a small amount, but that agricultural activity carries with it the need for more time spent laboring, which can easily translate into a need for a larger population. Some fixed upper limit exists to the time individuals can work in a day, whether fixed by simple physiological limits or by social custom. When labor requirements increase, as they must if the time required for work increases, a clear solution is to generate more people. Thus, the family should be expanded by either bringing more kin into the nuclear unit, defining the work process to include a larger number of people, increasing the family size by birth, or a variety of other mechanisms. The important point is that as the time devoted to agriculture increases, the number of people needed to help with the work also increases.

Thus, two forces are operating on the local population dynamics as agriculture is gradually incorporated. The maximum number of people that could be supported (the sustainable population density) increases, and the number of people needed to do the agricultural work (the population necessary to create the ecological niche) increases. The fundamental idea is the same as with other organisms, as described in the previous section, except that humans, because of their ability to plan ahead and understand what they do to their environment, generate the inverse dynamics, as illustrated in Figure 2.5.

A model of agricultural evolution is suggested by this simple relationship. If the actual size of the population (and here we are thinking of the population of the farming unit) is smaller than the necessary population, there will be a tendency to increase the population to meet labor needs. If the actual size is larger than the necessary population, there will be a desire for a smaller population (e.g., not to have to feed so many mouths). On the other hand, if the actual size of the population is larger than the sustainable size, there will be an absolute shortage of food each harvest. The experience of this shortage is most likely to have the effect of devoting even more time and energy to agricultural activities, which is to say, increase agricultural infrastructure. At the other extreme, if the actual population is smaller than the sustainable size, a surplus of food will exist. The most likely response to this eventuality is to store what is possible, but also to adjust the time spent in agriculture, to spend less time in production—no sense in producing food that will just rot.

Thus, we have, by simple responses of people to circumstances of work and scarcity, two simple dynamic rules: 1) When the population is larger than sustainable and smaller than necessary, increase population and increase agricultural infrastructure, and 2) when the population is smaller than sustainable and larger than necessary, decrease the population and decrease the agricultural infrastructure. This leads to an interesting dynamic pattern, as shown by the arrows in Figure 2.5. It is opposite to the pattern of all other organisms (Figures 2.1 and 2.2).

This simple model is similar to but in the end quite different from early attempts to tie the emergence of agriculture to population pressure. The argument was simply that as long as populations remained small the hunting and gathering lifestyle was sufficient to meet needs, but as populations got larger, more calories were needed, and thus, agriculture needed to be invented. This is essentially a Malthusian position, which assumes that there is an inexorable tendency for populations to increase and eventually outstrip their food supply. If we add the new ideas of niche construction to this position, we effectively reconstruct the basic ecological theory of niche construction (Figures 2.1 and 2.2), Malthusianism plus niche construction, a perfectly reasonable approach to all organisms except *Homo sapiens*. Conversely, the present approach follows Boserup's lead and ascribes a social dynamic to the tendency for a population to increase or decrease (concretely, to track the necessary population as set by the technological level, which is to say, the agricultural infrastructure).

The linear assumptions incorporated into Figure 2.5a (both functions are assumed to be simple lines) are not likely to actually occur in nature. In particular, for some forms of primitive agriculture, the initial investment into agricultural activity probably provides only marginal increases in the size of the sustainable population, whereas after the agricultural system becomes a more important component of subsistence activities, the number of people sustainably supportable probably curves upward. Furthermore, the necessary population probably increases quite rapidly as the agricultural infrastructure is initiated but increases more slowly with further improvements. Thus, the pattern is more likely to be something like that shown in Figure 2.5b, notably distinct in the form of the nonlinearities from the situation with all other animals (see Figure 2.2). The model now suggests an interesting dynamic. A break point still exists (indicated by the open circle), but now there is another equilibrium point, where the tendency to decrease population and time devoted to agriculture is balanced by the reverse tendencies. The population will either tend to the pure hunting and gathering form (the !Kung?) or to a stable form of casual cultivation (the Huaoroni?). The particular form of the nonlinearities is introduced here mainly in the context of early adoption of agriculture. Later in this chapter we deal with the more complicated situation of further intensifying the agroecosystem, but for now, we limit ourselves to the simple nonlinearities, as shown in Figure 2.5b.

This model has three qualitatively distinct outcomes. If the forces promoting early agriculture act to always promote a decrease in population and decreased time devoted to agriculture, as would happen in an environment of high productivity, the sustainable population curve would always be above the necessary population curve, as in Figure 2.6a. On the other hand, in Figure 2.6c, the sustainable population is always below the necessary population, and the tendency is always to move to the sedentary agriculture habit. This is likely the outcome in an environment of poor productivity in which food shortage is common. Naturally, the graph in Figure 2.6b is the intermediate situation in which a stable sustainable population is possible with a low level of agricultural activity, as was suggested in the graph of Figure 2.6b. The existence of alternative equilibrium situations under the same underlying conditions (Figure 2.6b) suggests that certain accidents of history, rather than underlying ecological or cultural forces, may sometimes be at the root of the question "why (or when) agriculture?" For example, suppose a migrating population reaches an unoccupied river valley characterized by the functions arranged as in Figure 2.6b. If the population is relatively large, it will move inexorably toward intensive agriculture, whereas if it is small, it will move toward a hunting and gathering existence. Such dynamics make it clear how some cultures, fully aware of agriculture, choose not to develop it (see the previous discussion), whereas others do.

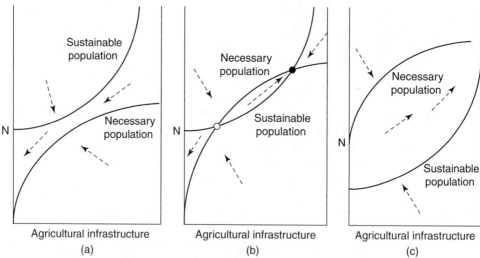

Figure 2-6 Same model as Figure 2.5 for more complex circumstances. In *a* and *c*, the nonlinearities are unimportant for the qualitative outcomes that would be identical even with simple linear functions; however in *b*, the two stable points (one at zero) are separated by an unstable point, identical to Figure 2.5b.

Agricultural Origins in Mesoamerica and the Middle East[10]

I choose these two sites first because probably more is known about them than any other sites in the world and second because they offer contrasting pathways to the evolution of sedentary agriculture, emphasizing the fact that there is not a single way that *Homo sapiens* carried out the agricultural revolution, or revolutions.

Ecological Background

The natural setting of the two sites is strikingly different (Figure 2.7). The area of concern in the Middle East begins in the Sinai peninsula, moves north through today's Israel and Lebanon, north to the southern half of Turkey, east through Iraq, and east and south through Iran, creating a crescent-like area known as the fertile crescent. The vegetation patterns in the area of the fertile crescent are complicated. Moving eastward from the Mediterranean, we encounter first typical Mediterranean vegetation with coastal scrublands; then grasslands; then sparsely vegetated desert; then gently rolling landscape originally covered with Savannah-like grasslands or open woodlands with not only many grass types but also oak, pine, pistachio, walnut, wild olive, and other trees; and then finally the flanks of the Zagros Mountains with their vegetation covering of oak/pistachio woodlands. The climate is typically Mediterranean, with cool wet winters and dry hot summers.

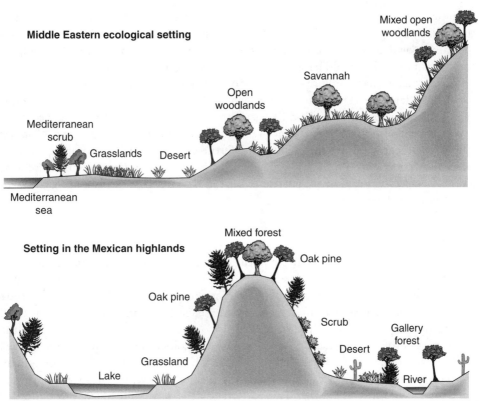

Figure 2-7 Diagrammatic representation of the ecological setting in the two examples of agricultural origins.

South Central Mexico could hardly be more different. The general topography is mountainous, with high flat valleys. Most of the valleys are dry with central rivers bordered by gallery forest, with the rest of the valley floor covered by desert scrub vegetation. The flanks of the mountains give way to scrubby xerophytic vegetation and, finally, at higher elevations oak/pine forests and at the highest elevation mixed forests with oak, pine, alder, ash, walnut, hackberry, liquidambar, and elm. The central valley of Mexico was distinct because it was mainly occupied by large shallow lakes with no water flow out of the valley, surrounded by grassland vegetation, in which the all-important *teosinte*, the forerunner of maize, was found. The climate was generally dry.

In addition to the climate and associated vegetation, the fauna available for hunting, fishing, and domestication is critical. In the Middle East, the western edge of the fertile crescent borders the Mediterranean sea, providing access to marine resources. Furthermore, the valley bottoms, especially associated with the Tigris and Euphrates rivers, provided fresh water aquatic resources. Animals for hunting were abundant. Wild goats and sheep of the hills and mountains were available first for hunting and eventually for domestication to form the basis of a herding society. In the Mesoamerican site, there was similarly a large fauna available for hunting, and the large lakes in the Central Valley especially provided abundant aquatic resources. Most important, there were no large grazing mammals that could have provided a source of domestication.

The key grasses that would first provide storable grain for hunting and gathering populations and later the genetic material on which domestication could operate were distinct in the two regions. In the Middle East, three key grass species were available: barley (*Hordeum spontaneum*), einkorn wheat (*Triticum boeoticum*), and emmer wheat (*Triticum dicoccoides*), occurring either throughout the fertile crescent or in significant quantities within subsections thereof. In Mesoamerica, the key grass species was teosinte (*Zea mays mexicana*), which occurred in abundance at least in the grasslands surrounding the lakes in the Central Valley.

Basic Settlement Patterns

The ecological differences between the fertile crescent and the Mesoamerican site led to distinct forms of agricultural evolution. In the Middle East, sedentary villages predated the evolution of agriculture, in contrast to earlier conventional wisdom that agriculture was a prerequisite for settled life. These settled villages were based on marine or fresh water resources, with no evidence of agriculture. At the same time, in the hills of the eastern part of the fertile crescent, are numerous archeological sites generally recognized as the remains of herding cultures. Thus, immediately before the development of agriculture here, two distinct lifestyles were extant, and certainly, these two lifestyles were aware of and interacted with one another. This pattern existed from about 12,000 through 9,000 before present (BP) and was followed by the gradual increase of agricultural products in the diet, first in the western part of the area and subsequently in the eastern part. Fully sedentary agriculture in the western part was established at least by 8,000 BP.

The preagricultural pattern in Mesoamerica can be characterized as a seasonal alternation between very small groups (microbands) and quite large groups (macrobands) of people, neither of which formed permanent settlements. The microbands occurred during the extensive dry seasons, presumably allowing for an easier mobility for the band to pursue scattered food sources. The macrobands resulted from the coming together of the microbands during the wet season when plant resources were more concentrated. This pattern is discernible in the archeological record for 2 millennia prior to 7,200 BP, a point in time when maize first appeared and the pattern began to change. Between 7,200 and 5,400 BP, the use of agricultural domesticates such as chili, gourds, avocado, tepary beans, and squash became more common, and the macrobands became larger. During this time, the famous American *metate*, used throughout the Americas for grinding maize, appeared in the record (Figure 2.8). From 4,300 to 3,500 BP, the basic pattern continued, with maize gradually becoming more important in the diet, but with the macrobands still breaking up into smaller units on a seasonal basis, presumably for hunting or pursuing scattered wild food in the dry season. This occurred even though agricultural products formed a

Figure 2-8 Metate and grinding stone, the first food-processing unit in the Americas, mainly used for grinding maize.

major part of the diet. Finally, after 3,500 BP, full-time agriculturalists were evident, living in communities of 100 to 300 inhabitants in permanent settlements.

Thus, we see the contrasting patterns of agricultural adoption. The Middle East simultaneously evolved sedentary villages and migratory herding prior to the adoption of agriculture (although casual cultivation was probably well known), leading to the full adoption of agriculture first in the already sedentary part of the population and later in the part of the population that practiced animal herding. The Mesoamerican pattern was a seasonal alteration of microbands and macrobands, gradually increasing numbers and probably spending more time in the macroband, along with the gradual increase of agricultural products in the diet, until the village of the macroband became the permanent agricultural settlement. A dramatic difference in the timing of the evolutionary process occurred. While the American people were restricted to hunting and gathering in 8,000 BP, the Middle Eastern people had already adopted a sedentary agricultural lifestyle. It would be another 4,000 years or so before the Americans finally settled down to permanent settlements and agriculture. Could chance have created such an enormous difference, or was some other factor(s) involved?

As evidence accumulates about the origin of agriculture in other parts of the world, we increasingly see a diversity of patterns and timings. While the underlying dual forces of the desire to feed everyone and the need for people to help in the resource acquisition process (i.e., sustainable population and necessary population) are likely the same in all cases, the rich details that cannot be discerned from the population dynamic model are as diverse as human cultures.

The Ecology of Extensive Food Production Systems

Throughout the contemporary world, nonindustrial agriculturalists still practice an agriculture that may be similar to that practiced by early agriculturalists. For example, fire is an important component of preindustrial agriculture and is also a common component recorded in the archeological record. Two forms of extensive agriculture still practiced by many indigenous people are worth considering: so-called slash and burn agriculture and nomadic herding.

Slash and Burn Agriculture

This form of extensive production begins with the felling of natural vegetation. The slashed vegetation is left to dry and is then burned, thus killing much of the fire-susceptible vegetation that could form a weed community. Crops are planted in the ash-covered ground and require little husbandry, at least in the first year. In the next cycle, either the next year or sooner, depending on the climate, the postharvest vegetation that has accumulated is burned, and crops are again planted. This burning and planting cycle eventually results in a combination of factors that reduces crop production each year. Soil fertility reduction and accumulation of pests, especially weeds, are thought to be the major factors. Crop production eventually declines to unacceptable levels, and the field is left fallow for some time. The number of consecutive years a particular piece of land can remain in production varies widely, depending on local ecological and cultural conditions (Figure 2.9).

The fallow time changes according to both socioeconomic and ecological criteria. Enough time is needed to replenish the soil, to deplete whatever disease agents have built up in the soil, and to develop the plant community that once again will permit fire to eliminate weeds—generally, fire-resistant plants will be eliminated by higher, woodier vegetation through the process of ecological succession. Also, the size of the relevant human population and its food requirements will dictate how much land must be in production at any point in time, which indirectly determines a limit on the length of the fallow. For example, if a total of 100 hectares is in the land pool, farming a single new hectare per year permits the potential of a 100-year fallow, whereas farming 10 hectares per year permits the potential of only a 10-year fallow.

In the contemporary world, one form of slash and burn differs substantially from more traditional slash and burn techniques.[11] Traditional methods of slash and burn are usually self-conscious of natural cycles, specifically adapted to local conditions and seemingly sustainable over the long term. In contrast, recent migrants into a new area sometimes practice what is formally slash and burn, but with less than perfect knowledge of local ecological forces and with techniques that are frequently unsustainable. It is thus useful to distinguish between slash and burn agriculture and migratory slash and burn agriculture. Classic slash and burn has a particular territory in which the farmer or farming community operates. The fallow lands are treated as part of the agricultural system, and only rarely is old growth vegetation cut. In contrast, migratory slash and burn cuts whatever vegetation

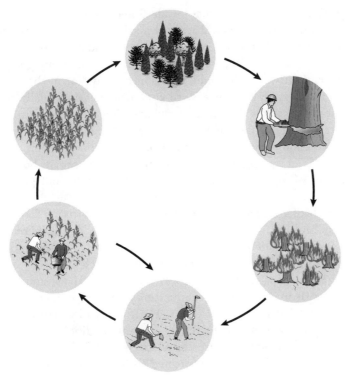

Figure 2-9 Diagrammatic representation of the slash and burn system.

is on the land, frequently old-growth forest, farms the land as long as it is productive, and then moves on to new land, with little notice of what had already been farmed.

A characteristic of classic slash and burn that is frequently not fully appreciated is the care and attention given to each phase of the agricultural cycle. For example, in experiments I was involved with on traditional slash and burn techniques in Mexico, I worked with several local farmers experienced in the techniques. After we had slashed the vegetation, we had to wait for it to dry sufficiently to be burnable. Each day the farmer would talk about the slash, obviously watching the drying process quite carefully. The point was that the slash had to be sufficiently dry before the rains came, for the first rainfall would significantly reduce the effectiveness of the burning. On the other hand, if the vegetation had been cut earlier, thus assuring a very dry vegetation to be burnt well before the rains came, the time may have been sufficient for potential weeds to resprout before planting. The process of cutting the vegetation and waiting for it to dry is far more fine tuned than one might have initially imagined.

Another fascinating example of high-level management of slash and burn agriculture is the tradition of *mal monte* and *buen monte* (bad weeds and good weeds) developed in the Maya areas of southern Mexico.[12] Farmers actively observe their fallow and have a detailed knowledge of the sorts of plants that invade. Some plants are "good plants" and others "bad plants." Sometimes farmers will actually weed out bad plants from the fallow and plant good ones. Exactly what makes some plants good and others bad is not necessarily part of the local knowledge. When asked why particular plants are good, one frequently gets the general answer, "They make the soil healthy," or in the case of wild banana (*Heliconia sp.*), for example, "it is easy to cut." Legumes are usually considered good plants, and one can speculate that the underlying ecological mechanism is nitrogen fixation. Spiny vines are usually regarded as bad, probably because they are annoying when cutting the vegetation during land preparation. The important point is that the fallow itself is not just land that is left idle for nature to take its course, but is thought of as a regenerative part of the agricultural cycle, with the possibility of being either less or more healthy.

An excellent example of a *buen monte* is *Thalia sp.*, a swampy herb that forms the basis of a particular slash and burn technique formerly common in southern Mexico.[13] Extensive swamps are dominated by this species, which has enlarged underground roots that serve for carbohydrate storage. The agricultural cycle begins by cutting the soft herbaceous vegetation as the water naturally drains from the swamp during the beginning of the dry season. Substantial residual moisture remains in the soil. After the vegetation is cut, it is left to dry for 1 or 2 weeks, followed by planting maize at a relatively high density in open holes created by a pointed "dibble" stick. When the maize germinates, the whole field is set afire, the relatively cool fire passing over the tops of the germinating plants, which remain protected from the fire within the planting hole. In this way, all of the vegetation that might compete with the corn is effectively eliminated. As the corn grows, the *Thalia* begins regenerating below it, using the energy stored in the underground roots, thus choking out any "*mal monte*" that may enter the system. By the time the maize is ready for harvesting, a waist-high carpet of *Thalia* already covers the ground below the corn, thus preparing the land for next year's cycle. The fallow period is thus less than a year and includes only the wet season time that the swamp is filled with water. It is obviously a special system in that it is based on a swamp soil that is inundated each year, probably importing much nutrient material in the aquatic sector.

Many other examples could be cited in which management details of the slash and burn cycle are based on a detailed knowledge of the underlying ecological system.[14] As much as any traditional agricultural system, the slash and burn system (but not necessarily the modern migratory system which superficially resembles it) is based on the idea that agricultural production is part of the natural world and needs to be attuned to the rules of that natural world. Consequently, normally the land in fallow should be regarded as part of the land base of the system.

Even though it is frequently cited as the most primitive of the traditional systems, slash and burn farmers are excellent experimental scientists, always trying out new techniques, new varieties, new crop combinations, and so forth. Even though it is steeped in tradition, that tradition is flexible and innovation is actually part of the tradition itself.

Nomadic Herding

Nomadic herding is the other general form of extensive agriculture. Based on the husbandry of livestock, this form of agriculture is actually older than settled agriculture (as described earlier). Its form is superficially quite simple. Herds of animals, mainly cattle, goats, and/or sheep, are grazed on a piece of land until either the characteristics of the vegetation or a change in season provides a clue to move on to different pastures; however, its details may be extremely complex. Human social interactions are frequently tied to the details of the grazing system and contact with settled agriculturalists is frequently regular. The Masai of East Africa are an excellent example of this form of agriculture.[15]

The range of the Masai incorporates the extreme northeast of Tanzania and the extreme southwest of Kenya, although the present-day range is considerably less extensive than at the turn of the century. The habitat is a typical East African mosaic of grasslands, woodlands, forests, stony and rocky outcrops, and isolated lakes, with strong seasonality. Extensive volcanic activity in the region, associated with a major East African fault, makes for great variability in elevations. Higher elevations are cold and wet in the rainy seasons, whereas lower elevations are hot and dry in the dry seasons, a fact that dictates some of the migratory habits of the Masai.

The Masai live in small settlements called boma, which contain small houses made of sticks and mud (Figures 2.10 and 2.11). Each boma may contain several family groups, each headed by a single male elder and containing his wives and children.

Daily activities revolve around the stock: cattle, goats, and sheep. In the morning, the women get up before sunrise and get the fire going (it is left to smolder all night) before they milk the cows. The process of milking involves getting the calves out of the corrals, allowing one teat to be suckled by a calf to get the milk flowing, and then milking from the other teats. For breakfast, milk is served in liquid form or is mixed with flour to make

Figure 2-10 Typical Masai boma in Kenya.

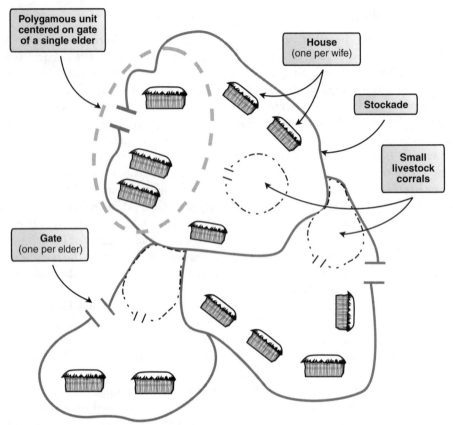

Figure 2-11 Layout of a typical Masai boma.

Adapted from: Homewood, K. M. & Rodgers, W. A. *Maasailand ecology: Pastoralist development and wildlife conservation in Ngorongoro, Tanzania (Cambridge studies in applied ecology and resource management).* Cambridge, UK: Cambridge University Press, 1991.

porridge for the elders and the children. After breakfast, the boys take the stock out of the boma to whatever grazing site is scheduled for that day. Typically, three different groups of livestock are pastured separately: calves and sick cattle, adult cattle, and small stock (goats and sheep). Frequently, the stock from various units within the boma, or even from different bomas in the general vicinity, is pooled for herding during the day. Depending on time of year and other conditions, the boys return with the livestock in the evening or sometimes even after dark, whence the cows are again milked by the women.

The social structure of the Masai is tightly attuned to the needs of the herding system. Although any attempt to summarize this structure in a paragraph is doomed to oversimplification, a quick overview is necessary to understand how the agroecosystem functions. There are five classes of people in the traditional Masai system: young boys, "warriors," elders (men), girls, and women. Women are effectively the backbone of the culture, responsible for feeding all residents in the boma and tending to the stock during those times when they are not foraging. Although they have little political power, they collectively ensure the survival of the culture not simply through reproduction and tending infants and younger children, but in tending the stock in the boma. Because the measure of wealth of a boma or a clan group is always measured in the number of animals they own, women's role is far more important than is usually acknowledged. Young girls are effectively the helpers of the women, contributing labor to the entire social enterprise while they rehearse for their adult roles. The male classes are based on "age-sets." Young boys, until about the age of 15 years, are responsible for the day-to-day herding of the stock. Warriors, more properly referred to as *murran*, are more or less the adolescent male class and range in age from about 15 to the late 20s. The *murran* live outside of the bomas in camps known as *manyatta* and spend much of their time moving from boma to boma, helping with the herding and interacting with the young girls, serving as their protectors and frequently their lovers. It is evident that the *murran* provide a certain social cohesion to the entire culture, as they are the major form of regular communication among far flung bomas. Masai traditions of comradeship and sharing are strongly enforced during the period in the *manyatta*, with the older *murran* acting as tutors for the younger ones. Somewhere during their 20s, a particular age-set of *murran* disbands, and the former *murran* now become elders. This is the time to begin taking wives and setting up one's own boma. As an elder, one has responsibility for one's own stock herd, the beginnings of which come as a dowry with the first wife. All social transactions, from taking new wives to arranging marriages for daughters, from trading stock to slaughtering stock, from where to place temporary bomas to when to migrate to the rainy season pastures, are the responsibility of the elder. A certain flexibility is incorporated within this structure, but the outlines are fairly general.

The dry season boma is semipermanent, usually located in higher elevations, although some are in lowland swampy areas. Each year the same boma is returned to, at least for many years running, and the dry season cycle begins. At first, there is significant forage, and it is not necessary to take the cattle too far from the boma each day. As the season progresses, however, it is necessary to take the cattle ever further to find adequate forage. Furthermore, as the dry season progresses it becomes increasingly more difficult to locate watering sources. It is not uncommon for the stock to go without water for several days running. The watering day, during these dry times, is devoted to nothing more than the trip to the watering hole, sometimes an arduous trek over inhospitable terrain. This occurs toward the end of the dry season, a time of hardship for everyone, and the time when most of the slaughtering of stock occurs.

The coming of the rains signals the time for migration to the rainy season pastures, typically lower in elevation. The dry season boma is usually located in an area that becomes wet and cold during the wet season or an area that floods when the rains come, uncomfortable for the humans at least. The important ecological fact, however, is that it is a time for which the dry season pasture areas can recuperate from intensive grazing.

This picture of migration from dry season site to wet season site, with a complex social structure based principally on the maintenance of animals and their herding, is typical of all migratory herding societies and likely reflects the outlines of early herding societies also, although the specifics vary enormously from culture to culture. From an environmental point of view, it is a system that can sustain itself in perpetuity as long as the environment and the culture are maintained in a relatively stable situation. Referring again to the basic ideas of sustainable and necessary populations, any tendency to intensify the system by, for example, establishing permanent gardens, is likely to be difficult because of the need to migrate to new pastures, either from seasonal driving forces or simply because of natural overgrazing. Any tendency to reduce the population is likely to be curtailed because of the need for herdsmen and women to tend the stock.

A final point to be made about extensive agricultural systems is the interaction among various types of systems. Although there is a tendency to look at each one as an integrated whole and a tendency toward celebration of how so many traditional systems are ecologically sustainable, it is also true that they are dynamic systems. Change is not uncommon, and even before the modern period of intensification, it is pure romanticism to think of any of these systems as Edenic and enduring. Indeed, most of them went extinct.

One of the major sources of change is interaction among systems. Although there is much to learn from the apparent ecological stability of many of these traditional extensive systems, it is also the case that contradictions frequently arise when they come into contact with one another. It is well known that contact is frequent. In the case of the Masai, for example, intertribal exchange is not unusual, where individuals take up the practice of agriculture and effectively join another tribe. This sort of interchange among systems is apparently continuous and common. Another form is sporadic and not nearly so subdued. For example, the range wars in North America were the result of contradictions between two systems: farming and ranching. Throughout history, similar confrontations have taken place and continue even today. Further analysis of this feature of agricultural evolution is, despite its importance, beyond the scope of this book.

Intensification

The Basics of Intensification

The same simple model used earlier can be applied to the process of change in extensive agriculture. For example, if the territory of a population and its basic technologies are fixed, about the only two things that can be modified in face of changing conditions are either the population density or the amount of time the land will be left in fallow (inversely related to the amount of land that will be cultivated at any time). The same variables used in the constructivist analysis of the origin of agriculture, the sustainable and necessary population, can be used here, assuming for the moment that intensification refers only to the reduction of fallow time and increase in amount of land actually under cultivation. The elementary dynamic consequences are unchanged. If the actual population is below the necessary one, there will be a tendency to increase population, whereas if the actual population is above the sustainable one, a shortfall in food will be experienced and there will be a tendency to increase the amount of production, which is identical to increasing the length of the planting cycle (reducing the length of the fallow period). These fairly obvious observations determine the way in which the agricultural system will evolve near its equilibrium point—it will be either stable, in which case a slight deviation from it will return, or unstable, in which a slight deviation from it will promote even more deviation. However, the ultimate fate of the system will depend not only on this local behavior, near the equilibrium point, but also on the way the equilibrium points them-selves are formed in a larger context, which is effectively the same as the way nonlineari-ties enter the system.

Although both sustainable and necessary population values increase as the agricultural technology increases, the previous assertion that "for some forms of primitive agriculture, the initial investment into agricultural activity probably provides only marginal increases in the size of the sustainable population, whereas after the agricultural system becomes a more important component of subsistence activities, the number of people sustainably supportable probably goes up nonlinearly" applies to the early adoption of agriculture and not necessarily to the future intensification thereof, which is to say the shortening of the fallow period. Other patterns are clearly imaginable. For example, if the background environment includes soil that rapidly loses its fertility under cultivation, conversion from fallow to production will occur more frequently, thus requiring further increases in labor, meaning that the necessary population curve has an accelerating pattern with respect to intensification (e.g., Figure 2.12b). If the background environment includes fallow vegetation that does not replenish the soil very rapidly, as the length of fallow decreases, the sustainable population falls off ever more rapidly, and a diminishing returns type of nonlinearity (e.g., Figure 2.12b). Thus, we can imagine the nonlinearities entering the process either as diminishing returns (initial fast rise followed by a leveling off) or an accelerating pattern (initial slow rise followed by a fast rise), depending on the underlying ecological conditions.

Figure 2.12 shows sample curves for the two cases that summarize the qualitative pos-sibilities this model offers (various situations can be envisioned simply by adjusting the positions and curvatures of the two curves). In Figure 2.12a, repeated from Figure 2.5, there is a general tendency toward a planting cycle of length L* and a population of P*. Any deviation from this point will be resisted by the basic dynamics elaborated previously, and the culture can be expected to stay at this point in perpetuity, at a constant popula-tion size and constant planting cycle. Any point below the lower unstable equilibrium point, however, will inevitably revert to the hunting and gathering mode. In Figure 2.12b, a society located at point P**, L** is balanced between two attractive positions, and any change in either the population size or the length of the planting cycle is likely to drive the system to one extreme or another, either a slash and burn system (with L* length of crop-ping cycle) or full intensification (i.e., no fallow, all land in cultivation all the time). When speaking of intensification, the arrangement in Figure 2.12b may usually be more likely, but that depends on exactly what sort of intensification is meant and what sort of eco-logical background is involved.

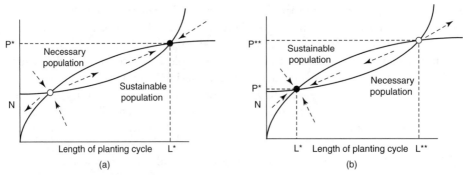

Figure 2-12 Sample curves for two qualitatively distinct patterns of dynamic change in the process of intensification, with characteristic nonlinearities added. (a) The situation in which increases in the length of planting cycle are initially labor intensive but, as less time is devoted to fallow, efficiencies of labor are adopted and the nonlinearity is a "diminishing returns" type. The sustainable population is initially not increased that much, but as less time is devoted to fallow, efficiencies of production allow for a larger increase in the sustainable population. This pattern is likely typical for the initial stages of agricultural adoption, as argued earlier, but probably unusual for intensification. (b) The alternative situation in which ever-increasing demands on labor accompany decreases in the fallow period and only modest increases in the sustainable population are achieved after an initial rapid increase with initial stages of intensification. This situation is likely more often the case in extensive agriculture in which the fallow is necessary to recuperate natural ecosystem function.

This simple view of the socioecological dynamics makes clear two underlying tendencies, the need to have sufficient labor for a given technology and the desire to produce enough but not too much food. It also shows how those dynamics could lead to the continual intensification of agriculture. It fails, however, to capture a crucial feature of the intensification process—as more land is cultivated and thus the fallow cycle is shortened, the ecological background that necessitated the fallow in the first place must be changed, or shortening the fallow cycle simply cannot work. That is, as intensification proceeds, some sort of technological change that obviates the original need for the fallow must come into being, or the process of intensification will most surely fail.

An expanded vision of the process of intensification is presented in Figure 2.13. Rather than imagining a gradual reduction in fallow period of a slash and burn system, we consider a population of flood plain farmers. Yearly flooding creates the conditions for planting crops as the floodwaters are receding, as indeed was the case in many early agricultural communities (e.g., the majority of farmland in ancient Egypt). Areas closest to the source of flooding would be first used (Figure 2.13a), but the yearly flood plain is a limited area. As more land is put into production, the population is forced further away from the river, using land that is flooded less frequently, perhaps using temporary dikes to reduce the flow of floodwaters back into the river (Figure 2.13b). As the population seeks yet more land, it will move further from the river and eventually be forced to tap feeder

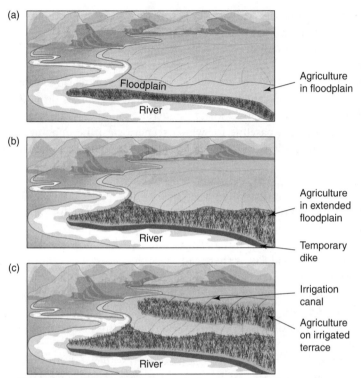

(a)

Floodplain

River

Agriculture
in floodplain

(b)

River

Agriculture
in extended
floodplain

Temporary
dike

(c)

Irrigation
canal

Agriculture
on irrigated
terrace

River

Figure 2-13 The three hypothesized stages in the evolution of irrigation systems.
(a) Using the annual floods and residual moisture from them. (b) Construction of temporary dikes to control water flow. (c) Intensive control of water using irrigation channels.

streams for irrigation (Figure 2.13c). Thus, the increased "intensity" of agriculture
(i.e., more land brought into production) is necessarily accompanied by technological
advances, from simple dikes to irrigation canals.

We can imagine the population dynamics of this situation by substituting "distance
from river" in place of "length of planting cycle" in the model presented in Figure 2.12.
The extreme right-hand side of the graph is accompanied by full-scale irrigation systems.
Exactly how difficult it might be to build dikes and irrigation canals determines the shape
of the necessary population curve. The situation in Figure 2.13a, for example, might be
the case where a significant amount of work is required to simply farm the flood plain or
build minor dikes, but adequate food for the population is easily provided without further expansion. In contrast, Figure 2.13b illustrates the situation in which simply gathering materials from the flood plain is one option that requires few hands but supports
quite a lot of people, whereas the construction of irrigation canals would require many
hands. If the population ever surpassed P** or the distance from the river ever surpassed
L**, a dynamic would be initiated in which ever more land was brought into production

by ever more people until the entire valley was under irrigation and cultivation (still not fully satisfying the population).

The model presented in Figure 2.12 is based on the simple ideas of land under cultivation expanding when a shortfall of food is experienced and numbers of people increasing when a shortage of labor is experienced.[16] Certainly the real world is more complicated. Although the complications themselves may be the key variables driving the process of intensification, nevertheless, as a sort of neutral model, considering only these two dynamic forces provides a baseline for which scenarios can be imagined and compared with actual situations. As with any model, however, it should not be taken too literally.

Examples of Early Intensification

As slash and burn agriculture intensifies, its main feature is shortening of the fallow period, going from extensive production (with a long fallow period) to intensive production (where the land is used without a rest or fallow period). The problem with intensification is that the ecological forces that made a fallow necessary do not disappear, which means that some technological advance must accompany intensification. Thus, the word intensification is nearly synonymous with technological advance, a notion that in recent years has taken on strong ideological connotations. The intimate relationship between irrigation and intensification is an excellent example of this connection (moving agriculture further away from the river requires development of irrigation technology).

Early farmers understood the connection between water and crop growth, and thus, water frequently was the focus of intensification efforts. For example, today's Hopi Indians in the southwestern United States still use the residual water stored at the bases of arroyos as cropping sites for maize cultivation, using varieties of maize specifically adapted to their generally xeric conditions. Although planting maize specifically where residual water accumulates is perhaps not much of a technological achievement, it is the likely precursor to the common practice of restricting the flow of temporary streams, thus creating pockets of residual moisture. The Anasazi of southwestern North America built rock dams across the intermittent streams of their desert home. Wadi agriculture of the Negev, probably from the famous Nabatean culture, was based on this idea, and even today the Negev is so filled with examples of small rock dams, some very ancient, that one is hard pressed to find an intermittent water course that lacks them (Figure 2.14). Thus, as agriculture intensified, one technology that early on was an aid in intensification was simply the construction of small dams or dikes that restricted water flow and created pockets of residual moisture where crops could be planted.

Moving from the utilization of small retaining structures to the digging of simple irrigation ditches would have been a small step. For example, the Sinagua people, relatives of the Anasazi, took advantage of the effluent of a flooded limestone sink, which was several meters above the water level of a nearby stream to divert irrigation channels to bring water to their fields.[17] The key to this system was that the limestone sink was perched above the nearby stream channel. Thus, the effluent, which normally flowed directly into the

Figure 2-14 Ancient retaining structures in desert washes in the Negev, Israel.

stream, could easily be diverted before it reached the stream. Even today, modern farmers use some of the old irrigation channels that the Sinagua constructed.

Ultimately, however, irrigation systems became extremely complicated, as in the case of the Moche civilization of ancient Peru[18] (by Moche I refer to a series of civilizations that occupied the Moche and nearby valleys of what is today northern Peru), occupying the extremely dry desert of northern Peru. This is one of the driest deserts in the world and is dissected at various points by rivers that drain the Andes. Tapping the rivers high in the mountains, the Moche built canals with increasingly sophisticated technology as the remarkably difficult environment periodically destroyed their previous engineering achievements. Earthquakes actually tipped canals upward so that the water would have flowed in the wrong direction. The periodic el Niño episodes washed away many canals, moving dunes-filled canals, and the highly erodible rocks of the area caused the inflows of canals to deepen continually, necessitating reengineering their slopes. In short, irrigation in this civilization was a difficult technology to adopt. Nevertheless, with extremely simple tools, the Moche people were able to do it.

Their increasingly sophisticated irrigation system began with an enormous canal known as the great trench. These canals were dug by work gangs of from 10 to 20 men using mainly stone tools, working out a labor tax in the apparently rigidly class-structured society. Wicker baskets were used to remove the dirt, and large boulders were removed by repeatedly setting fires below them and then dousing them with water, cracking them slowly from the outside and carrying off the pieces. These great trenches were dug in the mainly soft sand bases of the lower river courses and were apparently difficult to maintain because of the highly dynamic dunes of the lower basins.

The second phase of canal building was more sophisticated, drawing water from much higher in the river drainage. To irrigate as much land as possible, it was necessary to follow the contour as closely as possible. This phase was characterized by very gently sloping canals, which led to another problem. As the inlet eroded, the entrance to the canal became blocked, precisely because of the sophisticated engineering that enabled the gentle slope. The bottoms of the canals thus had to be continually deepened as the entrance to the canal eroded. Add to this the normal threats from El Niño and earthquakes. Thus, a continual building and rebuilding of canals occurred, leaving an archeological record of a complex network of canals of various ages. As these canals were continually eroded and damaged by the natural elements, the final stage was the massive construction of the intervalley canal, in which water from an entirely different river system was shunted to the necessary fields. This was a massive undertaking requiring an impressive array of engineering skills,[19] as can be appreciated by the enormous length of the canal. Note also that the need to irrigate the lands already in production required the building of this massive canal, but its building opened up yet new lands for agricultural development.

Irrigation channels were not the only route to the use of water in the intensification and technological developments in agriculture. A most interesting case is associated with the maintenance of the classic Mayan civilization in Mesoamerica. The importance of agricultural intensification can be seen in the problems associated with determining population density of the ancient Mayans with archeological methods. Because of the agricultural practices of the current Mayans, it had always been assumed that the ancient populations were slash and burn farmers, just like the majority of the contemporary communities. Given the slash and burn habit, it is a relatively easy task to compute the carrying capacity of the land. If there is a fallow period of 12 years, for example, and a cropping period of 3 years and a single family of five can be supported on a cropping land size of 1 hectare, it would take 5 hectares to support that family (after cropping the first hectare for 3 years it is left fallow—the second hectare is then cropped for 3 years, during which time the first hectare lies fallow, followed by the cropping of the third hectare for 3 years while the first hectare still remains fallow—for the first hectare to remain fallow for 12 years, the cropping cycle has to go through the fifth hectare). Using this sort of formula based on the actual productivity of contemporary Maya, anthropologists were able to come up with an estimate of the maximum that the ancient population could have been.

However, if populations are estimated by taking samples of house mounds and assuming some particular number of people per house, a direct census can be made of the ancient populations. Using this technique, archeologists estimated a population size far larger than the size estimated indirectly based on slash and burn agriculture.

The contradiction was resolved with further archeological information and new interpretations of old structures. Land forms that could not be explained were found throughout the ancient Mayan world. For example, in Belize, at an archeological site known as Pulltrouser swamp, large square shaped land masses were found arranged in rows such that they were unlikely formed by any natural force. Excavations at the site uncovered agricultural implements. At another site on the banks of the Rio Candelabra in southern Mexico, similar square shaped land forms were discovered in aerial photographs, and similar agricultural implements were found in excavations. These kinds of land forms were clearly the result of human intervention, and they certainly had something to do with agriculture. They are commonly found throughout the Mayan area of Mexico, Guatemala, Honduras, and Belize.

Further evidence came from old Spanish texts reporting on the existence of curious agricultural systems surrounding the city of Tenochtitlan, the central city of the Aztec civilization (Figure 2.15). Remnants of that system can be seen today in Xochimilco, a suburb of Mexico City. The Spanish called them floating gardens (although they do not float), and they represent a form of intensive agriculture that combines aquatic and terrestrial production. Terrestrial platforms are constructed, usually in a swampy area, such that a series of canals surrounds the platforms. Muck is dredged up from the bottom of the canals and spread out in a small area to form a seedbed. The drying, but still moist, muck is scored into small squares, and each square is planted to a crop. When the plant germinates, it is transferred to a permanent site on the platform where it grows and matures. Furthermore, aquatic plant material is collected from the canals and deposited as mulch on the platforms. In addition to the terrestrial production, fish and other aquatic organisms are produced in the canals.

Generally speaking, one of the principal problems of agriculture in tropical areas is the inability of many tropical soils to store nutrients (discussed further in Chapters 4 and 5). This system of production, known as *chinampas*, conveniently solves that problem by connecting the aquatic and terrestrial sectors. The nutrients leach out of the terrestrial sector just as they always do in these kinds of tropical soils, but instead of being lost from the system entirely, they are captured by the biological activity in the aquatic sector. Then the farmer harvests them in the form of muck and aquatic plants and recycles them

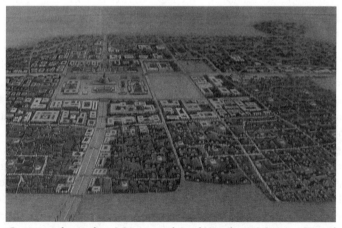

Figure 2-15   Section of mural in Museum of Anthropology, Mexico City, showing central Tenochtitlan (today's Mexico City) as it appeared to the Spanish on first encountering it. Note the extensive system of chinampas and their canals surrounding the central plaza and administrative area.

into the terrestrial sector again. The central problem of nutrient leaching from soils is thus resolved.

Throughout the Mayan region and beyond, remnants of chinampas-like production are now recognized, and there is general agreement this indeed was a major agricultural technology used by the Mayan civilization. It is a highly intensified form of production, dependent on a large and well-organized population. Canals must be physically maintained—collection of mulch and muck must be done on a strict time scale and requires large amounts of labor. It seems to be a high-yielding and sustainable system, but only with a highly organized society.

A further class of technological developments was an inevitable consequence of hydrological technology. Building an irrigation canal would hardly be efficient if the land below it remained so steep that water simply ran off the fields. The obvious solution to this problem was to build terraces, effectively leveling the land to maintain the water. Terracing was a common intensification technology in many parts of the world. Especially in mountainous regions, terracing enabled highly sloped land to come into agricultural production. In the Peruvian highlands, for example, near the Inca capital of Cusco, virtually every hillside is riddled with ancient terraces, from valley bottom to mountain top. It was certainly an incredibly intensive form of agriculture. Most important, terracing, much like irrigation, requires constant maintenance, and thus a large labor input. Most impressive in the region around Cusco are many terraces that are reinforced with stonework, an investment of tremendous quantities of human labor.

The combination of terracing and water management is common in Asian rice production. The bucolic rice terraces of Bali are testament to this ancient tradition. The rice cycle begins by draining the paddy and plowing the land, frequently using a flock of geese to eat the small plants turned up by the plow. Rice is sown in nurseries where it germinates and grows quickly to the seedling stage, whereupon it is transplanted into a newly prepared and flooded paddy. There are yet further complications that involve sociopolitical factors. The water in the entire basin must be managed centrally to insure that each terrace has water when needed. Thus, the sociopolitical or cultural system must insure water rights to all farmers concerned.[20] Indeed, this control of water in all paddy rice cultures has led to systems of centralized planning that Marx referred to as the "Asian system of production," contrasting starkly with the feudalism/capitalism axis of the European system.

In summary, early intensification of agriculture frequently involved water management and terracing. The basic population dynamics model of intensification carries with it the underlying assumption that as intensification proceeds, some form of technology must be invented and adopted to permit more land to be brought into production more frequently. The ecological and physical conditions that limited production to an interfallow period or to some smaller fraction of the land available had to be breached. This would be the pattern of intensification in all premodern agricultural systems. When we reach the modern era, new intensification technologies become important, and the rate of technological change and thus agricultural intensification increase dramatically.

Traditional Systems, the "New Husbandry," and the Dawn of Modernity

In tracing the evolution of agriculture during more recent times, it makes sense to start with the traditional systems that had evolved by the time we entered the modern age. Most evident were probably the paddy rice systems of Asia, which gave rise to extremely dense populations with a unique mode of sociopolitical organization (the "Asian mode of production") that derived from the ecological necessity of controlling water. However, many other systems existed on all continents except Antarctica and Australia. They were remarkably diverse. They ranged from simple slash and burn systems to extremely complicated ones such as the Mexican Chinampas, from dry land farming to complicated irrigation systems like those of the Moche, from those lacking draft animals to those with such animals, from almost all meat production to almost all vegetable production, and from rain forests to deserts in all areas of the globe.

Thus, as our time frame reaches the late stages of the European Middle Ages, the world was filled with an enormous diversity of agroecosystems. Nevertheless, one of those systems would become the stem that grew to dominate all of the others and transform into what we today know as industrial agriculture. Beginning in Flanders in the 18th century and migrating first to Britain and then France, the system became known to historians as the "new husbandry." This system eventually gave rise to the industrial system that dominates the world today.

Before the establishment of the new husbandry, the agroecosystems of Great Britain had been rather loosely organized affairs, with row cropping and common property pasture generating low soil fertility and the consequent need for large areas of fallow. Furthermore, pasture availability was strictly seasonal, generating a boom and bust cycle of meat availability—given the lack of forage availability in the winter, most animals were slaughtered in the fall, meaning that there was an overabundance of meat at one time and a scarcity for the rest of the year, plus a shortage of milk in the winter. The new system involved a host of particularities, but two of the most important were the introduction of the turnip and the land enclosure acts.

In 1730, Charles "Turnip" Townshend introduced the turnip to Great Britain from Flanders. The turnip was in a sense a miracle crop in the same way that potatoes would be later in Ireland. It could grow on almost any soils at all, was easy to cultivate, and provided a large yield in terms of potential calories. Rapidly it became adopted not only as a major food source for people but also as a winter food for animals, extending the time animals could be maintained alive during the winter.

The 18th century was also a time of growth in the industrial sector more generally. Chronic labor shortages in the urban factories stifled the growth of the new capitalist enterprises. As a partial solution to this recurrent crisis, the enclosure acts were promulgated, in which land that had been farmed in "common property" style was "enclosed" (i.e., privatized), and many former farmers found themselves to be owners of patches of land too small to make a living. This induced, as planned, a massive migration to the cities, thus creating the industrial reserve army that was need for the continued growth of industry.

Although the enclosures had the desired effect of creating an army of proletarians, its effect on the farming sector was severe and clearly could not have been sustained without major modification. Farmers who had normally operated with a common property attitude toward the land suddenly were faced with the prospect of a limited land base on which to grow crops and pasture animals. The solution to this problem was partly conditioned by the turnip—livestock could now be kept throughout the winter. Furthermore, the manure from that livestock could be recycled into the cropping system. Thus emerged, on these small enclosed farms, a system of rotation in which what had been previously a fallow field now grew turnips and in which manure concentrated in stables in the winter provided nutrients for spring crops.

This "new husbandry" involved a great many continually evolving techniques, many of which anticipated some of the later developments that we would see in the evolution of the industrial agroecosystem. The key ingredient, from an ecological point of view, however, was the integrated use of animals and rotations.[21] The most famous system was the so-called Norfolk four-course rotation in which turnip or some other root crop was followed by barley, which was followed by a legume (frequently clover) and finally wheat. Sheep and sometimes cattle were usually pastured within the system and maintained in stables during the winter. Most of Central Europe used some sort of three-field system usually with two cereals (e.g., oats and wheat) alternating with a legume. Key to the development of this new husbandry was a rise in demand for milk and meat, stimulated to ever greater extents by the growth of the factory system and concomitant growth in consumer demand (e.g., it has been estimated that by 1840 the average British citizen consumed 87 pounds of meat per year).[22]

The land was fully intensified with rotations obviating any need for fallow. The principles involved are among the principles used by today's organic farmers—rotating fields to avoid pest and disease buildup, incorporating animal manures, and the use of a legume as part of the rotation to replenish nitrogen. Indeed, it might be said that the evolution of agriculture through the new husbandry was an evolution using only what would be referred to today as organic methods—agriculture had been organic for thousands of years. The modern industrial system would come on the heels of the industrial revolution, but only as a later addition.

The industrial revolution marked the beginning of the "modern era." Its most significant effect was perhaps more social than technological, although certainly the application of technological solutions to problems of human existence was not insignificant. However, the change in work, the change in attitudes toward life in general, and the changes in general human social organization were truly revolutionary. Virtually all aspects of human existence came under the domination of capitalist rules. Dispersed artisans were replaced by workers organized in factories, products that were formerly valued based on how useful they were became valuable based on how much they could be traded for, and money took on a new significance, being an input into a process the output of which was more money. In short, the industrial revolution meant capitalist dominance of human activities. This transformation is discussed later in the text, but here some of its major features are outlined as they apply in a very general way to agriculture, especially focusing on how the agroecosystem was resistant to being incorporated into the sociopolitical changes emerging in the industrial urban centers.

A convenient way of visualizing the underlying basis of this urban industrial transformation is in the changing role of money, from an instrument of exchange to an input (discussed in more detail in the appendix to Chapter 7). Precapitalist economies were based on the principle that people produced goods (P = product) so as to sell them to get money (M), which they would use to buy goods that they did not produce. Thus, the cycle product-money-product (PMP) dominated. However, interspersed within this society were individuals who operated according to different rules, namely the merchants who used money to purchase products for the purpose of selling those products at a profit. This cycle, the money-product-money (MPM) cycle, was not dominant but existed in the margins of precapitalist societies (e.g., among traveling merchants). As this "mercantilism" came to dominate the economy, the dominant rules driving the economy began to change, and finally, with the full evolution of capitalism, the MPM system came to dominate. That is what we see today, and to some extent, it is part of the problem, as discussed in future chapters.

As the MPM system comes to dominate, products tend to be seen in terms of their value for exchange rather than any intrinsic utility they may have had. This has led to significant problems, both social and environmental, of which one of the most significant with regard to agroecosystems is speculation. An assumption of the MPM cycle is that the second M should be larger than the first M. Because the only value contained in P is its ability to fetch a larger price than that for which it was purchased, guessing what its future value might be is an obvious thing to do. However, others are also guessing, and purchasing the product depends on what others might be guessing. When the system functions well, large numbers of people evaluating a product provide a reasonable estimate as to what that product will be valued in the future, and according to traditional economic theory, the value will tend to a "natural" level. Yet sometimes a product is valued solely based on what is thought people will pay for it, ignoring any intrinsic worth that it may have. This tendency is part of the capitalist system and normally functions at least well enough that participants do not feel terribly cheated. But it also can get out of hand.

The archetypal example was the famous episode of buying and selling tulip bulbs in 17th century Amsterdam.[23] In 1634, particular varieties of tulips came to be highly prized. Foreseeing a burgeoning market for tulip bulbs, especially some of the more obscure varieties, merchants began bidding ever higher prices for these bulbs, obviously not based on any inherent value of the bulbs themselves, but on what some merchants thought other merchants would pay in the future. Prices rose dramatically. No one could see where the prices would stop rising, and with such an expectation, it made sense to pay as much as the market demanded, which of course made the market demand higher prices. Within a couple of years, some bulbs were auctioned for the price of a mansion, and fortunes were made buying and selling these virtually worthless items. Then, in 1637, it apparently dawned on someone that perhaps tulip bulbs were going to begin losing their value, thus generating a rush to sell. Within the year, tulip bulbs lost almost all of their value, and many people lost their fortunes. Speculative euphoria, as John Kenneth Galbraith called it, is a continuing characteristic of capitalist economies,[24] perhaps one of those famous seeds of destruction so frequently cited.

Part of establishing MPM as the principal economic mode is the commodification of as much human activity as is possible. This was especially important for agriculture and remains important today. If we assume that society's prime purpose is to make sure that the second M in the MPM cycle is larger than the first, in all spheres of human activity, it is obvious that we must make P tradable (the favored terminology in 2009 is "monetarize"). If some products exist as a consequence of human activities that are intrinsically unmarketable, they represent a barrier to this central goal. For example, a farmer who controls pests through a specific planting design rather than a purchased pesticide is not contributing to the profits of the pesticide manufacturer and thus to the general economy. No commodity is produced, thus there is nothing to buy and sell. Granted, the farmer as an entrepreneur may be able to gain more in his or her own transaction, but this assumes that he or she is a key participant in the MPM mode of action. The dominant participants, the industrial companies that manufacture the pesticide, get nothing.

This generalized path, so clearly followed in industry, was severely constrained in agriculture. Agriculture was limited by its articulation with natural processes. Although it was within human abilities to substitute all of the textile weavers with workers concentrated in large textile mills, all wheat producers could not be brought into a wheat factory. They remained scattered across the landscape, a condition necessitated by the nature of agriculture. An illustrative futuristic entitled "The farm beginning to approximate more closely to a factory" (see Figure 2.16) optimistically portrayed the penetration of factory ideals into agriculture. Despite hopeful projections of 19th century futurists, the penetration of capitalism into agriculture was distinctly less rapid than its penetration, or rather complete takeover, of industrial production.

Figure 2-16 The farm beginning to approximate more closely to a factory.

This penetration has been characterized as having two aspects: appropriationism and substitutionism.[25] On one hand, the inputs into agricultural production were gradually appropriated by capital (e.g., soil fertility became provisioning of fertilizer). On the other hand, the outputs of agriculture were gradually substituted with alternative products (e.g., ketchup rather than tomatoes). Indeed, the history of agriculture in the developed world during the past 2 centuries is a history of appropriation and substitution in this sense, such that agriculture today bears only a vague resemblance to agriculture of the 18th century in both technological and social terms. The distinction between agriculture and farming, as characterized by Lewontin ("Farming is growing peanuts on the land; agriculture is making peanut butter from petroleum"), would have made little sense at the beginning of the 19th century, yet today is a key conceptual device for understanding food and fiber production (as discussed in more detail in Chapter 8).

Endnotes

[1]Lewontin, 2001, 2003.
[2]Olding-Smee et al., 2003.
[3]Many uses of the word constructivism, from philosophy to sociology, exist. Here I use it only in the sense of an organism constructing the environment to which it must respond.
[4]Goldberg, 1990.
[5]Berkenbusch and Rowden, 2003; Flecker, 1996; Jones et al., 1997.
[6]Vandermeer, 2008.
[7]Lee, 1972a, 1972b.
[8]Rival, 2002.
[9]Boserup, 1965.
[10]Information in this section is largely derived from Flannery, 1973.
[11]Tinker et al., 1996.
[12]Chacon and Gliessman, 1982.
[13]The detailed description of this system was provided to me by Steven Gliessman.
[14]Useful references on slash and burn agriculture include the following: Uhl, 1987; Juo and Manu, 1996; Kleinman et al., 1995.
[15]Homewood and Rodgers, 1991.
[16]Boserup, 1965; Minc and Vandermeer, 1989.
[17]Pike, 1975.
[18]Billman, 2002.
[19]Ortloff, 1988.
[20]Lansing, 2007.
[21]Timmer, 1969.
[22]Grantham, 1978.
[23]Dash, 2001.
[24]Galbraith, 1994.
[25]Goodman, Sorj, and Wilkinson, 1987.

Chapter 3

Competition and Facilitation Among Plants: Intercropping, Weeds, Fire, and the Plow

Overview

Agriculture always involves biomass production from plants, whether harvesting the direct products of photosynthesis (e.g., leafy vegetables) or products several processing steps removed (e.g., grain-fed beef). Without exception, the products of photosynthesis involve plants interacting with other plants, whether in the common traditional pattern of combining a legume with a grass in intercropping or the perennial problem of controlling the negative impacts of unwanted plants (frequently called weeds) or the largely forgotten art of managing the mix of available food for pastured animals. Today much of the world, beholden as it is to the industrial system, ignores the complexity of these interactions, with monocultures having replaced most intercropping despite their well-known pitfalls, with herbicides replacing the plow and the machete even though they harm far more than only the weeds for which they were intended, and with feedlots robbing the world's grain stores, eliminating the need for any detailed understanding of the community structure of the complex assemblage of plants in natural pastures. Because the future is increasingly seen as incorporating now-devalued but previously widely appreciated techniques, the basic ecological principles involved in these sorts of issues will soon come to be seen as important, if not essential. Consequently, this chapter begins with a background in the theory of plant competition and facilitation and how it relates to intercropping, agroforestry, rotations, and especially weed management. Because early weed control was based on the fundamentally nonlinear ecological idea that crops should be given a "head start" in the competitive game with weeds, the plow was invented independently many times. The plow was undoubtedly one of those pivotal inventions that not only led to many other inventions, but also to a mindset that saw mechanization as a fundamental goal of agricultural development. Thus, this chapter ends with a historical survey of the developments in mechanization that eventually emerged from that pivotal innovation.

The Basic Process of Energy Capture in Photosynthesis

At its most basic level, ecology is all about energy—its capture, allocation, and dissipation. In the modern world, the capture and allocation of energy is automatically read as something along the lines of maintaining a strong military to secure the transport lanes for oil tankers and protect the oil fields themselves. The sort of energy we get by burning the fossil fuels that stored energy hundreds of millions of years ago is not really the place to begin. Rather, we begin with the question of how that fossilized material came to have all that energy stored in the first place, a question as basic to agriculture as it is to international politics.

Three nutrients—hydrogen, oxygen, and carbon—along with nitrogen, form a kind of gang of four that is essential for proper growth of plants. Because nitrogen is mainly absorbed directly through plant roots from the soil, we treat it in detail in a future chapter. Here the concern is with the other three of that quartet. Hydrogen, oxygen, and carbon, along with the energy that comes from the sun, make up the raw materials of the most important chemistry ever invented—photosynthesis, or the capture of usable energy from light.

It all begins when a chlorophyll molecule becomes excited by a photon. Actually, the chlorophyll molecules occur as small bundles attached to the membrane structures, the thylakoid membranes, within the chloroplast of the plant cell. A photon energizes an electron in one of the chlorophyll molecules, and that electron passes on its energy to successive electrons in successive chlorophyll molecules until it arrives at the so-called reception center of the bundle of chlorophyll molecules. At that point, it gives up this excited electron to enter into a transport chain, eventually losing the burst of energy that it captured from the photon to the elements of the transport chain, ultimately resulting in energy-rich ATP. This is the first step in the photosynthetic system. It must take place in the presence of light (where the photon comes from), and its only product is energy, which is to say ATP and other energy-storing molecules.

The next step is most conveniently thought of as a completely distinct step, the actual harvesting of the carbon dioxide. Within the chloroplast, the thylakoid membranes are embedded in the "stroma," which is the matrix that contains the membranes. Although the chlorophyll molecules are attached to the thylakoid membranes and thus the capture of the sun's energy occurs there, all of the biochemical machinery involved in the harvesting of carbon dioxide is located in the stroma. Thus, the thylakoid membranes keep feeding the energy for the biochemical processes into the stroma where the actual conversion of carbon dioxide occurs.

The first biochemical step in the CO_2 conversion is extremely important. One molecule of a sugar-phosphate, commonly referred to as RuBP (Ribulose 1,5-biphosphate), comes in contact with one molecule of carbon dioxide in the presence of a very special enzyme, rubisco (RuBP carboxylase), to form two molecules of another sugar-phosphate, PGA (3-phosphoglycerate). The accounting here is important. The RuBP molecule contains five atoms of carbon, whereas the PGA molecule contains three. Thus, the two PGA molecules together contain six carbon atoms, obviously five from the RuBP and one from the CO_2. Of course, the whole reaction is driven by the energy from the ATP generated in the nearby thylakoid membrane by the chlorophyll harvesting the energy from the photons that come from the sun.

Rubisco, a special enzyme, is probably the most abundant enzyme on earth, comprising fully 15% of all the protein mass in the chloroplasts of all green plants. It is also an unusual enzyme because it not only catalyzes the conversion of RuBP into PGA, but may, under conditions of excessive oxygen (relative to the amount of carbon dioxide), catalyze the transformation of RuBP into an entirely different compound, concomitantly *producing* more carbon dioxide, the reverse of its "normal" function of *harvesting* carbon dioxide. This process is called "photorespiration" and effectively interferes with what we normally regard as the function of photosynthesis, harvesting the CO_2 from the environment. The reason rubisco has this dual function is not completely understood, but probably has something to do with it being an evolutionary anachronism, having first evolved in an environment almost completely lacking in oxygen.

The PGA that results from rubisco catalyzing the reaction of CO_2 with RuBP enters the Calvin cycle, where it is converted into glucose, the ultimate product of photosynthesis, the raw material for other compounds, and the storehouse of energy.

This story of carbon dioxide harvesting by the conversion of a five-carbon molecule and carbon dioxide to two three-carbon molecules was more or less the standard story, thought to be universal until the 1960s. In 1966, two biochemists, M. D. Hatch and C. R. Slack, discovered that many tropical plants (maize, sugar cane, and others, especially tropical grasses) have evolved a different biochemical mechanism of harvesting the CO_2. This new system harvests the CO_2 before it reaches the site of rubisco activity. Rather than using RuBP to harvest the CO_2, this alternative cycle uses a three-carbon molecule, PEP (phosphoenol–pyruvate), and a different enzyme (i.e., not rubisco) for the initial harvesting of CO_2. This reaction creates a four-carbon compound that is stored in the mesophyll cells as malate (a four-carbon molecule). Then the malate is transferred to the bundle sheath cells, where it is split into pyruvate and CO_2. The CO_2 then enters into the regular Calvin cycle. Thus, the malate stores the CO_2 and delivers it in great quantities to the site of the Calvin cycle, making the likelihood of photorespiration very small (because the CO_2 overwhelms whatever oxygen is there).

Because the first product of CO_2 harvesting is a four-carbon compound rather than the three-carbon PGA, this type of photosynthesis is known as C_4 photosynthesis and is contrasted with the more primitive C_3 synthesis in which the CO_2 is brought directly from the outer environment to the site of the Calvin cycle.

A third mode of photosynthesis was discovered earlier than C_4 (in the late 1940s) even though it was eventually understood that C_4 modality was involved. This form was first discovered in the plant family Crassulaceae and thus became known as CAM (Crassulacean acid metabolism). In this form, the stomata open during the night (other plants have the stomata generally open during the day) and harvest the CO_2 using a C_4 technique and then break down the malate during the day when the stomata are closed. Once again, as it enters the Calvin cycle, it faces the possibility that rubisco will generate photorespiration, but because the malate continues supplying large quantities of CO_2, there is always a local abundance of CO_2 compared with oxygen. Although C4 photosynthesis separates the initial harvesting of CO_2 physically (mainly harvesting the CO_2 in the mesophyll cells and entering the bundle sheath cells for the Calvin cycle), the CAM method separates that initial harvesting in time, harvesting CO_2 at night and then entering the Calvin cycle during the day when the stomata are closed.

The role of stomata is crucial. The plant faces a fundamental contradiction regarding the functioning of stomata. On one hand, it is advantageous to have the stomata open as much as possible to insure the continual influx of CO_2 into the leaf. On the other, open stomata allow for water loss. This is especially problematical for C_3 plants because having closed stomata means that harvesting (within the mesophyll cells) of CO_2 insures the continual reduction in CO_2 relative to oxygen and the consequent switch of rubisco to an oxidizing agent and increased levels of photorespiration. Plants using the C_4 or CAM pathways do not face this problem. In coffee (a C_3 plant), for example, it has been shown that stomata stay open at low light (and presumably moist) conditions, whereas they close at higher light conditions, thus creating a deficit of CO_2 at higher light conditions, resulting in higher levels of photorespiration.

The ultimate response of photosynthesis to light can be visualized as the combination of two variables operating simultaneously: energized chlorophyll molecules and CO_2 concentration. As light increases, the amount of energized chlorophyll increases, but for C_3 plants, the relative concentration of CO_2 (within the leaf tissue) decreases. These two functions are illustrated in Figure 3.1 for both C_3 and C_4 plants.

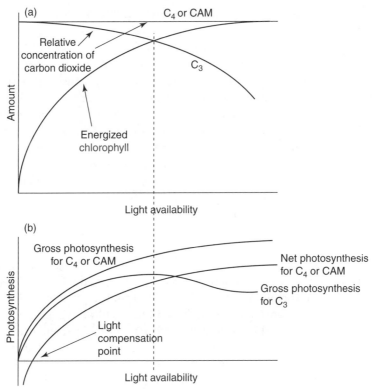

Figure 3-1 Basic processes involved in photosynthesis. (a) The two basic elements involved in the photosynthetic process, amount of energized chlorophyll and the relative concentration of carbon dioxide, as a function of light intensity. The CO_2 concentration profile changes as a function of light intensity, for a variety of reasons, in the case of C_3 plants. (b) The resultant photosynthesis production as a function of light availability. Gross photosynthesis accounts for only the carbon fixed, whereas net photosynthesis subtracts the necessary respiration from the equation.

Analyzing photosynthesis from this point of view is explicitly restricted to the process as it happens at the level of the single leaf. For most agricultural applications, we are not concerned with the leaf, but rather the whole plant, indeed with populations of plants. Once we begin thinking of populations of plants, the fundamental process of competition becomes important, and the availability of light is a consequence not only of the amount of solar radiation available, but of the positioning of the leaves of other plants, which is to say competition for light. Naturally, competition for resources occurs not only for light but also for nutrients from the soil. Before analyzing how this process relates to the structure of agroecosystems, we delve into some of the basic ideas associated with the way plants use resources in typical agroecosystem contexts.

The Theory of Plant Competition and Facilitation

Fundamental Ideas

Competition among species of plants is, at its base, no different from other cases of inter-specific competition. Indeed, the development of the competitive exclusion principle, Gause's principle, was a key event in the development of modern ecology and was inclusive of, but not exclusive to, plant competition. In recent years, the idea has received a great deal of criticism, and its original formulation has been shown to be highly dependent on some unreasonable assumptions. Nevertheless, as an historical event, one can hardly overplay the significance of the idea. The following presumes a familiarity with the basic principles of ecological competition. Appendix 3.A (at the end of this chapter) presents the basic theory of competition in a simplified form, specifically with reference to those aspects crucial to agroecosystems. If you do not already know what Gause's principle is, or what indeterminate competition is, proceed to Appendix 3.A and read it before proceeding with this chapter.

In 1944, a special meeting of the British Ecological Society was held in which the proposition that "no two species could live together permanently if their niches were too much alike" (Gause's principle) was hotly debated.[1] From their vast experiences, notable ecologists such as Arthur Tansley, David Lack, and Charles Elton argued that this was certainly a reasonable principle. Indeed, much of what followed in theoretical community ecology was conditioned by the seeming consensus that was achieved at this meeting. Today it is almost taken for granted that a major (frequently *the* major) problem in understanding any biological community is understanding how (or even whether) the niches of particular species are divided, tacitly assuming that they certainly must be divided if the species are living together.

Worth noting are the dissenters at that 1944 meeting. The great J. B. S. Haldane raised an issue that still has not been addressed—that of close species sharing diseases and the dynamic consequences thereof. Perhaps most significant, however, were the comments by a little-known (at least today) insect ecologist whose byline was Capt Diver. He noted that the realities of the world included enormous complexity that would almost automatically call into question the simplistic assumptions necessary to formulate the process of competition in the mathematical way of Lotka and Volterra or to correspond to the test tube experiments of Gause. In modern terminology, Diver's exception might be stated as follows: "To what extent is the competitive exclusion principle a consequence of the linear assumptions that go into the original equations and thus a model-bound and uninteresting principle?" Ayala[2] and later Gilpin and Justice[3] suggested an answer to this question with their experiments on two species of *Drosophila*. Much later, two very important theoretical papers answered with a resounding yes the modern form of Capt Diver's interrogation. First Levins[4] and later Armstrong and McGehee[5] showed with rigorous mathematical arguments that indeed the entire idea of competitive exclusion based on common resource use was based on the linear assumptions of the original Lotka Volterra equations, as further discussed in Chapter 7.

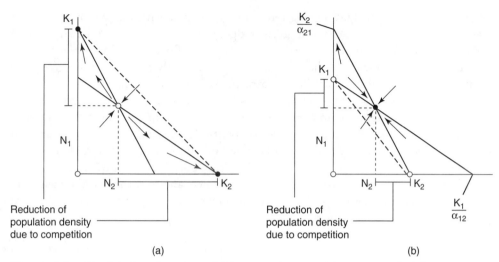

Figure 3-2 Graphical explanation of the competitive exclusion principle (if necessary, refer to the Appendix 3.A at the end of this chapter for a refresher in competition theory).

The competitive exclusion principle in its most elementary form has quite a simple graphical interpretation. In Figure 3.2, the two most important cases are illustrated: the cases of competitive exclusion in Figure 3.2a and of competitive coexistence in Figure 3.2b. An equilibrium point occurs where the isoclines cross, as always. If a line is drawn between the two carrying capacities (the dashed line between K_1 and K_2 in Figure 3.2), the position of the equilibrium point either above or below that line differentiates between coexistence and exclusion. In Figure 3.2a, the point falls below the line connecting the two carrying capacities, and exclusion will visit one of the species. In Figure 3.2b, the point falls above the line connecting the two carrying capacities, and both species will coexist in perpetuity.

Furthermore, the underlying reason for exclusion or coexistence can be readily seen on these graphs. The reduction from carrying capacity to the equilibrium point is illustrated on the two axes for both Figure 3.2a and 3.2b. In the case of exclusion (Figure 3.2a), the reduction has been relatively large, whereas in the case of coexistence, the reduction is relatively small. The principle then can be qualitatively stated that if competition is too large, exclusion will occur, and if competition is relatively weak, both species can live together in perpetuity, despite the fact that they are competing.

This idea, as indicated previously, is frequently related to the idea of the ecological niche, which itself has a long history in ecological thought. If two species have very similar niches (or very similar requirements for survival and reproduction), they will likely have large competition coefficients, and extinction will therefore occur. If their niches are distinct, they will likely have small competition coefficients, and coexistence will therefore occur. Fish do not compete with mice even though their ranges plotted on a map overlap.

Their niches are obviously different. Yet fish may compete with tadpoles because both may eat the algae in the lake, although the tadpoles may eat smaller items than the fish and therefore their niches are not exactly alike. However, if two species of mice eat exactly the same food, live in exactly the same habitat, have exactly the same nesting requirements, and so forth, they are likely to compete intensely, given that their niches are so similar, and one or the other species would likely be excluded from the region. Going from the fish/mouse competition (very low competition) to the fish/tadpole competition (intermediate competition) to the mouse/mouse competition, the competition coefficients go from zero to some high value, and the likelihood of competitive exclusion likewise goes from zero to some high value. To the extent the niches of two species are distinct, the likelihood that competitive exclusion will occur goes down, and to the extent the niches are the same, the likelihood goes up. This phenomenon is frequently summarized as "no two species can occupy the same niche."

A major experimental challenge to the competitive exclusion principle came from a classic experiment of Ayala, alluded to previously. Two species of fruit flies, *Drosophila serrata* and *Drosophila pseudoobscura*, were cultured together in bottles with simple medium. When cultured in monoculture (a single species), either species approached a constant population density, which is reasonably assumed to be that species' carrying capacity. When cultured together in the same bottle, both species persisted for a long enough period of time that it could be assumed they were in a state of stable coexistence. The three points—the carrying capacity of *D. melanogaster*, the carrying capacity of *D. simulans*, and the equilibrium point where the two species coexist in an apparent stable equilibrium—are graphed, along with what must be the isoclines, if the linear theory is true, in Figure 3.3.

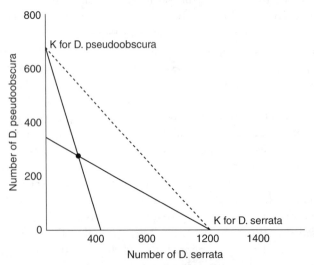

Figure 3-3   Results of Ayala's experiment. The equilibrium point is below the line connecting the carrying capacities, indicating an unstable solution.

The experiment clearly leads to an important contradiction. According to the theory, if the equilibrium point is below the line connecting the two carrying capacities, the system must be unstable. Yet here is a real system that is evidently stable, even though the theory applied to the data from the experiment says the opposite. Gilpin and Justice, as cited previously, resolved this contradiction with the simple observation of what the theory requires in a qualitative sense. They noted that the intercept of one isocline must be above the carrying capacity of the other species for both species to coexist indefinitely. Although it is impossible to make this happen if the carrying capacities and the equilibrium point are on the linear isoclines (as in Figure 3.3), the requirement for coexistence could be met if the isoclines were allowed to bend, which is to say, if we relax the linear assumptions of the classic theory. The isoclines themselves divide the space into the various sections where the two species increase or decrease (see Appendix 3.A if this is confusing). It is thus easy to see how bending isoclines can reconcile the theory with the experiment (Figure 3.4).

Figure 3.4 shows that the curved isoclines (nonlinear competition model) indeed can account for the stable system. Since this work, the idea of a simple linear approach to interspecific competition has been rejected as overly simplified. More complicated approaches were to develop later, yet even today, the theoretical aspects of interspecific competition are hotly debated. Some of this new, nonlinear approach (anticipated by Capt Diver in 1944!) is discussed in later chapters.

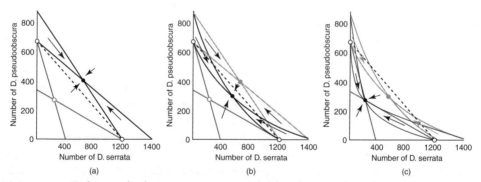

Figure 3-4   Relaxing the linear assumption, Gilpin and Justice (1972) drew isoclines that allowed coexistence of the two species, *Drosophila pseudoobscura* and *Drosophila serrata*. (a) Original nonviable isoclines shown in grey and theoretical isoclines shown with black lines, along with arrows indicating stabilizing vectors to the stable equilibrium point shown with a small black circle. (b) Theoretical stable isoclines (darker grey lines) bent to become nonlinear (black curves), taking with them the stable equilibrium point (because the isoclines define the dynamics). (c) Final state in which the theoretical stable isoclines have been bent enough to correspond to the original equilibrium point, but now making it a stable point through the curvilinear modification.

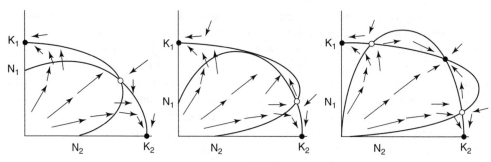

Figure 3-5 Another possible result of adding nonlinear elements to the classical competition equations. Here it is possible to generate a system with five alternative equilibrium points (right-hand graph).

Although not usually regarded as an important topic, the reverse of the Ayala/Gilpin conundrum is certainly possible. The isoclines might very well bend outward such that an equilibrium point that would normally signal coexistence actually represents an indeterminate state, as illustrated in Figure 3.5.

It is relatively easy to demonstrate that the sort of competition in which plants engage is very likely to generate such convex isoclines (see Appendix 3.A). The basic idea is that competition is likely to be quite symmetrical when plants are small, but as they become larger, because light comes from a single direction, the response of one species to being shaded by another will be highly nonlinear. Thus, the isocline, which may look linear when the biomass is very small, changes dramatically when one species becomes overtopped by another. In terms of weeds and crops competing with one another, such a nonlinear response is very likely, especially when dealing with annual crops. Thus, we can conclude that weed/crop competition is likely to be of the indeterminate variety when dealing with annual crops. If it is of the indeterminate variety, the most important issue is where the system begins—a small advantage for one species could translate into exclusion of the other species, a point to which we return.

The kinds of nonlinearities illustrated in Figure 3.5 are suggested by yet another framework in which competition theory has been formulated, that of the response/effect paradigm.[6] In a forest, for example, some trees cast very dense shade under which it is difficult if not impossible for other species to tolerate. Other species do not cast such a dense shade. Thus, from the point of view of a seedling seeking to grow in the understory, the first species "effects" a highly negative environment on it, whereas the second, lower shade-casting species does not. Clearly one species is a strong effect competitor in the sense that it strongly modifies (effects) the environment in which other species must survive. On the other hand, some species are capable of growing even in relatively dense shade, which is to say, they can grow or "respond" well to the effects of competition from others. Other species are not able to respond with positive growth to the shade created by the strong effect competitors. Taking a corn and bean intercrop as an example, one can say simply that the corn modifies (effects) the environment in which the bean must survive, and the bean responds to that modified environment. Reciprocally, the bean modifies (effects) the environment in which the corn must survive, and the corn responds to that modified environment.

This formulation of the process makes it clear that competition is not the only process that is operative. In the specific case of corn and beans, for example, the beans may *positively* effect an environment that has more nitrogen in the soil, thus "facilitating" the growth of the corn, whereas the corn, by creating a stalk that the bean may use as a trellis, "facilitates" the growth of the bean. It is thus clear that facilitation is the other side of the coin of competition, similar to the general idea of niche construction or ecological engineering (see Chapter 2).

The classic formulation treats competition as a complete abstraction. That is, we presume that two populations are competing, usually implying that there is some quantitative way of expressing the intensity of that competition and then infer consequences such as competitive exclusion, competitive equilibrium, or indeterminate competition (see Appendix 3.A for further explanation). It is as if we choose to determine that one person is a better athletic competitor than another without stipulating what sport is involved. Although it sometimes makes sense to think of "athleticism" as a phenomenon in itself, for many situations, we really need to know what is being contested during the competitive process, which is to say, what the mechanism of competition is. Thus, we commonly reduce the level of abstraction by saying that competition is "for" some resource. The word resource then becomes a new abstraction, and following the logic of reductionism, we then ask what the particular resource is that is under contention. As is always the case, the difference between the phenomenological and mechanistic approach is one of perspective. Nevertheless, in ecological competition, we usually think of a phenomenological approach when we speak of competition coefficients (quantitatively, how intense is the competition) and a mechanistic approach when we explicitly involve the resources under contention in the competitive process.

In thinking of mechanistic competition, there are generally two ways of formulating the process, depending on the underlying nature of the resource(s) involved. On one hand, some resources behave as biological objects (e.g., lions, zebras are resources), whereas on the other hand, resources are not autocatalytic (nitrogen may be removed and added to the environment independent of its own concentration). Thus, the differential equation we begin with to describe an autocatalytic resource (zebras) is the exponential equation

$$\frac{dR}{dt} = rR \tag{3.1}$$

while the one we use to begin with to describe a nonautocatalytic resource is

$$\frac{dR}{dt} = r \tag{3.2}$$

When speaking of plant competition, resources are usually of the second variety, and the parameter r is usually referred to as the resource supply rate rather than the intrinsic rate of natural increase (to which it is referred when in the exponential equation). In effect, when referring to the process of interspecific competition, the rate of increase of the zebra population from the point of view of lions is equivalent to the nitrogen supply rate from the point of view of a maize plant.

Intercropping: Applied Plant Competition

Especially in tropical areas of the world, it is common to see various types of crops grown in association with one another, as it apparently has been for quite some time (Figure 3.6). Indeed, the massive expansion of monocultures associated with the industrial system has little to do with the economic or ecological advisability of growing things in monoculture and almost entirely conditioned by the needs of the technical inputs that have become so much a part of the industrial system, as already discussed in Chapter 2. Nevertheless, little doubt exists that in the future, as systems become more ecologically sophisticated, intercropping[7] and agroforestry are likely to be more important components of overall productive systems. It is thus useful to explore the way in which plant competition and facilitation theory apply to them.

Exactly what are the ecological benefits thought to accrue from the practice of intercropping? The hypothesized ecological benefits have been divided into two categories: reduced competition (or the competitive production principle) and facilitation.[8] In the

Figure 3-6 Wall painting from ancient Mayan temple illustrating farmer sowing seeds below the "tentacles" of the storm deity. In back of him are two plants, one a corn plant the other a bean plant, presumably growing as an intercrop and illustrating the process of intercropping as an ancient tradition.

case of the competitive production principle, it is thought that two different species occupying the same space will use all of the necessary resources more efficiently than a single species occupying that same space, much as is sometimes believed to happen in natural ecosystems. Recall the basic ideas of interspecific competition. Plotting the density or biomass of one species against that of the other species and indicating where the carrying capacities are, we expect competitive exclusion if competition is strong enough to push the equilibrium point below the line connecting the two carrying capacities but competitive coexistence if the competition is sufficiently weak that the equilibrium point is above that line (see Figure 3.2). This basic idea of competitive exclusion has a counterpart in intercropping.

The classic criterion for deciding whether an intercrop is better than its associated monocultures is the land equivalent ratio (LER) (which is effectively equal to the relative yield total [RYT], a quantity used in some of the literature). Suppose we have a total of 1 hectare and we wish to produce maize and beans, would it be better to produce them together as an intercrop or divide the field into two parts and produce maize on one part and beans on the other? Presuming that the two crops compete with one another (the case of one facilitating the other is discussed later), we expect that the relative yield of each crop will be less than 1.0. That is, define the relative yield as P_i/M_i, where P_i is yield of crop i in the intercrop (P = polyculture) and M_i is its yield in monoculture; in both cases, the yield is expressed per unit area (kg/ha or biomass/ha, or some other relevant measure). Then the relative yield total is simply the sum of the two relative yields

$$RYT = LER = P_1/M_1 + P_2/M_2 \qquad (3.3)$$

The meaning of this measure is clear on some reflection. If the polyculture produces some specific amount in a fixed area (say an area of 1 ha), how much area would be required to produce that same amount in two separate monocultures? This idea is made clear in Figure 3.7.

Thus, LER > 1.0 is usually taken as the criterion as to whether the intercrop will perform better than the separate monocultures. It is certainly not the only possible criterion, and sometimes it can be misleading. Nevertheless, it is normally taken as the first step in analyzing an intercrop. If LER is less than 1.0, it is the case that the two monocultures would be better. However, if LER is greater than 1.0, yield performance will be better in polyculture than in monoculture, opening up the possibility that the intercrop would be beneficial (but not necessarily so, e.g., farmers may have no use for the bean yield in the first place).

Given that LER = 1.0 is the critical value in deciding whether an intercrop will be better than separate monocultures, we have

$$1.0 = P_1/M_1 + P_2/M_2 \qquad (3.4)$$

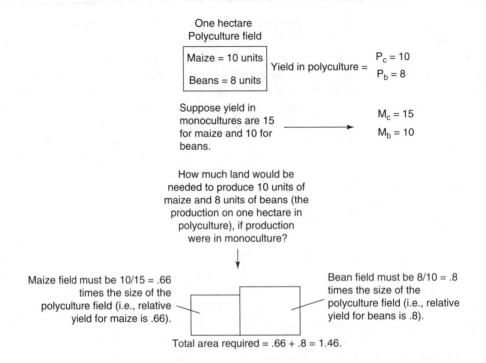

One hectare
Polyculture field

| Maize = 10 units | Yield in polyculture = | $P_c = 10$ |
| Beans = 8 units | | $P_b = 8$ |

Suppose yield in monocultures are 15 for maize and 10 for beans. ⟶ $M_c = 15$, $M_b = 10$

How much land would be needed to produce 10 units of maize and 8 units of beans (the production on one hectare in polyculture), if production were in monoculture?

Maize field must be 10/15 = .66 times the size of the polyculture field (i.e., relative yield for maize is .66).

Bean field must be 8/10 = .8 times the size of the polyculture field (i.e., relative yield for beans is .8).

Total area required = .66 + .8 = 1.46.

Thus, in order to produce in two monocultures that which would be produced in a polyculture of 1 ha, a field of 1.46 hectares would be necessary. Clearly it would be a more efficient use of the land to use the polyculture.

Figure 3-7 Explanation of the meaning of land equivalent ratio.

which can be rearranged as

$$P_1 = M_1 - [M_1 / M_2] P_2 \qquad (3.5)$$

If we plot Equation 3.4 on a graph of P_1 versus P_2, we see that it is a line connecting the two monocultures (Figure 3.8). Furthermore, we can easily see that LER > 1.0 represents the area above the line and LER < 1.0 the area below the line. Depending on the strength of competition (see explanation in Figure 3.2), the system will be found either above or below the line connecting the two monocultures. The parallel between Figure 3.8 and Figure 3.2 is not accidental. The basic ecological process of competition is in operation in either case, and the question as to whether the intercrop will yield better than the two separate monocultures (Figure 3.8) is parallel to the question as to whether the two species will coexist (Figure 3.2). This parallelism between the classic ecological principle of competitive coexistence and polyculture advantage suggests that this form of polyculture advantage be termed "competitive production."[9]

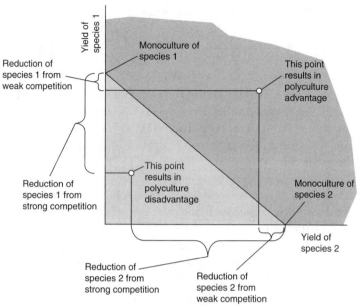

Figure 3-8 The graphical interpretation of the land equivalent ratio criterion (compare with Figure 3.2). Light shading indicates monoculture advantage (below the critical line), and dark shading (above the critical line) indicates intercrop advantage.

In any practical situation, it should be emphasized that computing LER could be misleading for various reasons. First, the monocultures that are used for comparisons must be optimal, which is to say the highest yield of all possible monocultures. Second, the LER is computed for one particular design of the intercropping system. Even if LER is less than 1.0 for that particular design, it may be greater than 1.0 for some other design. This design problem creates a practical difficulty for the rational design of intercrops. Almost an infinity of design possibilities exist in an intercrop of just two species (how many rows, what density with rows, should there be rows in the first place?). This problem can be dealt with conceptually within the construction of yield sets and decision functions[10] or the use of computer-aided modeling for exploring a large number of design options, but such fine details are beyond the scope of this text. For those who will someday be involved in the design of intercropping systems, simple solutions such as restricting the intercrop design to the two extremes of additive or replacement models will not suffice.[11]

The competitive production principle actually suggests that there is nothing particularly out of the ordinary about the crop combination, that it is simply whether the intensity of interspecific competition is relatively weak compared with the intensity of intraspecific competition. In many cases, it is well known that the more efficient use of niche space by two crop species (the essence of competitive production) is not what produces intercrop advantage, but rather some environmental modification of one crop that improves the conditions under which the other crop lives. Using the paradigm of response/effect, the intercrop can be represented as dialectically related to its environment, as pictured in Figure 3.9. The maize plant uses phosphorous from the environment and thus "has an effect on the environment" (or "constructs" a part of the niche of the bean plant) (see Figure 3.9). The bean is thus forced to respond to the environment that was changed by the maize plant—the distinction between effect and response in plant competition. With this response/effect framework, it is transparent how another ecological force may operate in intercropping situations (also in other situations, e.g., agroforestry). If one crop has an effect on the environment, that effect certainly could be salutary for the other crop, which then responds to an "improved" environment. In this case, the first crop "facilitates" the second crop, or returning to the conceptual framework introduced in Chapter 2, the first crop constructs a niche for the second crop. The theoretical details of facilitation are substantially more complicated than those of competition (ecologists generally do not pay much attention to facilitation) and are beyond the scope of this book. Details can be found elsewhere.[12]

The evidence for the competitive production principle is scant. For the most part, the enormous literature on multiple cropping[13] is not of use, as experiments do not generally

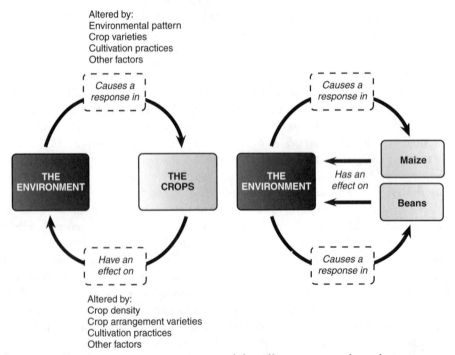

Figure 3-9 Diagrammatic representation of the effect–response formulation.

Adapted from: Vandermeer, J. H. *The Ecology of Intercropping Systems.* Cambridge, UK: Cambridge University Press, 1989.

accommodate the data required to answer the key question of whether the intercrop yields better than a combination of the two optimal monocultures. At one extreme, a review of experiments with mixtures of grasses[14] found only very weak evidence that interspecific competition, on average, was smaller than intraspecific competition (Figure 3.10a). Instead, the overall pattern of the data suggested something more akin to neutral competition in

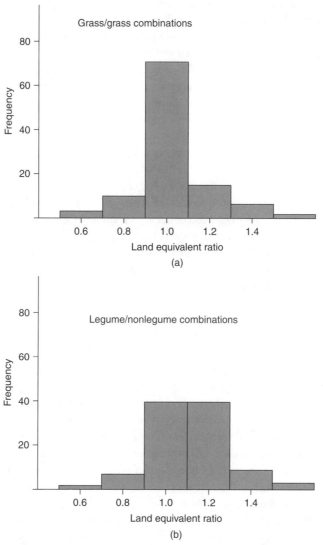

Figure 3-10 Distribution of land equivalent ratios of grass/grass combinations and legume/nonlegume combinations. Data are from B. R. Trenbath, *Adv Agron*, 26 (1974): 177–210 and B. R. Trenbath, *ASA Spec Publ*, 27(1976):129–169.

which interspecific competition and intraspecific competition are not significantly different. In many ways, this is not surprising; plants of similar stature use similar environmental factors—they basically do live in the same niche,[15] at least from the point of view of interspecific competition. On the other hand, if the competitive production principle is parallel to the competitive coexistence principle, it too may be overly dependent on the underlying assumptions of linearity. Perhaps in industrial agriculture, the environment has been simplified to such an extent that linear rules do in fact apply, and plants with the same "niche" may not be able to coexist (or demonstrate a yield advantage living together). On the other hand, in the agriculture of the future, with ecological complexity reintroduced into the equation, the nonlinearities of nature may once again dominate, making the competitive exclusion principle (and by implication the competitive production principle) invalid on first principles (i.e., multiple species sharing the same niche can coexist because of critical nonlinearities).[16] This is an intriguing avenue for future research.

At the other extreme, mixtures of legumes and nonlegumes frequently seem to provide evidence of intercrop advantage,[17] with one literature survey[18] demonstrating that legume/nonlegume combinations generate on average an LER of greater than 1.0 (see Figure 3.10b). Yet even here it is not really clear what mechanism is involved. On one hand, it could simply be a case of the competitive production principle if the legume and nonlegume are simply tapping different pools of nitrogen and thus reducing interspecific competition for that nutrient.[19] On the other hand, the legume could be facilitating the growth of the nonlegume by supplying it with extra nitrogen. Very few experiments have addressed this problem in detail, and the general literature is not amenable to differentiate between these two alternatives. It has been suggested that the legume may simply sequester nitrogen that would otherwise leach out of the soil and later release it to the nonlegume.[20]

The general conclusion seems to be that although theoretically it seems reasonable to suggest that two different crops could fit into an environment more efficiently than an equal biomass of the same crop, accumulated evidence offers some support for that idea, but not much.[21] It would seem to be equally logical, especially in light of recent theory in plant ecology, to suggest that different species of annual crops are more or less interchangeable ecologically, and the main advantage of intercropping is either from a socioeconomic point of view, perhaps from the long-term point of view of soil conservation or from some sort of special facilitative effect characteristic of the particular combination involved. Perhaps the competitive production principle will turn out to be the competitive neutrality principle, and the overall benefit of intercropping has to do more with some sort of effect/response-mediated facilitation or, taking a cue from the previous chapter, an interspecific constructed niche approach.

What has been frequently thought of as the alternative generalized mechanism for intercrop advantage, the facilitative effect,[22] although certainly not ubiquitous, seems well established in particular cases. For example, now a vast literature is available on how intercropping creates the sort of vegetative texture that may control specialist pests, both strictly empirical[23] and mechanistic.[24] In all of these cases, the second crop acts as a facilitator of the first crop by somehow controlling the potential pests, whether by the resource concentration hypothesis or the enemies hypothesis,[25] as further discussed in Chapter 6.

In the end, intercropping is not a panacea. After application of qualitative and quantitative ecological theory and substantial empirical work, several tentative conclusions suggest themselves. First, the crude idea that competitive production will inevitably lead to yield advantages for intercrops, similar to the crude application of Gause's principle, might best be abandoned. In light of Trenbath's studies (Figure 3.10) and in light of modern ideas of plant competition, it is likely that similar plants "occupy essentially the same niche" and thus yield advantage may not often emerge from such a simple idea.

Second, it certainly remains true that if crop types are very different, exploitation of various distinct niches becomes a distinct possibility. For most annual crop combinations, however, this seems unlikely. The exception, and it is an extremely important exception, is the case of a grass and a legume, in which two different niches associated with nitrogen supply may be involved.

Third, in most actual cases of dramatic yield advantage in an intercrop, the critical factor seems to be some sort of facilitative mechanism, be it nitrogen supply from a legume or protection from pests or modification of some physical factor. It would thus seem to be a good strategy to seek facilitative mechanisms when searching for intercropping designs.

Fourth, recent ecological theory has emphasized the limited nature of the competitive exclusion principle. Specifically, if the environment is variable or the competitive process is nonlinear, more than a single species can occupy a single niche. In modern industrial agriculture, environmental heterogeneity has likely been brought close to zero, and strong nonlinear competitive effects may have been reduced by simplification of the environment. Thus, yield advantages of intercrops are not to be expected. However, in more traditional circumstances or in alternative or ecological agriculture, the environmental variability may return, and/or the excessive simplification of the environment that generates linear competition rules may not be obtained. It is under such circumstances that yield advantages through the competitive production principle may actually happen.

Agroforestry: Niche Construction and Competition

In temperate and tropical ecosystems before the development of the industrial system, it was common for trees to be incorporated in some way into the agroecosystem (Figure 3.11). In the temperate zone, it was common, and in the tropics, it was almost universal.

With the environmental crisis looming in the 1970s and the realization that trees had been and still were incorporated into almost all tropical agroecosystems, interest in agroforestry burgeoned in policy institutes. Agroforestry would save the tropics from environmental devastation threatened by overpopulation (so the argument went). In 1977, the International Council for Research in Agroforestry (ICRAF) was established in Nairobi, Kenya, for the purpose of collecting and disseminating information about agroforestry

Figure 3-11 Examples of traditional use of trees in agriculture. (a) A mixed farm, with coffee dominating the understory, in Kerala, India. (b) Alley cropping system in the United States, with poplar trees combined with the annual crop Zea maiz. (c) Shaded cacao system in Bahia, Brazil.

and developing new methods for incorporating agroforestry into modern agricultural systems in the tropics. ICRAF developed a definition that seems to be universally recognized:

> Agroforestry is a collective name for all land-use systems and practices in which woody perennials are deliberately grown on the same land management unit as crops and/or animals. This can be either in some form of spatial arrangement or in a time sequence. To qualify as agroforestry, a given land-use system or practice must permit economic or ecological interactions between the woody and non-woody components.[26]

The actual practice of agroforestry is extremely diverse, ranging from the use of trees as boundary markers to their use for shade to the incorporation of trees and crops into the same complex mixed garden. Table 3.1 shows a classification of the known forms of agroforestry, based on the scheme of Young.[27]

Farmers' use of agroforestry systems is patently obvious to anyone who has traveled in the tropical areas of the world. Interviews with farmers reveal an enormous range of "reasons" for the practice ranging from the ecologically trivial such as boundary markers to the ecologically complex—farmers frequently make claims that sound almost spiritual about trees, but probably reflect a deep local knowledge about what trees actually do ecologically. The efforts of modern science to understand and propagate agroforestry systems have been mixed at best. Indeed, it could be argued that there is a rich potential for generating ecological understanding in these systems, which are probably still most completely understood by the farmers that employ them rather than the scientists that study them.

The intellectual organizational base for agroforestry seems to be converging on the basic idea of a balance between competition and facilitation,[28] normally focusing on the negative and potentially positive effects of the trees on the crops.[29] Initial enthusiasm for agroforestry seems to have assumed that trees must be bringing something positive to crops or pastures; otherwise, farmers would not use them. The competitive production principle, however, allows for a net negative effect of trees on the crops yet still provides a rationale for the joint production of trees and crops or pasture (see the previous section on intercropping). The joint production of trees and crops will clearly show a yield advantage if the two occupy different niches and the trees alone provide some sort of value to the farmer. Indeed, considering basic ecological principles, accordance with the competitive production principle seems far more likely with a combination of trees and crops than in the case of two annual crops.

By the nature of the tree/crop combination (in those systems in which trees and crops have a significant interaction), the scheme of response and effect competition is especially useful, as what is desired is a tree that is as weak an effect competitor as possible and a crop that is as good a response competitor as possible. In terms of niche construction, the tree clearly constructs a part of the niche in which the crop must live. Although it is normally thought of as a competitor with the crop and thus the goal would be to find the weakest effect competitor, it is also the case that the effect could be positive, as is commonly

TABLE 3-1 Classification of Agroforestry Practices

Mainly agrosilvicultural (trees with crops)

Rotational:

Shifting cultivation (trees ordinarily in the fallow system)

Improved tree fallow (planting valuable timber or other use trees in the fallow)

Taungya (crops are planted in areas in which tree plantations have just been initiated—a temporary situation)

Spatial mixed:

Trees on cropland (isolated useful trees, sometimes from previous systems)

Plantation crop combinations (e.g., shade coffee or cacao)

Multistory tree gardens (especially home gardens)

Spatial zoned:

Hedgerow intercropping (barrier hedges, alley cropping) (also silvopastoral)

Boundary planting (sometimes for nothing more than indicating property boundaries)

Trees on erosion-control structures

Windbreaks and shelterbelts (also silvopastoral)

Mainly or partly silvopastoral (trees with pastures and livestock)

Spatial mixed:

Trees on range land or pastures

Plantation crops with pastures

Spatial zoned:

Live fences (common in most tropical regions of the world)

Fodder banks

Tree component predominant (see also taungya)

Woodlots with multipurpose management

Reclamation forestry leading to multiple use

Other components present

Entomoforesty (trees with insects)

Aquaforestry (trees with fisheries)

thought of in the case of niche construction. Indeed, agroforestry could be thought of as a canonical case of interspecific niche construction in that the tree clearly is involved in the construction of the niche in which the crop exists. Recognizing this basic fact, Ong[30] suggested the following equation for assessing an agroforestry system:

$$I = F + P + L \pm M - C \tag{3.6}$$

I is the relative yield of the crop (in proportion to the yield in monoculture); F is the facilitative effect from the trees; C is the competition effect; M is a measure of above-ground changes in temperature, light, humidity, and wind speed; P is the consequences of changes in soil properties; and L is the reduction or avoidance of losses of nutrient or water. F, P, and L are clearly cases of positive niche construction, or facilitation, whereas C is negative construction, or effect competition, and M can be either positive or negative. Sanchez[31] used a simplified form of the Ong equation to evaluate many of the published reports on alley cropping, one of the more popular forms of experiments done with agroforestry. The form used by Sanchez was simply,

$$I = F - C \tag{3.7}$$

where the variables are as before. To use this approach in evaluating particular forms of agroforestry, however, it must be acknowledged that both F and C are not likely to be linear responses to planting designs. The amount of positive niche construction will obviously depend on how the planting density and arrangement of trees relates to the provisioning of the "niche" to the crop as well as the "responsiveness" of the crop to the constructed niche. Indeed, F and C themselves are likely related to one another in a nonlinear fashion such that I will be positive for some planting arrangements and negative for others.

An excellent example is the effect that shade trees have on water balance in coffee production.[32] Water is, of course, supplied by rain and removed from the system by evapotranspiration. The latter is effected by transpiration from shade trees and coffee, as well as direct evaporation from the ground. If we imagine a system that begins with completely unshaded coffee, the situation is that the coffee plants are using water at some rate and the water is being evaporated from the soil at another rate. If shade trees are added to the system, they have a negative effect in that they compete with the coffee for water, but they also change the microclimate at the soil surface such that evaporation is reduced. A linear mindset might say that either 1) shade is good because it reduces evaporation from the soil or 2) shade is bad because the shade trees use water that the coffee plants thus cannot use. A moment's reflection suggests, however, that both are true and that the overall response of the coffee to shade will most likely be a nonlinear affair with respect to water use, which indeed has been experimentally verified.

The nonlinear nature of both F and C in the previous simplified equation needs to be taken into account when analyzing agroforestry systems, especially when searching for an appropriate design of one. As with the case of intercropping, a strictly empirical approach may be so overwhelming as to be untenable. For example, in the case of an alley cropping system, consider just the possibility of simultaneously altering the width of the alleys (which is ultimately the same as the number of rows of trees) and/or the within-row density of either crop or tree and/or the spacing between rows of the crop. If we consider just three possible states of each of these variables, we have a total of 12 combinations that must be investigated empirically (three alley widths, three within-row density of trees, three within-row densities of crops, and three spacings between crop rows), and each of these must have the three treatments discussed previously (monocultural comparisons should be the same, i.e., the optimal monoculture, for all designs). Furthermore, many experimentalists would likely object (correctly) that fixing three states for each variable is not enough to estimate the functions involved, requiring yet more treatments. We rapidly run into the limits of empiricism, which encourages once again a more theoretical and/or mechanistic approach to the problem.[33]

If we presume that the relationship between F and C is a nonlinear one,[34] the basic form of the nonlinearity will likely be as indicated in Figure 3.12, where three distinct options are illustrated. The key nonlinearities arise from the way in which the niche constructive versus niche destructive elements enter into the system with different quantities and qualities of trees and crops. One can deduce from Figure 3.12 that for a particular combination of trees and crops (including pasture as if it were a crop) there may be no chance for an agroforestry advantage (system c), no chance for anything but an agroforestry advantage except at extremely high levels of tree intensity (system a), or agroforestry advantage at intermediate states of tree intensity (system b). It all depends on the combination of trees and crops and the planting design. Sometimes a particular combination of trees and crops will never be advantageous (system c), but under most circumstances, it is probably the case that some planting designs will yield an agroforesty advantage and others will not.

The particular way in which trees bring advantage to crops is varied and complex[35]: water balance, protection against pests, supplying nutrients, decreasing wind-born diseases, and many others. Discussing each of these possibilities is beyond the space limitations of a single volume. Nevertheless, one particular mechanism is worth mentioning, specifically that of pumping nutrients into the system. A problem in many tropical soils (as discussed in Chapters 4 and 5) is excessive leaching of some nutrients due to low cation exchange capacity (low ability of the soil to store nutrients for later use by the plants). Thus, it is generally thought that some nutrients may accumulate at very deep layers in the soil, effectively keeping them out of reach of crop roots. Thus, the agroecosystem exists in a zone in which many nutrients have been leached. If deep-rooting trees are put into the system, they offer the possibility of "pumping" those deep-seated nutrients back into the effective agroecosystem. Ironically, there is little political desire for deep-rooted trees even among those farmers and technical advisors that wish to promote agroforestry. Both farmer and advisor want quick results. Trees that grow fast usually have shallow root systems, whereas the deep-rooted species generally grow more slowly. Because both farmer and advisor want quick results, most modern agroforestry systems are based on trees with shallow root systems.

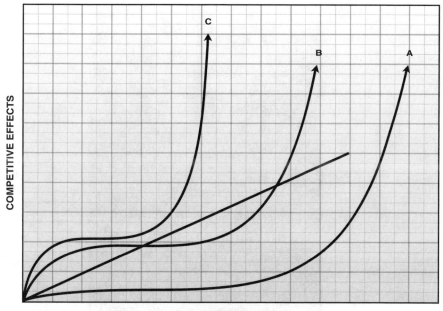

Figure 3-12 Theoretical relationship between the competitive and facilitative effect in an agroforestry system. The arrows indicate the increasing "intensity" of the tree component of the system, that is, increasing density or biomass of trees in the system. Different systems are indicated by different letters. (a) The agroforestry system is almost always beneficial because facilitation tends to be greater than competition. (c) The agroforestry system is always inferior to monoculture because the competitive effect is always greater than facilitative effect. (b) The intermediate situation in which not enough facilitative effect is felt at low tree intensity to offset the competition, but a window of agroforestry advantage occurs at intermediate levels of tree intensity.

Adapted from: Vandermeer, J. H., "Maximizing crop yield in alley crops," *Agroforestry Systems, 40* (1998):199–206.

Rotations: Competition in Space and Time

One of the most common situations in which plants are purposely brought into competition with one another in an agroecosystem is in the process of crop rotations, a key part of the traditional European three- and four-field system in the "new husbandry," and a key element in almost all movements toward ecologically based agriculture when dealing with annual crops. Why is it that sometimes particular crops will be rotated, whereas in other places or at other times, those same crops will be intercropped (Figure 3.13)? Ignoring for the moment the important sociopolitical forces that may be involved and thinking of the issue only from the point of view of ecological competition, rotations

(a)

(b)

Figure 3-13 Two corn/bean combinations. (a) Corn and beans simultaneously intercropped in southern Mexico such that the beans use the corn stalks as a trellis. (b) Corn (in front) bean (behind) rotational system in Cuba.

bring several important complications into the picture. First, because classic competition theory applied to plants normally implies a simultaneity in time, separating two or more species implies a lack of competition, at least that which is directly felt. Indeed, the entire idea of niche construction (or ecological engineering) becomes extremely important, and the balance between facilitation and competition a key idea. Second, the phenomenological approach normally applied to competition theory is inadequate to understand the most basic elements of competition in the case of rotations. Plants use nutrients, just as they do in simultaneous competition, but that use is strongly compartmentalized in time. These issues strongly imply that a more mechanistic approach to competition is desirable.

In rotational systems, 1) crop species compete for resources, and 2) this year's environment (niche) is constructed by last year's crop. To appreciate fully the competitive aspect of rotational systems, it is convenient to formulate the competitive process mechanistically (see Appendix 3.A), where there is a key resource being used by both crops. If a single resource is limiting for both crops, the rotational system generates a competitive situation that cannot be beneficial in the sense of overyielding (see the previous section about intercropping). That is, from the point of view of plant competition alone, it never makes sense to rotate crops if both crop species are limited by the same nutrient. In any given season, crop number one will tend to reduce the level of the critical resource to R1* (see Appendix 3.A), and crop number two will tend to reduce the level of that resource to R2*. Whichever R* is lower, that species will be able to survive, and the other will not have sufficient resource available—competitive exclusion, or in the context of agricultural production, underyielding. If both crops are desired, it would be best to not rotate at all. Thus, if a single resource is in short supply, it does not make sense to rotate.

As the competitive exclusion principle applies to ultimate survival of a plant (or animal) population, so the competitive production principle applies to rotational systems in the same way it applies to intercropping systems. If the critical resource for one species is different from the critical resource for the other (compensation occurs, or niche division), then a rotational system may provide advantages over a single monoculture of either of the species. This result, however, begs the question of why rotate rather than intercrop.

All of this is predicated on the simplifying linear assumptions commonly taken in competition theory. As discussed previously (also see Appendix 3.A), many of our most cherished ecological principles actually are consequences of this very special assumption. Thus, the result that rotational systems will never be advantageous if a single resource is involved in the competitive process can be seen as a clear example of such a problem.

The second important aspect of rotational systems is, as frequently stated, to avert soil-born pathogens from building up. From an ecological perspective, this observation places rotational systems directly within the framework of facilitation in an effect/response or niche construction context, precisely as previously described for agroforestry systems. For example, in a grass/legume rotational system, the grass constructs a soil environment free of legume diseases and vice versa. It seems likely, if for no other reason than farmers repeatedly state it, that facilitation of some sort is involved in successful rotational systems.

Plant Competition and Weed Control: From the Fire to the Plow

Plant Competition and Weed Control

Recalling the earlier introduction to nonlinear aspects of competition theory, when dealing with plants that compete with crops and have no use themselves, we have the definition of the weed problem in agriculture. As noted in the first section of this chapter, the kinds of nonlinearities pictured in Figure 3.5 are rather easy to imagine under the specifics of plant competition. A large plant responds to the ratio of a competitor's biomass to its own biomass, rather than directly proportional to the competitor's biomass. It makes little difference to a 5 cm maize plant seedling whether a bean plant is 6 cm tall or 50 cm tall with regard to light competition—living in the shade of a bean plant is not optimal no matter how tall the plant making the shade. But it matters a lot to a 20 cm maize plant. It is the ratio of the effector to the responder that matters, not the absolute size of the effector. The sorts of nonlinearities shown in Figure 3.4 are a consequence of this sort of competition (see Appendix 3.A for further explanation).

Given this nonlinear picture, it is evident that the point at which the process of competition is initiated is the critical issue. Whichever plant has the initial advantage is the one that will win in competition. Weed control, throughout history, has been based on this elementary fact of ecological plant competition theory. Slash and burn agriculture, as described in Chapter 2, can be considered in this light. From the initial burning of the slash that resulted from clearing the fallow to the repeated annual burning during the planting phase, the whole point of fire is to provide the crops with a head start over the other plants in the system.

Although fire was undoubtedly a critical factor in agriculture right from the beginning and remains an important tool throughout the tropical world even today, a far more efficient device rapidly replaced it in the old world. Invented many times independently, the plow is the most universal means for giving crops the advantage over other plants and has been cited as one of those pivotal innovations that gave rise to an avalanche of technological transformations.[36] Because of the presence of draft animals, the plow was an invention of the Old World, and its adoption in America came only through European expansion. Advances in metallurgy enabled the original wooden plow to be fitted with an iron plowshare, greatly increasing efficiency, but also signaling the further appropriation of this part of agriculture since making iron plowshares was clearly an industrial activity.

Universal adoption of the plow in agriculture was early and complete. Its significance was initially to provide the competitive edge to the crops over the weeds, as discussed previously; however, the plow also changed attitudes, even if the changes would be millennia in being realized. The changed attitudes were in the utilization of nonhuman power in the farming operation. Originally, oxen, mules, or horses were the source, and their use in farming commonplace. More important, the use of such nonhuman labor set the stage for the initiation of a cycle of mechanization that became the first of the four major "appropriations"[37] that eventually came to characterize the industrial system (mechanization, fertilizers, pesticides, and seeds). Its most significant birthright was perhaps not with the plow itself, but rather with the harvest operation.

In 1833, Obed Hussey received a patent on a mechanical reaper, and in the following year, Cyrus McCormick patented a similar machine. McCormic's machine surprisingly did not come into use until well after the Civil War had begun, despite the fact that it was well-known some 20 years earlier. It was not some subtle cynicism that farmers had about "new-fangled devices" that stood in the way of its adoption—it was in fact a rational assessment of economic facts that caused them to remain with a technology that required five people per acre when they were well aware of this new technology that required only one.

The all-important factor was availability of labor.[38] Westward expansion of agriculture, under way at the time the Civil War broke out, required a supply of harvest labor that was easily met by the comparatively large population of potential workers. The many men who were looking to establish homesteads or in some other way articulate with the agrarian economy provided a ready source of cheap harvest labor. With a steady supply of cheap labor, it made no sense to invest large sums of money in an automatic harvester, and the McCormick reaper remained an underused "new-fangled" device for almost 2 decades. Thus, the device that revolutionized agriculture and even strongly contributed to the making of the industrial revolution in the United States was ignored for a significant time, not because of unreasonable or backward farmers, but because of logical and essentially correct economic calculations.

With the outbreak of the Civil War, labor became short, and McCormick's reaper was widely adopted, initiating a change in the farming mind set. "Machines in the fields" was to be the future of agriculture, and a host of automation devices followed the automatic reaper, ranging from corn harvesters to threshers to automatic seeders and many others. Recall Samuel Copeland's "Agriculture, Ancient and Modern" with the barn housing a variety of threshing and other machines, all driven by belts that attached to a central drive shaft that was driven by a stationary steam engine, such that the barn was made to have an industrial look (Figure 2.16 in Chapter 2). At this point, however, a major contradiction existed in the attempt to push the industrial capitalist system into agriculture—energy. The belts and pullies that made the barn look like a factory in Copeland's drawing did not encompass the fields where the crops were grown. As noted by Goodman et al.[39]

> Perhaps the most striking indication of the difference between rural and industrial activities is the contrasting development of their energy base in the nineteenth century. While the steam engine provided the motive power in manufacturing, mechanization in agriculture continued to draw its power from the horse or mule. In manufacturing, "nature" is broken down by processing and introduced into the machine as a raw material input, which thus can be adapted to the speed of machine production. In contrast, nature in agricultural production cannot be reduced to an input; indeed, it is the "factory" itself. Consequently, instead of restructuring the production process, mechanization

effectively represented an implement adapted to the spatial and temporal characteristics of agriculture. Rather than the Copernican revolution of manufacturing, whereby nature must circulate around the machine, nature in agriculture maintains its predominance and it is the machine which must circulate.

Because of this characteristic of agriculture, the steam engine, so revolutionary for manufacturing and the fundamental energy basis of the entire industrial revolution, was only effectively applied in agriculture to threshing, mainly because the product, wheat stalks, could be brought to a central location. Thus, whereas the horse team was replaced by the steam engine for certain processes, the majority of agricultural activities could not conveniently use steam engines.

Despite substantial technological know-how of the late 19th century, steam power ultimately saw limited application in agriculture. This set the stage for the next great event. The internal combustion engine made all of the automated devices that had already been invented even more efficient. In 1892, the first self-propelled gasoline tractor was introduced. Initially, the acceptance of this new device was low, but with the burgeoning of the automobile at the beginning of the 20th century, tractors rapidly became the norm in farming in the United States and Europe. Although it took some 50 years for full adoption, the replacement of horse and mule power by traction power derived from the internal combustion engine was clearly a turning point in agricultural history. It paved the way for ever more sophisticated mechanized harvesting activities, beginning with hard grains such as wheat and corn, proceeding to the more difficult problem of cotton, and finally to the seemingly insoluble problem of perishable fruits and vegetables. It also, in each wave, created social upheavals because of its effects on labor relations and sometimes on the very fabric of societies.

What the Plow Has Wrought—Mechanization as a Principle

Cotton was, as is oft repeated, king.[40] The southern colonies of British America derived their income from cotton and tobacco. Their political power in the new United States was based on the textile mills they supplied with cotton, and their secession from the Union was based on a threat to their labor supply in cotton production and restrictions on their ability to market that same cotton overseas.

Slavery was notoriously incompatible with the growth of capitalism, and the Civil War of the United States effectively eliminated this form of labor exploitation. However, an equally pernicious form of exploitation emerged with the establishment of Jim Crow laws and the system of sharecropping. This system would remain in existence for almost a century, but its economic inefficiency was well known. With the success of automated harvesting of hard-seeded grains behind them, researchers turned to the question of mechanical harvesting of cotton. But a variety of seemingly intractable problems had to be faced. The main problem that eventually emerged was that cotton ripened unevenly through the later part of its growing cycle. Thus, although machines were developed in the late 19th and early 20th century, very little adoption occurred because they resulted in far lower yields per acre than sharecropping. The only reason for adoption was to eliminate the "labor problems" associated with this form of agriculture labor.

Thus, this was the birth of a new mentality in the agricultural sector—one that would have far-reaching consequences. Production goals began to shift from yield per acre to yield per labor time. With labor accounted for as a simple cost of production, as in industrial production, an obvious contradiction was lurking between seeking highest yields per acre or highest yields per person-hour invested. Clearly, the two goals could be perfectly compatible, but what about the cases where they are not? What about the adoption of a mechanization (or any other) technology that implied a reduced yield per acre but an increased yield per person hour? Although straight-forward economic calculations are elementary and obvious, farmers (like others) are not the *Homo economicus* of elementary economics textbooks, which means that a host of social, cultural, and political factors enters the equation in the real world. Adoption of mechanization is thus like any other technology—it depends on the underlying conditions of production: ecological, social, political, cultural. In the particular case of adoption of mechanical harvesting of cotton, the legacy of slavery and intransigent racism were certainly more than marginal components of the process. The maintenance of the old system of production thus incorporated far more than only simple economic calculations, as it always does.

The technological package that eventually ushered in cotton mechanization was developed piecemeal.[41] The cotton gin enabled the harvesting of less than perfect bolls and set the stage for harvest machines. But the machine was not yet available. When the machine became available, an efficient power source was not available to make the machine attractive. When the internal combustion engine came into use, the machine looked more attractive, and by the mid 1940s, a self-propelled mechanical harvester was available. The problem of unequal ripening remained, however, and the large amount of foliage on the cotton plants made the machine inefficient. The establishment of the National Cotton Mechanization Project in 1946 through an act of Congress provided the necessary research for genetic changes to create varieties in which the bolls ripened simultaneously and higher up from the ground to facilitate the needs of the mechanical harvester. Finally, in the late 1940s and early 1950s, herbicides came into wide use (discussed later) so that the unwanted foliage could be eliminated before harvesting.

Throughout this process, innovations evolved as their need emerged, resulting in a gradual adoption of mechanization. The ultimate effect could not have been more spectacular. The northward migration of hundreds of thousands of African Americans filled the labor requirements of expanding industries in northern cities, resulting in an urban racial crisis as of yet not completely resolved.[42]

Although mechanization of the cotton harvest presented significant technological challenges, the history of the mechanical tomato harvester was even worse. Ironically, it was this technological difficulty that made it not only dramatically different from cotton[43] but ushered in an entirely new mode of agricultural research—the "technological package."

The entrance of the United States into the Second World War was a watershed for California tomato producers. As in all productive activities, agriculture was put on a war emergency basis, and California was targeted for providing the Allied effort with tomato sauce, among other products. The forced incarceration of Japanese Americans helped consolidate large land holdings for the establishment of large tracts of vegetable production, including tomatoes. The shortage of labor, however, conditioned partly by the conscription needs of the war effort, but also in part by the requirements of a low-paid labor force, encouraged the U.S. Congress to pass legislation allowing for the seasonal migration of thousands of migrant workers from Mexico, to pick the tomatoes each year. This was the infamous Bracero law.

Even before the war, two visionary agricultural scientists, Jack Hanna, a plant geneticist, and Coby Lorenzen, an agricultural engineer, recognized the ultimate need for automatic tomato harvesting as a means to eliminate the problems with labor. Indeed, Hanna anticipated the emergence of the United Farm Workers of America.[44] In those days, the mechanization of cotton was still a dream, and the mechanization of a soft fruit, such as tomatoes, was thought to be technologically impossible. Thus, when Hanna announced that he would begin work on the development of an automatic tomato harvester, he faced a great deal of skepticism. An important component of this research was that the mechanization process was conceived at the outset as a combination of genetics and mechanical engineering. In contrast to what had been happening in the program to mechanize the cotton harvest, the tomato harvester would be initially conceived as a machine that did not exist, to harvest a tomato that did not exist—the first coordinated development of a "technological package" in modern agricultural history.

But the research was not to bring to fruition a workable harvesting system soon enough. With the end of World War II, the theoretical justification for the Bracero program vanished. Yet the large producers of California needed a cheap and docile labor force. Appropriate lobbying achieved the result of extending the Bracero program, despite the lack of theoretical justification and in the face of a coalition of church and union interests that continued protesting against it. Furthermore, the unionizing efforts of Cesar Chavez and what would become the United Farmworkers of America got under way in the postwar years. During the late 1940s and early 1950s, the need for a mechanical harvester thus became ever more evident.

Finally, in 1958, the work begun in 1940 was completed, and the first experimental mechanical harvester and the compatible tomato were introduced to California farmers (Figure 3.14). Industrial production of the machine began in the early 1960s, and the Bracero program was eliminated in 1964. The adoption of this new technology proceeded at a remarkably rapid rate, much faster than the earlier harvest mechanizations of wheat and cotton, and that speed of development is largely attributable to the new philosophy of the technological package.

From Obed Hussey and the wheat harvester through the massive transformation of the United States by the mechanical harvesting of cotton to the sophisticated machinery needed

Figure 3-14 Early prototype of the mechanical tomato harvesting machine from the University of California at Davis in cooperation with the Blackwelder machine company.

for soft fruit harvesting, appropriation of harvesting technology by the industrial capitalist sector can thus be seen as a primary force resulting in great changes in the rural labor process. From the beginning of the industrial revolution and continuing to today, the tendency in all industrialized countries has been qualitatively the same. Agricultural workers and farmers have been converted to industrial workers. The consequent magnitude of change in rural society can hardly be overstated. This change has come about through the formal anarchism of the capitalist mode of production and has created the conditions under which new forms of agriculture must evolve. It brings up the question of what kind of rural society is desirable—a question that needs to be faced by all concerned with the promotion of sustainability as a key goal for agroecosystems (a point to be revisited in Chapter 8).

Herbicides: "Chemical Mechanization" of Weed Control

We now come full circle, so to speak, in that the development of herbicides can be seen as the mechanization of the weed control process. As noted previously, weeds and crops are almost certainly related to one another in a competitively indeterminate fashion. Thus, from the beginning of agriculture, the imperative existed to try and start the season by giving the crops an advantage in the competition space (formally, try to start the process of competition such that the system as a whole would be in the basin of attraction of the crop's carrying capacity). Fire was the first way to manage this process, usually coupled with manual weeding to steer the system toward the crop's basin of attraction. Mechanical cultivation, facilitated by the culture of mechanization, was better than fire since it could be applied even as the crop was growing. Nevertheless, the significance of mechanical cultivation was mainly in its effect of reducing labor requirements, much as mechanical harvesting.

Near the end of World War II, however, a new idea came to center stage. The U.S. and British militaries had begun experimenting with herbicides during the war. During the insurrection in the Malay Peninsula, British forces used two herbicides, 2,4,D and cacodylic acid, to destroy food crops in areas known to be strongholds of the extensive left-wing revolutionary movement. For much of the history of herbicide application, both military and agriculture, these two herbicides and some relatives were a mainstay of vegetation destruction. During wartime, by far the most important use of herbicides until the 1980s, they were used mainly by U.S. and British forces for the destruction of crops, cacodylic acid to destroy narrow-leaf crops and 2,4,D to destroy broad-leaf crops. When the transfer came to agriculture, the same general strategy was employed, applying narrow-leaf plant poisons to broad-leaf crops and broad-leaf plant poisons to narrow-leaf crops. The negative implications for the use of polycultures, the vast majority of which were grass/legume combinations (and thus broad-leaf and narrow-leaf crops grown side by side), are obvious.

As the United States was increasing its attack on Vietnam, the Air Force began experimenting with various cocktails to defoliate forests. The problem was that the Vietnamese nationalists were using hit-and-run tactics that were extremely effective against the U.S. military. Attacking from forested cover and then disappearing into that cover were effective strategies for the guerrillas. Defoliating the forests into which the guerillas would escape became a major program, known as Operation Ranch Hand, and a cocktail made up of mainly 2,4,D and a related compound, 2,4,5,T, came to be known as "agent orange." Sprayed extensively over thousands of hectares of Vietnamese forests, it did untold damage to those forested habitats and resulted in events that are still playing themselves out. One of the components of agent orange, 2,4,5,T, as a conseqence of its manufacture, contains Dioxin, one of the most potent carcinogens known. U.S. veterans of the Vietnam War continue to suffer from what is now admitted to be the consequences of exposure to Dioxin, including elevated rates of several kinds of cancers.[45]

In the 1970s, a miracle herbicide, atrazene, was developed. It had two important properties: It was extremely cheap to produce and had a low breakdown rate in the environment. Atrazene rapidly became the most widespread herbicide in grain production in the United States and was exported around the world to control broad-leaf weeds. Recently, a raging controversy has gathered steam surrounding atrazene's apparent effect as a hormone mimic in nature. For totally unanticipated chemical reasons, atrazene acts biologically in a fashion similar to estrogen. Consequently, feminization of wild animals, especially noted in amphibians and reptiles, has sounded an alarm that has yet to be fully explored. The effects on humans may also be consequential, as noted in the little read but important book, "Our Stolen Future."[46] For example, the decline in sperm counts in industrialized countries is strongly correlated with the spread of atrazene and related compounds. Although correlation does not imply causation, it does cry out for an explanation.

The latest chapter in herbicide use is still playing itself out today with the controversy over Monsanto's Roundup. The active ingredient in Roundup is glyphosate, which is a critical enzyme inhibitor, apparently effective against plants with minimal effect on animals. It is translocated into the plant tissue and is thus an incredibly effective plant biocide, like fire in that it kills all plant material, but better than fire because it kills all of the tissue, including residual rootstock. Combined with genes inserted into crops to resist the effect of glyphosate, the technological package known as "roundup-ready" crops is the latest in the unending march to rid agriculture of the need for labor. With roundup-ready crops, one can plant the crop and then spray roundup over the entire field, killing all plants except the crop itself. These products are extremely attractive, especially to larger producers whose main goal is to reduce the cost of labor. Recent concern over roundup-ready crops emerged from studies at the University of Pittsburgh demonstrating that roundup directly kills amphibians.[47] Atrazene and related compounds have an indirect effect, through feminization and carcinogenicity. Roundup has a direct effect, killing the frogs that it meets and whatever else eventually becomes known. Its immense popularity in some farming sectors has environmentalists the world over extremely concerned.

Appendix 3.A: The Ecology of Competition

The Three Key Concepts

Warning: If you are frightened by the presence of equations or think that you are somehow "bad" at math, read this section very carefully. If you do not have "math phobia," you may skip this "three key concepts" section and proceed to the section entitled "exponential population growth." This section is for those who have been convinced that they cannot do math. It is an attempt to introduce three basic concepts in such a way that you realize they are simpler than they perhaps seem to you. The intention is to back up and state some of the obvious so that a bit of the less obvious can be brought to light. The three concepts are 1) equations, 2) logarithms, and 3) derivatives. Understanding these three concepts, if you have become rusty in your math skills, will "set you free." It will also permit you to read the rest of this appendix.

Equations

The basic idea of population ecology is that changes in numbers and/or biomass over time and in space can be understood more easily if elementary mathematics is applied to the basic processes that lead to those changes in numbers or biomass. Consider, as an example, the changes in a population of bacterial cells in a small glass dish filled with some sort of appropriate medium. In the beginning (which we shall call time zero), there is a single cell, and 1 minute later there are two cells. The following picture shows the bacterial population for the first 3 minutes (time = 0, time = 1, and time = 2).

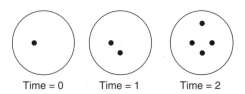

Time = 0   Time = 1   Time = 2

Thus, for time 0, 1, and 2, the bacterial population is 1, 2, and 4. The population for the next times (times 3, 4, 5, and 6) is shown in the next figure.

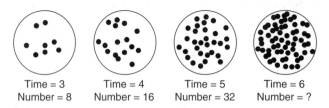

Time = 3   Time = 4   Time = 5   Time = 6
Number = 8   Number = 16   Number = 32   Number = ?

For times 3, 4, and 5, the bacterial population is 8, 16, and 32. Now here is a problem that I fail to understand. If I give you the task of finding the number of bacteria at time 6, the math phobe will, if he or she is true to principle, count the little dots in the picture. Yet I have never met a math phobe who actually does that. They always cheat! That cheating (i.e., not actually counting the little dots) involves the use of mathematics, whether the math phobe wishes to admit it or not. That is, the true math phobe will refuse to use any sort of mathematics and will insist on spending his or her time physically counting the little dots in the last picture. Amazingly, however, all math phobes I know cheat, and simply multiply 32 by 2 to get the answer of 64. Of course, the truly militant math phobe will *claim* that he or she counted the dots, but secretly, he or she just did the simple multiplication.

Hysterical blindness is a remarkable disease. Victims really cannot see anything. Yet photons stimulate the sensors on the retina, and the proper signals are sent to the brain, so, in a sense, you could say that they really do "see," yet their experience is one of blindness. If confronted by an episode that normally elicits an automatic response, like jumping out of the way of a bus that suddenly appears in the street or flinching at a fist thrown in their direction, they do indeed engage in the expected automatic response. All of this suggests that the neurology is in place, but something in the brain says, "I refuse to acknowledge those visual signals." It seems to me that math phobes are similar. Although they may *claim* to spend the time to count all 64 dots, secretly, they multiply 32 by 2. Perhaps—and I am assuming that if you are reading this at all, you are a math phobe—by noting that you in fact do use mathematics all the time, you can begin the process of discovering what you have long known subconsciously: that mathematics is a useful tool that saves you the time of having to count all those things all the time.

Let us begin by analyzing the particular example at hand. You did not actually count all the dots—you noted that the number of dots at time 4 was twice the number at time 3, and the number at time 5 was twice the number at time 4. Thus, you reasonably concluded that the number at time 6 would be twice the number at time 5. To go from time 3 to time 4, you multiplied 8 by 2 and got 16. To go from time 4 to time 5, you multiplied 16 by 2, and to go from time 5 to time 6, you multiplied 32 by 2. In other words, to go from one time to the next, you multiplied the number by 2.

Now remain calm. We are going to write those statements in a slightly different form. The number at time 4 is equal to twice the number at time 3. To make that sentence a little easier to read, we could have spaced it out a little like this:

The number at time 4 is equal to twice the number at time 3.

Mathematics is really nothing more than making things more efficient, saving you time. So, instead of "is equal to," which is a phrase we use a lot, we use an arbitrary symbol, "=". From the wasteful "is equal to" (11 characters, including spaces) to the compact and beautiful "=," we have increased our efficiency by 92% (excuse me, *ninety-two percent*). Thus, we begin with "the number at time 4 is equal to twice the number at time 3," and we clarify that sentence with

The number at time 4   is equal to twice the   number at time 3

Finally,

The number at time 4   =   twice the number at time 3

This dramatically reduces the amount of writing we have to do. Why stop there? What if we were to make the phrase "the number at time 4" more efficient? Why not say, for example, "N(4)" or "N_4"? If we are going to replace "number at time 4" with "N(4)," we might as well replace "number at time 3" with "N(3)". Exactly how we write it is completely arbitrary, so, N(4) versus N_4 is only a matter of convention and convenience. Here I use the parentheses, but not for any particular reason. Thus, in pursuit of more efficiency, we can change

The number at time 4   =   twice the number at time 3

to

$$N(4)   =   \text{twice } N(3)$$

and then we can replace the word twice with the number 2; thus, we get

$$N(4)   =   2 \, N(3)$$

In pursuit of efficiency, we just eliminate all those spaces and get

$$N(4)   =   2N(3)$$

Since N(4) = 16, and N(3) = 8, this equation gives us 16 = 2(8).

This entire argument could be applied to the transition between time 5 and time 6 as well. Thus, we could have

$$N(6) = 2N(5)$$

Or between 1 and 2, in which case we have

$$N(2) = 2N(1)$$

Now it may have occurred to you that to get the number 6 you might simply add 1 to 5, or to get the number 2 you might simply add 1 to 1. In pursuit of efficiency, rather than write

$$N(1) = 2N(0)$$
$$N(2) = 2N(1)$$
$$N(3) = 2N(2)$$
$$N(4) = 2N(3)$$
$$N(5) = 2N(4)$$
$$N(6) = 2N(5)$$

It really would be more efficient to write

$$N(t + 1) = 2N(t)$$

which would save five entire lines.

This entire example was constructed under the assumption that at each time interval each bacterial cell doubled, which is to say, from one cell, we get two. What if at each time interval each bacterial cell produced two other cells, such that we get three cells from each single cell in one time interval? The sequence through time then would be $1 - 3 - 9 - 27 - 81$ and so forth. Which means that instead of $N(t + 1) = 2N(t)$, we would have

$$N(t + 1) = 3N(t)$$

Furthermore, depending on how we chose our time interval for sampling the bacterial population, it could be that each bacterial cell produced an *average* of 1.5 (or any other number) cells, in which case the equation would be

$$N(t + 1) = 1.5N(t)$$

This means that we could put any number we please in place of the number that multiplies the previous population or

$$N(t + 1) = RN(t) \qquad (3.A1)$$

where R stands for any number at all. R (or you could use some other symbol if you wanted) is called the "finite rate of increase" (finite, as later we shall have cause to talk of the infinitesimal).

You have now come to the end of the equation section of the math phobe section. If you wish, you can use Equation 3.A1 to understand population dynamics. But if you insist that you are blind, or rather that you just cannot understand mathematics, go ahead and count the dots. But really, math just makes it a little easier.

Logarithms

The other scary parts (logarithms and derivatives) are on the one hand conceptually simple, but on the other hand can get a bit dicey when you actually use them. We begin with logarithms. First, you must realize that logarithms reduce the number of computations you need to do, making things much simpler. For example, what is 100,000 times 100? Now, the true math phobe, if true to his or her principles, will put 100,000 beans in 100 bowls and then count them all. However, using equations will make life much simpler and save you a lot of time. The equation is pretty simple: $10 \times 10 = 100$ (okay so far?). $10 \times 10 \times 10 = 1,000$. Finally, $10 \times 10 \times 10 \times 10 = 10,000$. So one ten is 10. Two tens are 100. Three tens are 1,000. Four tens are 10,000. Five tens are 100,000. The actual number is a 1 (a one) with a certain number of zeros after it. Why not call that number of zeros something? Let us call it "proportion number" so that the proportion number for 10 is 1, the proportion number for 100 is 2, the proportion number for 1,000 is 3, and so forth. We could have called it something else, but if we call it proportion number and then translate that into Greek, it becomes *logos arithmes,* which we then shorten to logarithm—it's really quite arbitrary, what we call it.

Now, what we have done is taken the exponent of the number and called it the logarithm. Thus, the log (short for logarithm) of 10 is 1 because 10^1 is 1. The log of 100 is 2 because 10^2 is 100, and the log of 10,000 is 4 because 10^4 is 10,000. If we generalize a bit, the log of 10^x is x. If we return to the original question, "what is 10,000 times 100?" we can rephrase that question as "what is 10^4 times 10^2?" It turns out to be 10^6, which points out the incredibly useful aspect of logarithms—to multiply two numbers, add their logs (and, if you recall, to divide one number by another, simply subtract the log of the second from the log of the first).

Why are these examples using 10 only? After all, there are many many ways of representing the number 100 (i.e., not just 10^2). For example, $10 + 90$, or $50 + 50$, or $3^2 + 91$, or $2.71828^{4.60517}$. They are, after all, all precisely equal to one another. However, 2.71828 is indeed a very special number. It is usually called Euler's constant and has its origin in some fairly simple—but difficult to explain at this level—calculus. Fortunately, it is completely unnecessary to understand from whence 2.71828 comes.

A logarithm is actually an exponent (the log of 1,000 is 3 because $1,000 = 10^3$). Why use 10, however? Why not use the totally obvious choice of 2.71828? For reasons that need not enter the discussion here (later in this appendix it will become a bit clearer), all you have to accept is that there is a logical (mathematical) reason for defining a type of logarithm based on exponents of 2.71828.

The number whose exponent you are going to call a logarithm is referred to as the base of the logarithm. Thus, $\log(1,000) = 3$ is to be read "the log to the base 10 of 1,000 is 3." When the base is 10, we frequently refer to the log as a *common* log. However, the log to the base 2.71828 is something else and could be read "the log to the base 2.71828 of 100 is 4.60517." When the base is Euler's constant (2.71828), however, we frequently, by convention only, refer to the log as a *natural* log. Common logs (base 10) are most frequently written as "log," whereas natural logs are most frequently written as "ln." Thus, we have $\log(100) = 2$ and $\ln(100) = 4.60517$.

Derivatives

On an auto trip from Chicago to Detroit, you casually look down at the speedometer and find you are going 76 miles per hour. Because the speed limit is 70, you responsibly slow down to 70. Again, as a true math phobe, you do not concern yourself with what that number means on the dashboard and simply release the pressure from the "accelerator" to be a responsible citizen. You certainly do not have to know anything about calculus to drive a car. However, you probably have no trouble understanding this—that the "speed" of the car was 76 or 70 miles per hour. Yet, if you stop and think about it speed is not as obvious a concept as it first appears.

If no speedometer existed on the dashboard, how would you measure the speed? The most obvious thing to do would be to look at how many miles would be traveled in a particular period. Thus, starting at mile 100 (i.e., 100 miles from Chicago), we set our stopwatch to zero and start clocking. Then at mile 170, we stop the watch. The speed is something like the amount of space traveled (i.e., 70 miles) divided by the number of hours it took to travel that distance. If it took us exactly 1 hour to travel that distance, we say that our speed was 70 miles per hour. But, at any point during that 1-hour time interval, had we looked at the speedometer, it (the speedometer) would have known that the car was going 70 miles per hour. How could that be when we (humans) had to wait a full hour and then figure out, afterwards, how far we had traveled? How could the car have known that before we had all of the data necessary to calculate the speed? The answer is that the car (or rather its speedometer) knows calculus.

Actually, we had used that fundamental concept of an equation to calculate the speed anyway. We had subtracted 100 miles (the starting point) from 170 miles (the end point) and then divided that by the time elapsed. In other words, we applied this equation:

speed   =   (position at end − position at beginning) divided by (time elapsed)

If we let speed be symbolized by v (velocity) and position at time t by $x(t)$ and time elapsed by Δt, we could write

$$v = \frac{x(t + \Delta t) - x(t)}{\Delta t} \tag{3.A2}$$

which is precisely the same thing as speed = (position at end minus position at beginning) divided by (time elapsed), and for this example $x(t + \Delta t) = 170$, $x(t) = 100$, and $\Delta t = 1$.

The problem with this method is that we may have been traveling at 90 miles an hour for the first half hour and 60 miles an hour for the second half hour and still come to the conclusion that the speed was 70 miles an hour. If we had been traveling 90 and then suddenly reduced our speed to 60, the speedometer would have read 70 for only a fraction of a second as our speed was being reduced. What is it that the speedometer is measuring—obviously not the average speed over an hour period?

The speedometer actually measures the speed when Δt is equal to (not exactly right, but bear with me for a moment) zero. Now, you certainly know that mathematicians will not allow us to divide anything by zero (sort of like getting infinity, which is a very difficult topic). If we cannot let Δt equal zero and apply Equation 3.A2, what can we do? We could quite easily let Δt be a little smaller. Rather than a full hour, we could record the distance traveled in half an hour and divide that distance by 0.5 or in a quarter of an hour and divide that distance by 0.25 or in a tenth and divide by 0.1 and on and on. And precisely what do we mean by "on and on"? That is the concept of a "limit." The speedometer is calculating the velocity (using Equation 3.A2) for a very small value of Δt and then for an even smaller value if Δt and then for an even smaller value. If I say "even smaller value" an infinite number of times, that is the same as saying that the speedometer "takes the limit" of the velocity as Δt approaches zero.

Mathematicians use the symbol "lim" to denote "take the limit of" and usually put a little arrow below it to indicate "as something approaches zero" (or some other value). In our case, the true velocity at some particular point between Chicago and Detroit (what the speedometer actually records) is

$$\lim_{\Delta t \to 0} (v)$$

If we let $[x(t + \Delta t) - x(t)]$ be symbolized by Δx (the Greek symbol Δ is frequently used to denote a "difference"), Equation 3.A2 turns into

$$v = \frac{\Delta x}{\Delta t}$$

We have already noted that the speedometer is actually recording the limit of the ratio of the distance divided by the time interval (limit as the time interval approaches zero). Thus, the velocity recorded on the speedometer is precisely that v that is approached as Δt approaches zero (we always have to say "as it approaches" rather than "when it becomes," since the equation really does not make any sense if we let its denominator *equal zero*). Thus, the last equation is modified (to see precisely what the speedometer sees) and becomes

$$\lim_{\Delta t \to 0} (v) = \lim_{\Delta t \to 0} \left(\frac{\Delta x}{\Delta t} \right) = \frac{dx}{dt}$$

The simple equation, $\lim_{\Delta t \to 0} (\Delta x / \Delta t) = dx/dt$, might be the most important equation in all of mathematics (value judgment, obviously). The symbol dx/dt is the derivative (pronounced "derivative of x with respect to t") and is the limit of the change in x as a function of the change in t or, if you wish to stick with the speedometer, the recorded value on the speedometer at a particular point in time. In other words, whenever you see a derivative, just think "rate of change."

Exponential Population Growth

In Figure 3.A1, we show data for the growth of a single population of an arbitrary plant population, represented as its total biomass.

One way of devising an equation to describe these data is to look at the change in numbers over time. For example, going from time 1 to time 2, the change was 0. From time 2 to 3, the change was 1. From time 9 to 10, the change was 6, and so forth. Thus, for times 1 through 10, the changes in numbers are 0, 1, 1, 1, 1, 2, 3, 5, and 6. Now we can ask, what is the ratio of the change to the number that originally existed, which is sort of like asking how many grams of biomass were added per individual gram of biomass (or, if we were representing the data as numbers of individuals rather than biomass, it would be the number of individuals added per capita)? For example, we have, for the change from time 8 to time 9, 5 extra grams that came from 10 original grams, and thus, the grams added per gram is $5/10 = 0.5$. Calculating this number, the grams added per gram, for each time interval we have 0, 1, 0.5, 0.33, 0.25, 0.4, 0.42, 0.5, and 0.4, which are highly variable, but in the end, they seem to stabilize at about 0.4.

Thus, we have computed the change per time at each individual time unit. We could easily suggest that a good model for these data would be the number at one time unit in the future equals the number now plus the added amount per that number. Thus,

biomass next week = biomass this week plus 0.4 times the biomass this week.

If we let the biomass next week be symbolized as $B(t + 1)$ and the biomass this week symbolized as $B(t)$, the above word equation simply becomes

$$B(t + 1) = B(t) + 0.4\,B(t)$$

which could also be written

$$B(t + 1) = 1.4\,B(t) \tag{3.A3}$$

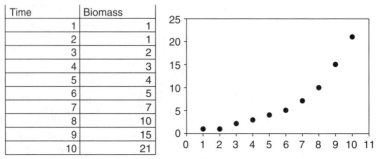

Time	Biomass
1	1
2	1
3	2
4	3
5	4
6	5
7	7
8	10
9	15
10	21

Figure 3-A1 Exemplary data for an arbitrary plant population. These artificial data represent biomass of the whole population over 10 time units.

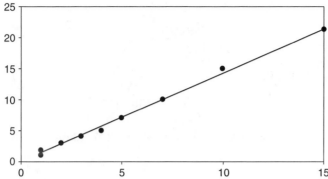

Figure 3-A2 B(t + 1) plotted against B(t), with the best fit linear regression providing an estimate of 1.43, which is very close to the expected.

If we use this equation as a model of the system, we can plot B(t + 1) versus B(t) and expect a linear relationship with a slope of about 1.4, as is done in Figure 3.A2.

To understand the rate of growth of a population at a particular point in time (much like reading a velocity on a speedometer), we need to make some simple changes in Equation 3.A3. First, take natural logs of both sides, to obtain

$$\ln[B(t+1)] = \ln(1.4) + \ln[B(t)]$$

By subtracting $\ln[B(t)]$ from both sides, we obtain

$$\ln[B(t+1)] - \ln[B(t)] = \ln(1.4)$$

Now, if we let the left hand side of that equation be the change over some time interval that is not equal to 1, we can write

$$\frac{\ln[B(t+\Delta t)] - \ln[B(t)]}{\Delta t} = \ln(1.4) \tag{3.A4}$$

(if you let $\Delta t = 1$, you get the original equation back) and then let the time interval become very small, to the limit of Δt approaching 0, we have, by the basic rules of calculus

$$\lim_{\Delta t \to 0} \left\{ \frac{\ln[B(t+\Delta t)] - \ln[B(t)]}{\Delta t} \right\} = \frac{d\ln(B)}{dt} \tag{3.A5}$$

And if you recall one of the basic equations of calculus (if you don't, just take it on faith to be true):

$$\frac{d\ln y}{dx} = \frac{1}{y}\frac{dy}{dx} \tag{3.A6}$$

We combine Equation 3.A4 with 3.A5 to get

$$\frac{d\ln(B)}{dt} = \ln(1.4)$$

and then apply the rule stipulated in Equation 3.A6 to get

$$\frac{1}{B}\frac{dB}{dt} = 0.336$$

and finally, multiplying both sides of the equation by B, we obtain

$$\frac{dB}{dt} = 0.336B$$

where it is clear that the 0.336 represents the per gram increase, which can be generalized to give the general model

$$\frac{dB}{dt} = gB \tag{3.A7}$$

where we can call the constant g, the growth rate of the plant population.

The Logistic Equation

The general form of Equation 3.A7 is the classic exponential equation and is frequently taken as a model of the growth of a plant population during its early phases of growth. However, Equation 3.A7 carries with it the consequence that it grows without limit, which no population can actually do. Thus, something must be added to account for lowered growth rate as the population accumulates more biomass.

Consider another artificial data set, presented in Figure 3.A3. If we were to presume that the exponential equation applied to these data, we would be saying that the actual per gram of growth per unit time is constant (that is what Equation 3.A7 says) and does not vary with the total biomass of the population at any given time. Thus, if we compute the per gram of growth for these data and plot that per gram biomass against the actual biomass, we would expect a flat curve, a nonsignificant regression between per gram growth and biomass. In Figure 3.A4, we compute those data and plot them for the artificial data of Figure 3.A3.

Clearly, a strictly constant per unit biomass rate of growth does not exist, but rather, there is a linear decline in that rate as a function of the biomass itself. This means that the constant g (which is what is plotted on the y axis of Figure 3.A4) is not really a constant, but rather a decreasing linear function of the biomass (which is what is plotted on the x axis of Figure 3.A4) or

$$g = a - bB(t)$$

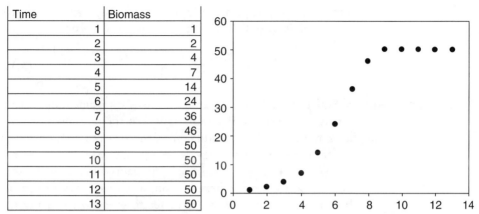

Time	Biomass
1	1
2	2
3	4
4	7
5	14
6	24
7	36
8	46
9	50
10	50
11	50
12	50
13	50

Figure 3-A3 Artificial data set representing longer term growth of a plant population that stabilizes its biomass at about 50 units.

which we can substitute into Equation 3.A5 to give us

$$\frac{dB}{dt} = aB - bB^2 \tag{3.A8}$$

where we have left the parenthetical t out of the equation to make it look simpler (remember that B does in fact vary with time, and if we were to be totally formal about it, we would write B(t) rather than B). Equation 3.A8 is known as the logistic equation. Another way of looking at this equation is that the rate of growth (g in Equation 3.A7) varies with the relative size of the population, relative to its maximum. Thus, a plant that is 50% of its maximum would be growing at a rate of $g = .5$ K, where K refers to the maximum

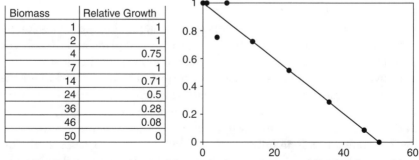

Biomass	Relative Growth
1	1
2	1
4	0.75
7	1
14	0.71
24	0.5
36	0.28
46	0.08
50	0

Figure 3-A4 Relative growth rate (change in biomass per unit biomass) versus biomass.

size the population can attain. At any given time, the growth rate of the plant will be proportional to its relative size, which is $(K - B)/K$, whence Equation 3.A7 becomes

$$\frac{dB}{dt} = gB\frac{K - B}{K}$$

(3.A9)

which is precisely equal to Equation 3.A8 with the appropriate rearrangement of the constants. Many ecologists prefer to work with Equation 3.A9 even though Equation 3.A8 may seem more straight forward to a mathematician, as the constants (K and g) in Equation 3.A9 have very clear biological meanings (K being the maximum size of the population, its carrying capacity, and g being its initial exponential-like growth rate). Furthermore, as will be demonstrated in the next section, when we speak of competition between two individuals, the formalities are easier to grasp when using the form 3.A9.

The Classic Competition Equations

We now ask what happens if we have two populations growing in competition with one another—struggling to acquire the same resources, be they nutrients or light. A quick look at Equation 3.A9 suggests a simple way of incorporating a second population into the system. If the overall rate of growth of the population (which is the derivative dB/dt) is "discounted" by the percentage of its maximum total size it has already achieved, will not a further discounting be introduced in a similar way by the introduction of another population? We could simply subscript the variables such that B_1 refers to the first population and B_2 refers to the second population and expand Equation 3.A7 to read,

$$\frac{dB_1}{dt} = gB_1\frac{K_1 - B_1 - \alpha_{12}B_2}{K_1}$$

(3.A10a)

and a similar equation for the other species

$$\frac{dB_2}{dt} = gB_2\frac{K_2 - B_2 - \alpha_{21}B_1}{K_2}$$

(3.A10b)

where the term α_{12} refers to the competition coefficient, relating the depression of plant growth of the first plant due to a unit biomass of the second plant, and the subscripts on the K indicate that the maximum size of each population may be different.

When analyzing the dynamics of a system, we normally begin by asking where the equilibrium points are located, and what is their nature, which means, are they stable or not? Stability here must be properly understood in its classic formulation. A stable point is one in which the equations representing the populations will act to return a perturbed system to its previous state, whereas an unstable system will act to reinforce any deviation from its initial state. A physical metaphor is usually useful. A stable point is like a marble at the bottom of a bowl. If it is disturbed by being pushed up a little bit on the edge of the bowl, it immediately moves back to the middle. An unstable point is as if the bowl is turned over and the marble balanced on the top of the upturned bowl. If nothing disturbs the marble, it remains stationary right at the top of that bowl—clearly an equilibrium state. The slightest deviation from that balanced position causes the marble to careen down the side of the bowl. The marble in the properly oriented bowl is a "stable" equilibrium point, whereas the marble balanced on the top of the overturned bowl is an "unstable" equilibrium point.

We seek to understand how Equations 3.A10a and b determine the equilibrium states of the two populations under consideration. First, we can compute the equilibrium points simply by tending to the definition of what equilibrium means, that the system remains unchanged in perpetuity—neither B changes. Mathematically, this means that the derivatives of the two equations are equal to zero. More important, we really want to know under what circumstances B_1 will either increase or decrease its biomass and likewise for B_2. We are interested not in just $dB/dt = 0$, but also in $dB/dt > 0$ and $dB/dt < 0$.

Taking Equation 3.A10a and setting the derivative equal to zero (with the ultimate aim of figuring out what values of B_1 and B_2 will give us that zero change value), we write

$$0 = gB_1 \frac{K - B_1 - \alpha_{12}B_2}{K}$$

which can be arranged to give

$$B_1 = K - \alpha_{12}B_2 \tag{3.A11}$$

the graph of which is simply a line on a graph of B_1 versus B_2. However, we are more interested in the question of when B_1 will increase or decrease, which is to say when dB/dt is greater or less than zero. So we can write,

$$\frac{dB_1}{dt} > 0 \rightarrow B_1 < K - \alpha_{12}B_1 \tag{3.A12a}$$

and, of course

$$\frac{dB_1}{dt} < 0 \rightarrow B_1 > K - \alpha_{12}B_1 \tag{3.A12b}$$

Relations 3.A12a and b can be represented graphically as a series of vectors in the geometric space of B_1 versus B_2. Equation 3.A11 is known as the isocline of the system and obviously represents the set of points (B_1 and B_2) for which B_1 does not change (since we derived it by setting $dB_1/dt = 0$). But we are more interested in those vectors that tell us how the system will change when it is not *exactly* at the set of points stipulated by the isocline. In particular, in Figure 3.A5, we present a graphical representation of Equation 3.A12a.

Figure 3-A5   Isocline and dynamic vectors for
Equation 3.A10a.

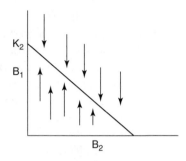

Doing the same process for Equation 3.A12b, we present its graphical representation in
Figure 3.A6. We are interested in how these two isoclines work together, and thus we need
to overlay Figures 3.A5 and 3.A6 on one another and look at the combined action of the
dynamic vectors of the two. In Figure 3.A7, we first lay the two graphs on top of one another
(left panel) and then clean up the vectors a bit (middle panel) and finally indicate the sum-
mation of the vectors (right panel), where the original (partial) vectors are shown in grey.

The overall sense of the dynamics of the system is easily seen in this picture. The vectors are tending toward the intersection of the two isoclines, meaning that the intersection is a stable equilibrium point and that the expectation of the competitive process is
that both plants will come to lie at an intermediate value of what would have been their
ultimate stature if they had grown alone.

Figure 3-A6   Isocline and dynamic vectors for
Equation 3.A10b.

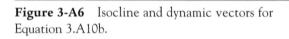

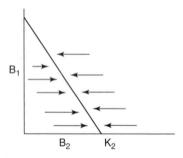

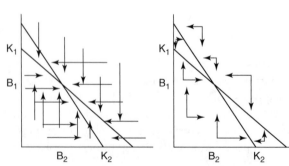

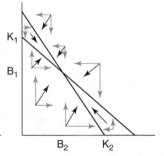

Figure 3-A7   Combining the vectors for the two isoclines.

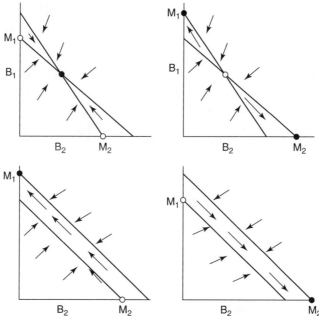

Figure 3-A8 The four possible outcomes of competition between two competing populations. Stable equilibria are illustrated with black circles and unstable equilibria with open circles.

The situation displayed in Figure 3.A7, however, is only one of four possibilities. Basically, it is possible to ask what the qualitatively distinct ways in which two lines can be drawn on a graph are. That question corresponds to what is actually a rather complicated analysis if you were to do it more analytically. There are actually only four possibilities, as shown in Figure 3.A8.

Thus, we see four distinct qualitative outcomes of competition. As reported before, the two populations may simply share the resources, both coming to coexist at a less than maximum size. However, it may be that one population will always dominate the other, as is the case in the bottom two graphs of Figure 3.A8. The most interesting case is when one population will always dominate the other, but which one dominates depends on where the starting point is in the growth cycle. If one gains only a small advantage over the other, it will dominate the process, but the reverse could be the case also. This is the situation in the upper right panel of Figure 3.A8, in which the equilibrium point is unstable (actually it is formally known as a saddle point, but for our purposes, it is only necessary to recognize it as unstable), and all trajectories of the system go either to K_1 or K_2. This situation is referred to as the indeterminate case, since the actual outcome of competition (which plant wins) depends on the starting point.

In all of the above there was a tacit assumption of "two-sided" competition going on between the two plants. That is, to write the basic equations in the first place, we assumed that the response of one population to the other was proportional to the biomass of that competing plant. However, plants are frequently characterized by what is known as "one-sided" competition, especially when the resource in question is light. A plant that shades another plant obviously exerts a far greater effect than if it only stood side by side. In other words, rather than the competitive effect being proportional to the biomass of the competitor, that effect is proportional to the biomass of the competitor relative to the biomass of the plant being affected. Thus, the basic Equations 3.A10a and b must be modified as follows:

$$\frac{dB_1}{dt} = gB_1 \frac{K_1 - B_1 - \beta_{12}\frac{B_2}{B_1}}{K_1} \tag{3.A13a}$$

$$\frac{dB_2}{dt} = gB_2 \frac{K_2 - B_2 - \beta_{21}\frac{B_1}{B_2}}{K_2} \tag{3.A13b}$$

which give rise to the isoclines

$$B_2 = \frac{K_1 B_1 - B_1^2}{\beta_{12}}$$

$$B_1 = \frac{K_2 B_2 - B_2^2}{\beta_{21}}$$

which are obviously quadratic equations in the variables B_1 and B_2 and thus provide an entirely new set of possible competitive outcomes, the most important of which is illustrated in Figure 3.A9.

In Figure 3.A9, there are three stable possibilities, depending on where the competitive process begins. If both populations start out growing with individuals that are about the

---

Figure 3-A9 Arrangement of two populations competing in a one-sided competitive arrangement.

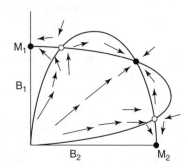

same size at the same time, they are likely to reach a stable equilibrium in which both persist, sharing resources as in the cohabitation state in the two-sided case above (Figure 3.A7). If one population begins with a slight edge, however, it is possible that it could dominate the competitive process, and the other two stable outcomes are a total dominance by either B_1 or B_2, depending on which one had the original edge.

The "Mechanism" of Competition

All of the previous development has assumed that the "phenomenon" of competition could be incapsulated in a simple constant, frequently called the "competition coefficient" (either the **a** of Equation 3.A10 or the **b** of Equation 3.A13). However, what is normally behind the competitive process is a competition for something, some critical resource that both populations seek. Explicitly incorporating the resource or resources into the development of competition theory is frequently referred to as a mechanistic approach, the "mechanism" of competition being the contention for some critical resources. At its most elementary level, we modify the exponential equation (Equation 3.A7) so that the parameter g is broken up into its component parts

$$\frac{dB}{dt} = gB = (b - m)B \tag{3.A14}$$

where b is the birth rate and m is the mortality rate. Suppose that the birth rate is a function of a particular resource that itself is supplied at a particular rate r and removed by the plant population at a rate a. The equation for that resource is, at its most elementary

$$\frac{dR}{dt} = r - aRB \tag{3.A15a}$$

and the birth rate of the plant population can be approximately assumed to be proportional to the amount of resource available, which means that Equation 3.A14 becomes

$$\frac{dB}{dt} = gB = (abR - m)B \tag{3.A15b}$$

where the parameter b refers to the conversion efficiency of R to B.

Equations 3.A15a and b represent the simplest way to incorporate explicitly (i.e., mechanistically) a resource into the basic exponential equation. The equilibrium value of R is easily calculated (set the derivative equal to zero in equation 3.A15b) as

$$R^* = m/ab$$

whereas the equilibrium value of B is (from Equation 3.A15a)

$$B^* = \frac{r}{aR^*} = \frac{rab}{am} = rb/m$$

whence we see the important result that the equilibrium value of the population is directly proportional to the supply rate of the resource (r).

This idea becomes especially important when considering competition between two species, B_1 and B_2. Modifying Equation 3.A15b

$$\frac{dB_1}{dt} = (a_1 b_1 R - m_1) B_1 \qquad (3.A16a)$$

$$\frac{dB_2}{dt} = (a_2 b_2 R - m_2) B_2 \qquad (3.A16b)$$

From Equations 3.A16a and b, we can ask what will be the final amount of resource available when each of the populations reaches equilibrium independently of one another. Thus,

$$\frac{dB_1}{dt} > 0 \Rightarrow R > \frac{m_1}{a_1 b_1}$$

and

$$\frac{dB_2}{dt} > 0 \Rightarrow R > \frac{m_2}{a_2 b_2}$$

Thus, B_1 will tend to drive R to the point $(m_1/a_1 b_1)$, whereas B_2 will tend to drive R to the point $(m_2/a_2 b_2)$, and unless $(m_1/a_1 b_1) = (m_2/a_2 b_2)$, one of the populations will exist in an environment in which it must decline. That is, one of the species will drive the resource to a point that is lower than the critical amount of resource necessary to sustain the second species. Because equilibrium values are frequently stipulated with a "star" (*) superscript, and in this case $Ri^* = m_i/a_i b_i$, this criterion for exclusion has been referred to as the R* criterion.[48] Thus, for example, if $(m_1/a_1 b_1) < (m_2/a_2 b_2)$, B_1 will drive the resource to the point $(m_1/a_1 b_1)$, which means that $R < (m_2/a_2 b_2)$, which means that B_2 will be unable to survive because of a low resource supply. This result is effectively the same result we saw before with the competitive exclusion principle. With a single resource (effectively equivalent to the same niche for each species), only one of the two species can survive. Although beyond the scope of this text, the point made earlier about nonlinearities possibly changing the outcome of competition is equally true in this mechanistic approach to the problem. Several authors[49] have made this point in a variety of dynamic modeling frameworks. While the competitive exclusion principle certainly remains a bedrock of thinking about population interactions, its conclusions must always be taken tentatively since they depend so strongly on simplifying linear assumptions.

The classic way of avoiding competitive exclusion in the mechanistic framework is to add an additional resource to the system. Thus, we have R_1 and R_2, and we presume

$$\frac{dR_1}{dt} = r_1 - (a_{11}B_1 - a_{12}B_2)R_1 \qquad (3.A17a)$$

$$\frac{dR_2}{dt} = r_2 - (a_{21}B_1 - a_{22}B_2)R_2 \qquad (3.A17b)$$

$$\frac{dB_1}{dt} = (a_{11}b_{11}R_1 + a_{21}b_{21}R_2 - m_1)B_1 \qquad (3.A17c)$$

$$\frac{dB_2}{dt} = (a_{12}b_{12}R_1 + a_{22}b_{22}R_2 - m_2)B_2 \qquad (3.A17d)$$

We retain the fundamental ideas of competition for each of the resources taken separately. The plant species that can reduce the level of the resource below that which would be required for the other species to survive would be the one to win the competition. However, now we have two resources and must face the possibility that one species could win on one resource, but the other species wins on the other resource, which sets up the possibility of coexistence. From Equations 3.A17c and d we can ask the simple question, when will dB/dt be greater than zero, whence we obtain

$$\frac{dB_1}{dt} > 0 \Rightarrow R_1 < \frac{m}{a_{11}b_{11}} - \frac{a_{21}b_{21}}{a_{11}b_{11}} R_2 = R_1^* - \frac{a_{21}b_{21}}{a_{11}b_{11}} R_2 \qquad (3.A18a)$$

$$\frac{dB_2}{dt} > 0 \Rightarrow R_2 < \frac{m}{a_{22}b_{22}} - \frac{a_{12}b_{12}}{a_{22}b_{22}} R_1 = R_2^* - \frac{a_{12}b_{12}}{a_{22}b_{22}} R_1 \qquad (3.A18b)$$

Endnotes

[1]Anonymous, 1944.
[2]Ayala, 1969.
[3]Gilpin and Justice, 1973.
[4]Levins, 1979.
[5]Armstrong and McGehee, 1980.
[6]Goldberg, 1990.
[7]Francis, 1986; Vandermeer, 1989.
[8]Vandermeer, 1989
[9]Vandermeer, 1981, 1989.
[10]Vandermeer 1989.
[11]Vandermeer, 1986; Garcia-Barrios et al., 2001.
[12]Vandermeer, 1984, 1989.
[13]See bibliographies in Francis, 1986; Vandermeer, 1989.
[14]Trenbath, 1976.

[15]Silvertown and Law, 1987; Silvertown, 2004.

[16]Levins, 1979.

[17]Haynes, 1980; Nair et al., 1979.

[18]Trenbath, 1976.

[19]Snaydon and Harris, 1979.

[20]Agamuthu and Broughton, 1985.

[21]Recent debate on this issue has emerged in the theoretical ecology literature. For a review, see Huston et al., 2000.

[22]Vandermeer, 1984.

[23]Risch et al., 1983; Andow, 1991.

[24]Kareiva, 1986; Root, 1973.

[25]Root, 1973.

[26]Lundgren, 1987.

[27]Young, 1989.

[28]Sanchez, 1995; Ong, 1994.

[29]Garcia-Barrios and Ong, 2004.

[30]Ong, 1994.

[31]Sanchez, 1995.

[32]Lin, 2007.

[33]Garcia-Barrios and Ong, 2004.

[34]Vandermeer, 1998.

[35]See the special volume of Agroforestry Systems, 2004.

[36]Johnston, 2003.

[37]Here, as elsewhere in the text, Goodman et al.'s (1987) schema of inputs to the farming operation being "appropriated" by capital as part if the industrialization process and outputs contributing to the "substitution" of value-added commodities on the market is used.

[38]Peet, 1969; Irwin, 2001.

[39]Goodman et al., 1987.

[40]Smith and Cothren, 1999.

[41]Fite, 1984.

[42]Collins, 1997.

[43]Friedland and Barton, 1981; Vandermeer, 1986.

[44]This point, as well as most of the insights in this section, come from the "Interviews with persons involved in the developers of the tomato harvester, the compatible processing tomato and the new agricultural systems that evolved," Sheilds library, UC Davis.

[45]U.S. Department of Veteran Affairs, 2009.

[46]Colborn et al., 1997.

[47]Relyea and Mills, 2001; Relyea, 2005.

[48]This criterion is commonly associated with Tilman and thus is frequently called "Tilman's R*." To be fair, as has been pointed out before (Abrams and Wilson, 2004), the idea is actually a rather old idea stemming back at least to Volterra in 1931.

[49]Levins, 1979; Armstrong and McGehee, 1980.

Chapter 4

Soils and the Emergence of the Industrial Approach

Overview

Evolution of the new husbandry in Europe was partly fueled by recognition of the need for healthy soil. The use of animals to produce manure was not just an unforeseen consequence of allowing animals to graze on the stubble left after harvest but a calculated strategy for reinvigorating the soil. These early farmers had evolved a system that effectively solved two problems: the disposal of animal waste and the fertilization of the soil (many years later, Wendell Berry would comment that modern animal confinement systems actually represent the construction of two problems from an elegant solution). Certainly, by modern standards, it was a good strategy.

Nevertheless, the way soil would be treated in agriculture changed dramatically in 1840 with a key conceptual breakthrough. In a Victorian Britain fascinated with an impressive collection of technological achievements attributed to science, the chemist Justus von Liebig articulated what has come to be called Liebig's law of the minimum, in which it was recognized that nutrients were required in specific proportions and thus a single nutrient in short supply would prevent plant growth—if your wheat plants needed more phosphorus than was available, the amount of nitrogen in the soil was irrelevant. The nutrient that was limiting plant growth was at its minimum with respect to that growth: thus the "law of the minimum."

This law led to the idea of providing the soil with specific nutrients, the limiting ones, so as to increase plant growth and eliminate the specific factor that was limiting it. Thus, in the mid 19th century, an incipient fertilizer industry arose, producing superphosphate with a simple chemical procedure and importing phosphate rock, guano, sodium nitrate, and potassium directly from natural sources. Mainly through the process of producing superphosphate, the fertilizer industry had become the British chemical industry's largest customer by the dawn of the 20th century.[1] Then, fueled by the need for nitrate explosives during the world wars, the industry grew ever larger and became one of the major components of the chemicoindustrial complex we see today. To appreciate its growth and development fully, we first must understand the basics of soil ecology as related to nutrient storage and plant nutrition. Some of this introduction relies on knowledge of basic chemistry. For readers who need a refresher on some of the basics, a short primer on relevant aspects of inorganic chemistry is presented in an appendix to this chapter (see Appendix 4.A).

The Basics of Soil Science—The Nature and Origin of Soils

What is soil and where does it come from? Soil is not a "thing" but rather a system, composed of five interacting factors: mineral matter, dead organic matter, water, air, and living organisms. It is best to think of the soil as a spongiform (structured like a sponge) mineral matrix, in which the spaces are filled with either air or water, with the water containing colloids and dissolved chemicals, plus a host of living organisms. The water with its colloids and solutes is referred to as the soil solution. Biological organisms act to hold the mineral matrix together, modify the dissolved chemical constituents of the soil solution, and ultimately, after death, form a major part of the collection of colloids in the soil solution.

Soil derives from three interacting factors: parent material, time, and vegetation. This way of looking at the genesis of soil is derived from the now classic equation of Jenny,[2] namely

$$S = f(cl,o,r,p,t)$$

where S is soil state, which is a function of cl (climate), o (vegetation type), r (topography), p (parent material), and t (time). Because climate determines vegetation type so strongly and topography is generally related to parent material, it is convenient to consider only the three main formative forces: parent material, time, and vegetation.

Before soil, there must be the material from which it will be made. The water, air, living organisms, and dead organic matter arrive only after the mineral matter matrix is established. This matrix is known as parent material and comes in a variety of forms, cycling over geologic time at a scale intermediate between rock cycles and ecological cycles.

The rock cycle begins with the formation of igneous rock, either the intrusive type in which molten rock cools beneath the earth's surface under pressure or the extrusive type in which it is formed by cooling in the air, typically in a lava flow. The igneous rock then is weathered by chemical, physical, and biological forces to become a parent material that eventually becomes a soil, which in turn is eventually covered by other lava flows, oceans, landslides, and other geological forces. The covered soil may then be subjected to intense pressure by the overburden, in which case it becomes a sedimentary rock (e.g., sandstone or shale). Through erosion, sedimentary rocks are exposed and begin the process of weathering, being broken, moved by water and ice, mixed with other rocks of various types, and occasionally covered again. Through pressure, the mixture of rocks may be fused to form metamorphic rock, which may again be exposed by erosion of the covering layers, starting the weathering cycle anew. During all of this erosion, burial, and reexposure, at any point, the rock may be submerged beneath a subducting tectonic plate and melt, forming lava, which when cooled becomes igneous rock. The cycle is illustrated in Figure 4.1.

Obviously, the soil cycle is embedded in the rock cycle, with soils forming and being buried at unpredictable intervals over geological time. Similarly, although not depicted in Figure 4.1, ecological cycles are embedded within the soil cycle. Within this hierarchical arrangement, a roughly logarithmic scale difference separates the rock from the soil from the ecological cycle. The rock cycle operates over millions of years, the soil over thousands of years, and the ecological over tens or hundreds of years.

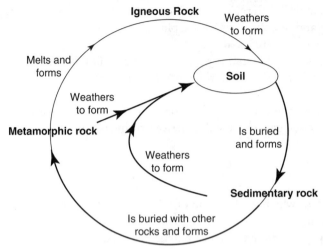

Figure 4-1 The basic rock cycle, with bold arrows illustrating the central position of the soil cycle.

Parent Materials

Parent materials are generally classified into the tripartite classification of 1) residual, 2) transported, and 3) cumulose, with the transported category accounting for most agricultural soils in the world. Residual refers to the weathering of the parent materials directly where they congealed, as in the weathering of an old lava flow. Transported parent materials are those whose geological origin was in a locality different from the point at which they are undergoing the process of soil formation, having been transported there by wind or water. Cumulose parent material is organic material deposited directly by plants.

Residual parent matter is usually classified according to the type of rock: igneous, metamorphic, or sedimentary. Of the igneous rocks, a variable chemistry makes any sort of classification that might be useful for soil science difficult, but frequently, a dual classification is employed. When high in silicon, aluminum, potassium, and sodium, igneous rocks are classified as silicic, whereas a high concentration of magnesium, iron (ferrous), and calcium is referred to as mafic. These two chemical classes are frequently crossed with physical classes, based on rates of cooling, rapid cooling (which generally gives rise to softer material) versus slow cooling (which generally gives rise to harder material). In Table 4.1, the four main classes of igneous rocks, relevant for soil parent material formation, are presented. Generally, fast-cooled igneous rocks weather more rapidly than slow-cooled rocks, and the type of secondary clay molecules formed (as discussed later) depends on whether the rock is silicic or mafic.

TABLE 4-1   Types of Igneous Rocks Resulting from Crossing the Chemical Characteristics of Igneous Rocks with Their Mode of Cooling

	Silicic	Mafic
Fast cooling	Rhyolite	Basalt
Slow cooling	Granite	Gabbro

Sedimentary rocks are mainly of three types: limestone, sandstone, and shale. The type depends on the sediments that had been covered in the process of their formation (see Figure 4.1). Limestone results mainly from marine deposits in which a variety of organisms uses calcium carbonate as a skeletal material. This calcium-based skeletal material settles to the bottom or is directly incorporated into rocky structures by coral reefs. Limestone originated in coral reefs is sometimes not regarded as sedimentary rock, but that resulting from accumulated marine sediments clearly falls in the sedimentary category. As limestone weathers, it leaves behind mainly the "impurities" it contained within its basic calcium carbonate structure, since the weathering process results mainly in volatiles such as CO_2 and ionics such as CaOH, which are largely carried off by water during the weathering process. Thus, depending on the exact composition of the limestone, the quantity of actual mineral material left for soil formation from the weathering process is highly variable.

Sandstone and shale result from the burial of sand and silt, respectively, and subsequent cementing action of pressure from above. They typically decompose into finer material quite rapidly under almost any exposed weather conditions. Sandstone presents a parent material that is usually similar chemically to silicic igneous rock, whereas the chemical composition of shale is highly variable.

When the residual parent material is metamorphic rock, we encounter an obvious difficulty in attempting any sort of summary statement, since the content is highly variable, and any characteristics are going to be largely the characteristics of the rock types from which it was formed.

Transported parent material presents a qualitatively distinct beginning for soil formation. Although residual material must be weathered into small particles, a process that may take hundreds of years, transported material typically arrives on site in a small particle form. Soil genesis based on transported materials is thus much more rapid than that based on residual materials. Classification of transported material is usually made based on the mode of transport: water, wind, or ice.

Flooding rivers usually carry an enormous amount of sediments with them, and when the floods recede, these sediments become alluvial deposits, forming particularly rich parent material. Especially in the humid tropics, one can frequently make an approximate map of soils of alluvial origin simply by looking for the long-term agricultural sites.

Alluvial plains represent some of the most attractive agricultural land in almost all areas of the world and have normally been the first to see the conversion to agriculture.

Lacustrine deposits, the result of sediments settling out of lakes, can also form quite rich parent material, but with particle size generally much smaller than alluvial deposits. The consequences of such small particle size are frequently poor structural quality, as explained later.

Eolian deposits, from wind transport, form some of the most important agricultural soils in the world. Most large grain-producing areas have soil based on material that was transported by wind, formally called loess, much of it deposited during the Pleistocene. Typical areas include the Mississippi valley and corn belt of North America, the plains of Germany, the Pampas of Argentina, and Shensi province of China, where loess deposits over 100 meters deep have been reported. A special form of eolian deposit is formed by pyroclastic volcanic eruptions, in which deep layers of volcanic ash form the parent material. Depending on the specific chemistry of the ash, this parent material usually contains most of the nutrients necessary for plant growth (typically lacking only nitrogen). Soils formed from volcanic ash, especially in tropical areas, are usually very rich, comparatively speaking. Generally, the nature of soils derived from eolian deposits varies enormously, depending not only on the physical and chemical characteristics of the deposit, but also on the time elapsed since the deposit was made.

Cumulose parent material derives from living plant matter and is the basis of one particular soil type, histosols. Although these soil types are not very common worldwide, they nonetheless can be extremely important agricultural soils. For example, the so-called muck soils of California, the basis of a multimillion-dollar vegetable business, are derived from this parent material.

Weathering

After the parent material is in place, the processes of weathering begins, slowly converting the parent material into the system called soil. These processes are commonly classified into physical, chemical, and biological, even though they operate simultaneously and in constant cause and effect relation to one another.

Physical weathering is most important when the parent material is residual. In high latitudes, seasonal freezing and thawing are key, water entering any available space and breaking parts of the rock on freezing. Similar forces in tropical areas are wetting/drying and heating/cooling, although their effects are not nearly as dramatic as freezing and thawing. As wind blows particles, their continual physical contact with parent material produces an effective grinding of that material. Finally, one of the most important of the weathering forces results from biological action. Plants disperse into any available opening in the parent material, and the force of growing plant roots is equivalent to, if not sometimes greater than, freezing water. As discussed later, biological action also has effects on chemical weathering and other soil formation processes, but its physical effect alone is always substantial.

Chemical weathering begins immediately after the parent material is deposited. One of the first types of chemical weathering is hydrolysis, due to the almost ubiquitous availability of water. For example, the mineral orthodase ($KAlSi_3O_3$) reacts with water to produce acid silicate ($HAlSi_3O_3$) and potassium hydroxide, thus releasing potassium ions into the soil solution. Potassium ions, in turn, are not only important for plant nutrition, but also are involved in the chemical construction of secondary clays, an extremely important component of the soil system, as discussed later. As a general phenomenon, hydrolysis is essential to chemical weathering.

With biological activity, carbonic acid is inevitably released into the soil solution. Carbonic acid then can act to promote carbonation of many of the original minerals of the parent material. For example, calcite ($CaCO_3$) may be acted on by carbonic acid to produce calcium bicarbonate [$Ca(HCO_3)_2$], a source of calcium ions, important for both plant nutrition and formation of secondary clay minerals.

It is important to note that the process of weathering is a continual process in soil development. It is not simply a question of parent material being established, weathering happening, and a soil emerging. All soils are always dynamic, continually under the influences of weathering, not just at a stage we can recognize as their "origin." Indeed, the practice of agriculture can dramatically modify the weathering process, either "improving" the quality of the soil for agriculture or, unfortunately more frequently, degrading it.

Time and Vegetation

In combination with the forces described in the previous section, the time passed since the original deposition of the parent material is of crucial importance. From an ecological perspective, primary succession involves the initial colonization of organisms onto parent material and the subsequent sequence of organisms that develop, eventually culminating in the so-called climax community. Primary succession is thus the biological side of soil formation. As time passes, various stages of the successional process pass, and with each new phase, the soil develops further. Thus, for example, the plant community invading the ash deposits from a volcanic eruption on the pacific coast of Nicaragua is an early successional plant community, and the soil on which it lives is barely more than original parent material. Although it is clearly a soil and contains all of the necessary ingredients normally contained in a soil system, one can easily recognize the parent material, volcanic ash. In contrast, the forest on the Caribbean coast of Nicaragua is late successional vegetation, and the soil on which it lives is very highly developed. It is difficult if not impossible to recognize what the parent material may have been, although from other evidence it is clear that this too was a soil originally derived from volcanic ash. The first example is a soil of about 100 years of age, whereas the second is at least 10,000 years old.

Closely related to the effect of time is the type of vegetation that develops. For example, because of the high probability of fire associated with various physical forces acting in consort, the area of North America in the southern great lakes region has seen the development of extensive pine forests. The soils under these pine forests tend to be quite acid, a consequence of the leachate from the pine needles. A nearby soil in an area protected from fires will generally develop a beech/maple forest and give rise to a soil with far higher pH. The age of both soils is about 10,000 years, yet their chemical reactivity is quite different, mainly because of the vegetation that had developed on them, which in turn was a consequence of a climate that either did or did not allow for periodic fires.

Of course, it is not possible to disentangle completely the effects of parent material, climate, time, and vegetation, since they all act dialectically and simultaneously. We cannot, for example, ask what would happen if a tropical rain forest were to develop in a cold northern climate.

The Main Types of Soils

The history of classifying soils is long. Probably when agriculture first evolved, the first farmers noted that some soils were good and others bad. Historical records indicate that at least the Egyptians, Sumerians, Ancient Greeks, and Romans all had soil classification systems of one sort or another, and Darwin clearly noted distinct classes of soil in his famous treatise on earthworms.[3] But the modern idea of constructing systems of classification for soils was born with the ideas of a Russian soil scientist, V. V. Dokuchaev, in 1883. Dokuchaev and his coworkers were the first to state unambiguously the relationship between climate, vegetation, and soil characteristics, a basis for many subsequent classification schemes. The system most used today is that invented by the United States Department of Agriculture, a system that has now gone through a variety of revisions.[4] Although the U.S. Department of Agriculture system is based on measurable parameters and not on presumptions about either parent material or developmental history, its specific categories are, not surprisingly, related to all of the soil formation processes. It is thus convenient to use this system of classification to understand how soil development processes have created the worldwide distribution of soils that we see today.

The system uses 12 major categories, the soil "orders," within which a variety of subcategories are embedded. For our purposes, we shall ignore the subcategories. The 12 soil orders are sufficiently descriptive. They can be conveniently categorized into 1) those soils more or less defined by the parent material that went into forming them (histosols, gelisols, vertisols, and andisols), 2) very young and poorly developed soils (entisols and inceptisols), and 3) soils highly correlated with particular vegetation types (aridisols, mollisols, alfasols, spodosols, ultisols, and oxisols). The classification system, in fact, uses measurable physical attributes in its definitions, a fact that makes this system highly objective. Nevertheless, correlations associated with the formative processes help to understand the general patterns of origin and give some indications as to the potential limitations in agroecosystems.

Figure 4-2   Harvesting peat cubes for fuel in England.

Classification Based on Parent Material

Of the four soils defined by their parent material, none is so completely thus defined as the histosols. These are organic soils. They are typically formed in wetlands where incompletely decayed dead plant matter accumulates and forms the soil. So laden with organic matter, these soils are sometimes mined for fuel. The peat bogs of Scotland and Ireland, for example, represent thousands of years of accumulation of organic matter (Figure 4.2). Highly compressed and only partially decomposed, this material differs only quantitatively from coal or petroleum, and indeed had been used for centuries for home heating and cooking in the northern British Isles and other parts of Europe. The famous "muck" soils of North America are of this sort. The highly profitable vegetable industry of California, for example, is based on this soil order. Generally, histosols can be very productive when converted to agriculture (which usually entails draining the wetlands where they are found), although they also can present difficult management problems. For example, because they are mainly organic matter they decompose—they "disappear" as they are being used.

Gelisols are soils that are perched on top of permafrost. They are extensive in polar and subpolar regions and typically have a restricted horizon structure. For all practical purposes, they are irrelevant for agriculture.

Vertisols are one of the most easily recognized soils in the world, at least when they are dry. They are characterized by shrinking and swelling clays (discussed below), such that

Figure 4-3 Typical vertisol cracks during dry season.

when they dry they form characteristic cracks (Figure 4.3). Wind erosion causes small particles to fall into these cracks, and the subsequent swelling of the clays in the wet season causes further miniaturization of the particles due to a sort of "vertical compaction." Not surprisingly, these soils occur in areas where there is a significant dry season, although the importance of fine structured parent material cannot be overstated. These soils frequently pose difficult problems for agriculture since the changes from wet to dry involve dramatic physical modification. When a dry vertisol begins to take up water, at first the rate of water penetration is very rapid as the water falls quickly into the cracks. Just as quickly, the clays begin to swell, the soil effectively closes, and puddles begin to form. These physical changes can have drastic effects on some plant roots and even make road building tricky. In northwestern Nicaragua, for example, there are extensive deposits of vertisols, and one can approximately map them by looking at the condition of the highways. An extremely deteriorated highway frequently indicates the presence of an underlying vertisol, which shifts around so much during its wetting and drying phases that it causes the highway to buckle and crack.

Because of the physical limitations imposed by vertisols, they are frequently used for low-intensity activities such as extensive cattle ranching, although less heavy (i.e., with lower clay content) vertisols are also used for producing cotton in the United States, maize, millet, and sorghum in Ethiopia, the Sudan, and India. In scattered areas around the tropical world, heavy vertisols are also used for paddy rice production because their puddling behavior is convenient for the construction of paddies.

Andisols are based on relatively recently deposited volcanic materials. Deep layers of volcanic ash form the basis for some of the most productive areas of the tropics, including the western seaboard of meso-America, the island of Java, and some areas of Ecuador and Colombia. Also, the highly productive wheat-growing soils of northwestern North America are andisols, as are some of the medium productive soils of Japan.

Andisols typically contain a colloidal fraction that makes them highly variable in plant nutrient availability, even though the nutrients may exist in abundance in the soil solution. This paradox is because the availability of nutrients in ionic form in the soil solution is dependent on the pH of the soil, largely because of the unique way in which the soil colloids are formed. In this aspect, andisol colloids act like the colloids in the much older and more highly weathered soils of tropical rain forest areas.

Young Soils

There are two soil orders whose characteristics are, effectively, that they have few recognizable characteristics. Entisols are barely weathered parent material, with very little horizon formation. Inceptisols are soils with only slightly more formation. Entisols can represent the extremes of productivity, from highly fertile recent alluvium on the flood plains of the world's major rivers to the barren sand dunes bordering the world's lakes and oceans. Entisols by definition form whenever new parent material becomes stabilized and therefore can form in any climate and associated with any vegetation.

Whatever one can say about entisols, the same applies to inceptisols, except they are somewhat more advanced in development. There is thus a continuum from parent material to entisol to inceptisol along the normal pathway of soil formation from any parent material except organic, volcanic, or extreme swelling and shrinking clays (which constitute histosols, andisols, and vertisols, respectively).

Classification Correlated with Vegetation

The remaining six soil orders are specifically associated with particular vegetation types, although exceptions exist. Beginning with the driest conditions, desert vegetation usually gives rise to aridisols. These soils are usually not suitable for agriculture, but not because of their inherent structure or chemical constituencies. They exist in areas where water is scarce. If irrigation is available, they can be among the most productive soils in the world. Indeed, some of the earliest advanced civilizations, including Egyptian, Harrapan, and Sumerian, had extensive areas of aridisols under irrigation. On the other hand, under irrigation, aridisols are susceptible to the accumulation of salts in the upper layers. The Hohokam civilization of southwestern North America, for example, traces its rise to the invention of irrigation on aridisols, as much as its decline to the salinization of those same aridisols.

Less zeric conditions usually favor the development of grasslands, which depend on biological and climatic conditions that allow fire to be a formative element. Temperate grasslands usually give rise to the most productive soils in the world, the mollisols. Mollisols are characterized by a large amount of organic material in the upper layers of the soil, the result of an accumulation of cycles of burning and deposition of dead plant material over long periods. The three major extensive areas of mollisols in the world are the famous corn belt of North America, the extensive grasslands of Argentina, and the enormous belt of grasslands extending from eastern Europe to northern China.

These soils typically have a large reserve of organic matter that slowly decomposes and releases nutrients to the soil solution. Because of this, they have been especially important in industrial agriculture. What aridisols were to the first advanced civilizations, mollisols were to the development of modern industrialized societies, based on petroleum and cheap food products made from cereal crops, most of which are grown on mollisols.

Initiating agriculture on native mollisols, however, was not a simple activity. Indeed, the westward expansion of Europeans in North America was stalled for many years at the edge of the "prairie" because of the physical problems presented by mollisols. The eastern third of the continent is dominated by alfasols (discussed later), which were covered with forest. To initiate a farm on alfasols, all that was necessary was to cut down the trees, burn the trunks, and till the soil. Once westward expansion reached the central grasslands, the accumulation of extensive root layers of the grasses formed a rather impenetrable mat known as sod, which made the ax and match unsuitable tools for farm initiation. Not until the extensive use of the moldboard plow by migrating teams of oxen driven by "plowboys" were these extensive deposits of mollisols available for agricultural production. So strong was this layer of sod that it was frequently used as construction material (Figure 4.4).

As we move to more mesic conditions in the temperate zone, hardwood forest tends to dominate the vegetation, and the soils usually formed under this vegetation type are alfasols. These soils do not have the rich organic top layer of mollisols, but they are rich in exchangeable bases (Ca^{2+}, Mg^{2+}, etc.). They are the basis of agriculture in northeastern North America, much of western Europe, and the extensive croplands throughout Asia north of the mollisol belt.

Figure 4-4 Sod house in late 18th century North American prairie. Thousands of such houses dotted the prairies of North America at this time.

Temperate alfasols have developed under conditions of extreme nutrient pulsing associated with deciduousness of the vegetation. Each spring the partially decomposed organic material from the fall leaf drop enters the soil as a pulse, and by the following fall the organic matter is fully decomposed, and the released nutrients are reabsorbed by the vegetation in a relatively closed cycle. Because of the relative abundance of organic matter and usually an excellent clay mineral structure, these soils probably have excellent ecological buffering capacity. For example, in the middle of a clearing made by a fallen tree, it might be expected that the normal cycle of summer absorption of the minerals released in the spring nutrient pulse would be disrupted since the tree that formerly held that spot is no longer capable of absorbing those nutrients. However, the favorable colloidal collection in the soil solution is capable of banking those nutrients, effectively holding them in reserve for when the vegetational succession proceeds toward the climax state (as explained in the section on soil colloids). This ecological buffering capacity is probably responsible for the success of these soils after agricultural conversion. With agriculture, as in the case of a clearing formed from a fallen tree, we would expect that the normal nutrient cycling system might be interrupted due to the trees having been removed, an expectation that is not realized because of the high ecological buffering capacity of temperate alfasols. This point is especially important when compared with tropical ultisols, as discussed later.

Another major vegetation formation that produces alfasols is the tropical savannah or grassland. Unlike temperate grasslands whose annual burning continuously deposits organic matter that never completely decomposes because of the relatively cold conditions, tropical savannahs present a hotter environment which leads to more complete decomposition every year, and thus, the typical rich upper layer of mollisols is never formed. Thus, the soils that form under tropical grasslands and savannahs are usually alfasols. Tropical alfasols have distinct characteristics from temperate alfasols, frequently lacking the characteristic ecological buffering capacity.

When climatic conditions are appropriate for the development of conifer-dominated forests, water leaching from fallen leaves (needles) tends to be highly acidified. This continual acid input into the soil leads to a characteristic soil development with high acid reaction and a distinct surface layer of organic material and aluminum oxides (and sometimes iron oxides) in the subsurface. These soils are the spodosols. They are usually very poor for agricultural production, as the extreme acidity has resulted in the complete leaching of most cations. Frequently, spodosols are associated with sandy soils and fire, the sandy nature permitting very dry conditions to develop in dry periods, thus inviting fires to spread through the vegetation and promoting dominance by fire resistant species such as pines. On the other hand, they do not develop if dry conditions persist during the entire year. Leaching of aluminum and iron oxides is a consequence of relatively high rainfall, at least during part of the year. These soils most characteristically develop in temperate latitudes, where most conifer-dominated forests are located.

With higher rainfall, typically in hotter conditions than those that give rise to alfasols, are found the ultisols, soils that may be thought of as highly weathered alfasols. These soils are formed under forests, ranging from southern hardwood temperate forests to tropical rain forests. Those formed under tropical rain forests are especially acid with low nutrient holding capacity and typically poor for agricultural purposes. Unlike the temperate alfasols that have strong ecological buffering capacity, tropical ultisols have the kinds of soil colloids that are not capable of storing nutrients well. Thus, when the natural vegetation is removed, much of the ecosystem's nutrient pool is rapidly leached out of the soil. Although poor in natural fertility, ultisols have been managed with large fertilizer input in southeastern North America with highly productive results.

Oxisols are the classic poor tropical rain forest soils. They effectively represent an extremely weathered ultisol, in which the soil colloidal properties are incapable of storing nutrients. They are usually very acid with large accumulations of iron and aluminum oxide clays. Furthermore, the depth of weathering in oxisols is normally much greater than for any other soil type. Although oxisols are not necessarily low in organic matter under normal conditions of forest cover, if the forest is cleared, the hot wet conditions cause rapid decomposition of remaining organic matter, and the soil can be left with a virtual absence of it. Couple this with a clay fraction dominated by hydrous oxides of aluminum and iron (as discussed later), and the result is a soil with extremely low natural fertility. In many areas of oxisols (and tropical ultisols as well), the poor nutrient holding capacity has apparently resulted in plants evolving specific mechanisms for obtaining nutrients, one of which is to forage near or even on the surface of the ground for nutrients. Consequently, many rain forests have root mats that exist literally above the surface of the soil, absorbing nutrients before they even enter the mineral soil matrix, a process sometimes referred to as direct nutrient cycling.

These then are the 12 soil orders, a summary of which appears in Table 4.2.

Physical Properties

Various soil physical features are important for ecological interactions. These features are a consequence of the soil's basic physical nature under the influence of chemical and biological modifiers. At the most elementary level, consider a soil made up of uniformly round particles of exactly the same size, a bunch of marbles in a beaker (Figure 4.5). Even this simple model suggests an important physical feature. The soil, in this case symbolized by the space taken up by the marbles in the beaker, contains both a solid area composed of the marbles and the spaces between the marbles. At the most basic level of definition, the soil system is the solid area plus the spaces between the solids, plus everything that goes on in the space between the solids—physical, chemical, and biological. Typically, the space between the solids is occupied by either air or water, and that part occupied by water is known as the soil solution.

TABLE 4-2 Summary of the 12 Soil Orders				
I. Young soils				
With very little horizon development	Entisols			
With somewhat better horizon development	Inceptisols			
II. Soils defined by parent material				
Nature of Parent Material	Soil Order			
Organic	Histosols			
Volcanic	Andisols			
Fine swelling and shrinking clay	Vertisols			
Arbitrary, but on top of permafrost	Gelisols			
III. Soils correlated with vegetation				
Vegetation type	Desert	Grassland	Hardwood forest	Coniferous forest
Temperate	Aridisols	Mollisols	Alfasols Ultisols	Spodosols
Tropical	Aridisols	Alfasols	Ultisols Oxisols	Spodosols

Basic Physical Measurements

In Figure 4.5, the soil fills the beaker—the size of the beaker is the volume of the soil. We can learn the volume of the space between the particles by filling the beaker with water (with the marbles in place) and then decanting the water and measuring its volume. Then the volume occupied by the particles themselves is simply the volume of the beaker minus the volume of the decanted water (the volume of the space between the particles). The particle density of the soil is simply the mass of the particles (estimated by the weight of the marbles in the beaker) divided by the volume occupied by those particles, usually given in terms of megagrams per cubic meter. The volume of the soil is the volume of the particles plus the volume of the spaces between them, but the particle density is calculated with only the volume of the solids. In contrast, the bulk density is the mass of the particles divided by the volume of the soil, which is equal to the mass of a unit of dry soil, again given in megagrams per cubic meter. The ratio between bulk density and pore density ($\times100$) is the percent pore space.

Bulk density, particle density, and percent pore space are interrelated concepts that have important consequences for the agroecosystem. At the theoretical level, these

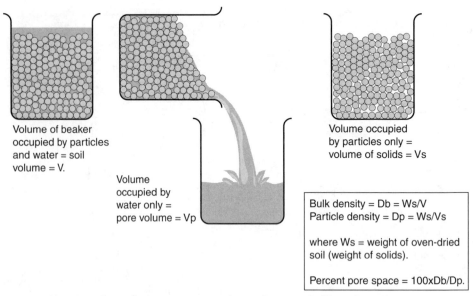

Volume of beaker
occupied by particles
and water = soil
volume = V.

Volume
occupied by
water only =
pore volume = Vp

Volume occupied
by particles only =
volume of solids = Vs

Bulk density = Db = Ws/V
Particle density = Dp = Ws/Vs

where Ws = weight of oven-dried
soil (weight of solids).

Percent pore space = 100xDb/Dp.

Figure 4-5   Metaphorical representation of some basic soil physical attributes.

concepts are obvious, but in practice, there are many complicating factors. First, particle sizes are never uniform, as was illustrated in Figure 4.5, but rather always exhibit a range of sizes. The relative distribution of particle sizes is the *soil texture*, as discussed later. Furthermore, the nature of the biological forces operative in the soil can change the effective nature of the soil particles, a concept giving rise to *soil structure*, as discussed later. Finally, soil organisms such as earthworms and termites can change the nature of the pore space by constructing tunnels throughout the soil, thus changing the distribution of pore sizes in the soil, a phenomenon discussed in Chapter 5.

Soil Texture

The distribution of soil particle sizes is soil texture. The classification of solid soil material, according to the International Society of Soil Science, is gravel (>2.0 mm), sand (between 0.02 and 2.0 mm), silt (between 0.002 and 0.02 mm), and clay (<0.002 mm). (In the classification system used by the U.S. Department of Agriculture, the dividing line between sand and silt is placed at 0.05 mm, instead of 0.02 mm.)

The largest particles normally considered part of the soil are those in the sand category. They are usually irregularly shaped, and because of their very large size, when packed together, they leave quite a lot of air space between them. For this reason, soils dominated by sand tend to dry out rapidly. Sand particles are dominated by quartz (SiO_2), although other minerals occur also, such as feldspars (aluminosilicates) and micas (iron and aluminum silicates).

Silt particles are intermediate in size between sand and clay. They are irregular in shape and usually dominated by quartz. Frequently, silt particles are coated with a layer of clay, making them somewhat sticky and able to absorb water, although this quality stems from the clay coating, not from the silt particle itself.

The smallest-sized particles are the clays. The smaller clays are usually held in colloidal suspension in the soil solution, and because of their small size, their surface area per unit mass is enormous. For example, you can fit about 1,000,000 clay particles in the area occupied by a single sand particle, and each of those clay particles has about 100 times the surface to volume ratio as the sand particle! Because so many features of soil physics and chemistry are consequences of surface area phenomena, the colloidal clays are thus extremely important components of the soil. They are extensively discussed later.

Classifying soils according to texture is effectively determining what fraction of each of the three categories exists in a soil. A sand soil, a silt soil, or a clay soil are obvious categories. Loam refers to a soil that contains all three in significant quantities, with an adjectival distinction for various grades (e.g., sandy loam, clay loam). Generally, a loam soil is usually best for agricultural production because too much clay makes the soil difficult to work as well as difficult for plants to take up water (as discussed later), and too much sand does not allow for much water retention.

The U.S. Department of Agriculture developed a classification scheme for textural classes that seems to be quite generally accepted. It is illustrated in Figure 4.6, along with the graphical procedure for determining to which textural class a soil belongs, from a knowledge of the proportional representation of the three particle size classes. It is only

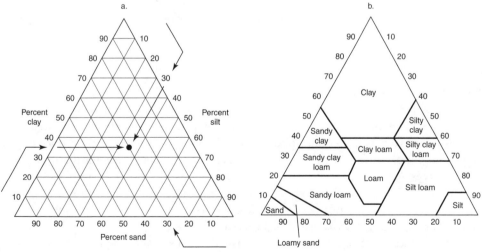

Figure 4-6 The U.S. Department of Agriculture textural classification scheme. The point on the left graph is of a soil containing 35% clay, 35% silt, and 30% sand.

with great difficulty that one can change a soil's texture. Adding large quantities of sand, for example, will change the texture and is frequently done for greenhouse operations. At the level of field production, however, the expense of hauling large quantities of sand is prohibitive for only small changes that are possible in texture.

Soil Structure

Basic soil particles rarely exist in the soil as distinct elements, but rather are held together by the stickiness of clays and clay skins, the cementing action of organic matter, and the chemical properties of colloids. The unit that is thus held together is referred to as an aggregate or ped. The nature of the peds and their interaction with one another is called soil structure and, unlike soil texture, is easily modified, sometimes unintentionally by agricultural practices. Maintaining or improving soil structure thus refers to the maintenance or improvement of the ped structure of the soil, a topic to which we return in Chapter 5.

Chemical Properties

Plant nutrients are made available through the contact of root hairs with soil solution. Exactly how the nutrient materials wind up in the rhizosphere along with how they come to be absorbed by the root hairs is fundamentally a chemical story. Key to understanding this chemistry is a basic feeling for the way soil colloids function in the chemical dynamics of the soil solution, more or less as a bank into which nutrients can be deposited and stored until they are needed for plant nutrition. We thus include here a rather lengthy discussion of soil colloids, beginning with a brief survey of the types that exist, followed by a qualitative description of how they function in the chemistry of the soil. We then discuss in detail the formation and function of silicate clays, one of the most important classes of soil colloids.

Types and Properties of Soil Colloids

Soil colloids are characterized mainly by their extremely small size, on the lower end of the particle sizes that are known as clays in soil textural analysis. Although their absolute size is very small, as molecules, they are huge, frequently with an orderly crystalline structure, but also occurring as rather amorphous material. Most important, their surface area is very large compared with their weight, a product of such a small size. Furthermore, they carry charges on their surface that ultimately determine much of the chemistry of the soil solution. In addition to their very large surface area, many of the colloids also have "internal" surfaces, parts of the crystal that are effectively open to the soil solution and charged as much as the actual external surface itself.

These surfaces contain both positive and negative charges, although the overwhelming number of charged sites are negative. These sites are the points on which cations (or anions if the charge is positive) can be adsorbed. Adsorbed cations cannot be easily leached out of the system, resulting in the "banking" function of colloids, alluded to previously. However, when arriving in the rhizosphere, the colloids may exchange some of the adsorbed nutrient ions for other ions made available on the surface of the root hair, thus making the nutrient ion available for uptake by the plant root. Consequently, both for the storage of nutrients (and the ecological buffering capacity that implies) and for making nutrients available to plants, soil colloids play a crucial role.

The four basic types of soil colloids are layered silicate clays, hydrous oxides of aluminum and iron, amorphic clays, and organic colloids. Layered silicate clays and organic colloids are the most important for plant nutrition generally and will be discussed in detail. Amorphic clays are important in some soils, especially andisols, and present opportunities and problems that are quite special. Hydrous oxides of iron and aluminum are mainly the product of excessive weathering of parent material and layered silicate clays and indicate very poor soil quality when they dominate. Unfortunately, they are extremely common in tropical soils, presenting daunting problems for tropical agriculture.

Layered Silicate Clays

The word clay is used in at least three different forms in the soils literature: 1) a soil textural class (as described previously, also see Figure 4.6), which is a categorical use, 2) a particle size class, which is a physical usage, or 3) a particular type of mineral, which is a chemical usage. When referring specifically to electrostatic properties, crystalline structure, and the like, the chemical usage is clearly intended.

The basic structural units of layered silicate clays are tetrahedral crystals of silicon and oxygen and octahedral crystals of aluminum (or magnesium) and oxygen and hydroxide. The fundamental structure of the tetrahedra and octahedra are illustrated in Figure 4.7. Both crystal types have unsatisfied negative charges on all of the exterior atoms and thus attract one another strongly. The consequence is the formation of large sheets of crystals, arranged such that a sheet of octahedral crystals is sandwiched between two sheets of tetrahedral crystals, as illustrated in Figure 4.8. This basic arrangement forms a "platelet" in which the individual atoms of the three sheets (tetrahedral-octahedral-tetrahedral) present a distinct layering of atoms beginning with a plane of oxygens, followed by a plane of silicon and then an O, OH plane, followed by an Al, Mg plane, followed by another O, OH plane and then another Si plane, and ending with an oxygen plane, as shown in Figure 4.8.

An individual platelet may be very large in its planar extent but is always as thin as indicated by the three-layered structure. However, platelets may be linked together by adsorbed cations and water, forming much larger structures. The nature of this interlayer can be very important in determining the reactive nature of particular clay types.

As these clays crystallize, depending on what ions are available in the solution in which the crystallization is occurring, the normal position of a particular ion may be occupied by some other ion of a similar size. This is known as isomorphic substitution. It is the main source of surface charges on clays, the substituted ion either providing an extra positive charge or one less positive charge than the ion that normally sits in that position. For

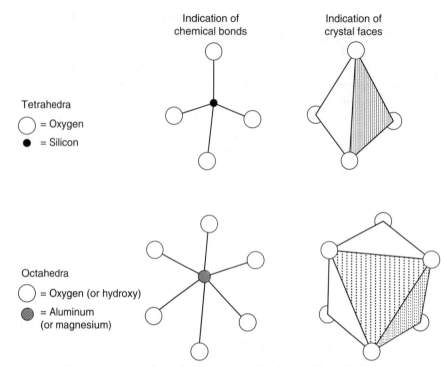

Figure 4-7   Basic structure of tetrahedra and octahedra in clay colloids.

example, if aluminum, which is only slightly larger than silicon, comes to lie at the center of one of the tetrahedra in the tetrahedral plane, the normal four-plus charge of silicon that, in conjunction with the oxygens surrounding it, contributes to the electrical neutrality of the clay particle as a whole, is now without one of the plus charges that position normally contributes to the crystal. Thus a single negative charge must remain unsatisfied in that crystal, because aluminum is plus three and silicon is plus four. If this sort of isomorphic substitution occurs in thousands of places in a single clay particle, it results in thousands of negative charges on the surface of the crystal. Because there are hundreds of thousands or even millions of "normal" ions involved in creating the basic crystalline structure, the basic identity of the clay is not really different, even though there is a large number of isomorphic substitutions.

Classification of silicate clays is usually based on the relative proportions of tetrahedral versus octahedral sheets they contain. There are some with the basic structure of a single tetrahedral sheet to each layer of octahedral sheet, the 1:1 clays. Although many clay minerals take this form, kaolinite is, for all practical purposes, the only one in soils. When several of the hydroxyl ions are replaced by oxygen ions, whole sheets of kaolinite are sandwiched together through hydrogen bonding. Other ions, and notably

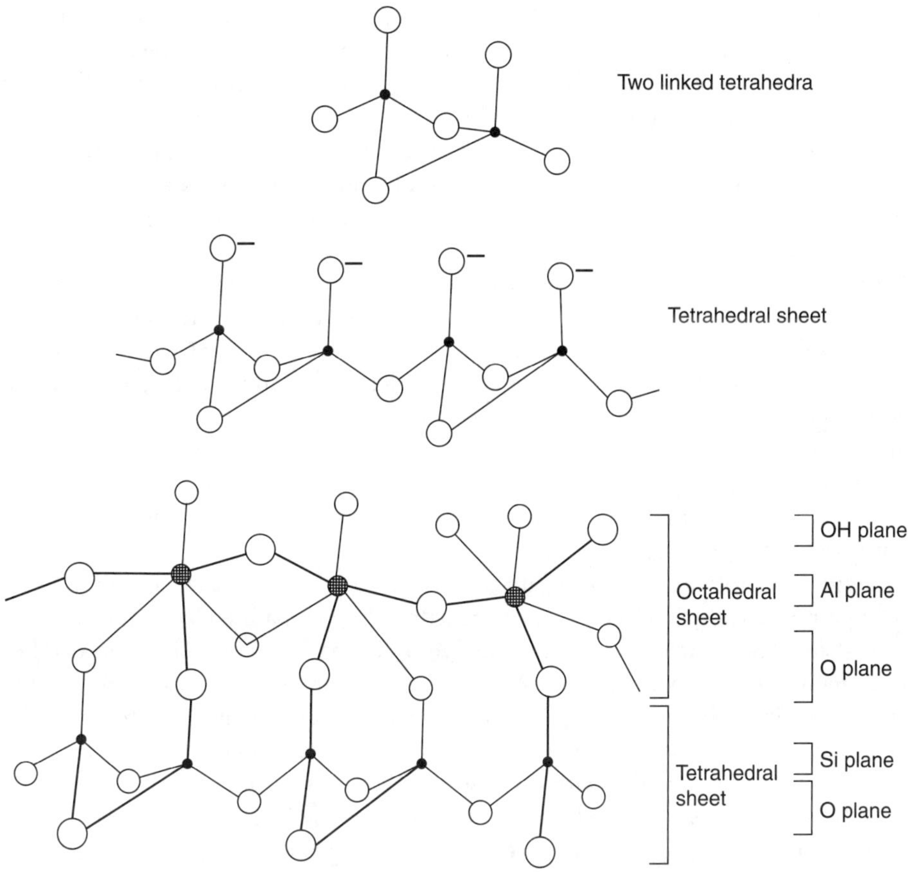

Figure 4-8 Constructing the basic sheets of layered silicon clays.

water, are not able to penetrate between these layers, and thus, kaolinite does not change form very much with wetting and drying. Kaolinite also has a very low surface charge.

Other clays are characterized by having two tetrahedral sheets enclosing an octahedral sheet and, with multiple platelets bound together by other ions, the 2:1 clays. One of the most important in this class is illite, characterized by the fact that many (usually about 20%) of the silicons in the tetrahedra are occupied by aluminum, resulting in a large negative charge. Potassium ions, if available, satisfy this charge and bind the layers tightly together. As with kaolinite, illite does not change morphology during wetting and drying, largely because of the tightness of the potassium binding.

Another important 2:1 clay is montmorillonite, a member of a larger family referred to as smectite clays. Of the smectites, montmorillonite is the most common in soils. These clay minerals are characterized by dramatic changes in morphology with wetting and drying. The platelets of montmorillonite are weakly attracted to one another either directly or through exchangeable cations. Because the attraction of platelets to one another is weak, there can be dramatic morphological changes upon wetting and drying. Water makes its way into the interplatelet space, forcing the patelets apart, the effect of which is for the clay to swell dramatically. Also, because of the weak attraction between platelets, a great deal of "internal surface area" is available for exchangeable cations, in addition to the normal surface area.

Vermiculites are another class of 2:1 clays, similar to montmorillonite, with the basic difference that many of the tetrahedral silicons are replaced with aluminum, much as in the case of illite. The substitution in vermiculites, however, has not been large enough to cause so much negative surface charge as to attract large numbers of potassium ions as binding agents, and vermiculite is effectively intermediate between illite and montmorillonite. As might be expected, vermiculite is also intermediate between illite and montmorillonite regarding morphological changes with wetting and drying. However, because of all the substitutions of aluminum for silicon, the surface charge is typically larger than montmorillonite, although not so large as to attract potassium ions.

The final category of layered silicate clays is chlorite, with a rather complicated structure of pairs of tetrahedral/octahedral/tetrahedral sheets separated by an additional octahedral sheet. The basic structural unit includes alternating tetrahedral and octahedral sheets, giving rise to the designation 2:1:1 (two tetrahedral sheets to each one octahedral to each one octahedral) or 2:2 type clays. Furthermore, the octahedral sheets are dominated by magnesium rather than aluminum. Chlorites have little internal surface area, and most charges are satisfied with little isomorphic substitution.

The fundamental structural features of most of these layered silicate clays are illustrated in Figure 4.9. Note that these types are ideal types. Most layered silicate clays are really mixtures of all of them, with for example, montmorillonite-type structure dominating but with vermiculite features, or some other combination.

Other Types of Soil Colloids

Hydrous oxides of iron and aluminum are very highly weathered clays, commonly found in tropical rain forest areas. Imagine, for example, a kaolinite crystal in which all the silicon had been weathered away, leaving only aluminum and oxygen. The crystal that forms may retain some sheet-like structure by virtue of the octahedral connections, but it is more likely to begin bending and folding as the extremely thin single octahedral unit lacks the same structural cohesion as the octahedral/tetrahedral combination. Thus, this aluminum oxide tends to have an amorphous form, as do the similarly structured iron oxides, in which the aluminum has been replaced by iron. The most common of these families of amorphous clays are gibbsite $[Al(OH)_3]$ and geothite $[FeO(OH)]$.

Because of the instability of its sheet structure, most of the surface charges become satisfied with the bending and folding of the basic crystal. Thus, the number of negative

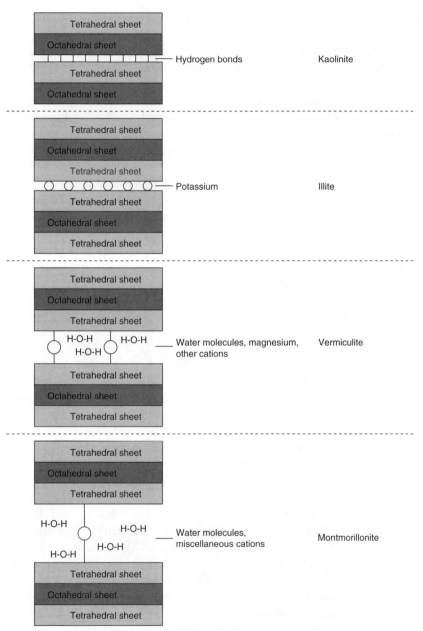

Figure 4-9 Basic structural types of silicate clays.

charges on the surface of the crystal is small. Furthermore, the balance of positive and negative charges on the surface can change dramatically depending on the pH of the soil. They thus may change from a small net negative charge to a small net positive charge. The significance of this fact is discussed in the following section.

Organic colloids, or humus, are highly decomposed organic matter, the formation of which will be discussed in detail in Chapter 5. As colloids, they have much the same chemical function as the other colloids, although their chemical properties have a different origin. With many negative charges on their surfaces, they can be exceedingly important with respect to plant nutrient dynamics and soil storage capacity. On the other hand, much like allophane and iron/aluminum hydrous oxides, the charge can be variable. The source of negative charges on the surface of humus stems from partially dissociated enolic ($-C = C-OH$), carboxyl ($-C00H$), and phenolic (phenolic ring-OH) groups. The bulk of the humic colloid can be of a great range of sizes and basic composition, but the surface charges are usually the result of these partially dissociated groups.

Allophane is an extremely important clay, although its properties are less well known than the more orderly layered silicate clays. It is the main clay that forms from volcanic ash deposits and, like the iron and aluminum oxides, has an amorphic form and a relatively low surface charge. It is, as in the case of organic colloids and aluminum/iron oxides, variable in charge.

Variable Charge

The source of variable charge in soil colloids is a result of a distinct chemical mechanism whereby surface charges become established. Unlike the large layered silicate clays such as smectite and vermiculite in which surface charge is primarily caused by isomorphic substitution in the body of the crystal, in allophane, aluminum, and iron oxides, as well as organic colloids, much of the surface charge is due to hydroxy groups on the edges and surfaces of the particle. As the pH increases (fewer protons in the soil solution), the hydrogen disassociates from the hydroxyl groups, leaving a negative charge. For example, if we begin with Al-OH and add OH^- (which is to say increase the pH), we end up with $Al-O^-$ and water. That is, a negative charge has been added to the surface of the crystal because the hydrogen that had formed part of the hydroxyl has been brought into the soil solution. In general, therefore, an increasing pH results in larger numbers of negative charges on the surface of the clay. Because the reaction is reversible, lowering the pH will have the opposite effect, eliminating many of the negative charges.

An additional mechanism whereby pH can change the charges on the surface of clays is associated with aluminum hydroxyl ions. These ions $[AL(OH)_2^+]$ combine with hydroxyl ions to form the insoluble $Al(OH)_3$, thus removing the aluminum hydroxy ion from the surface of the clay particle, where it had been neutralizing a negative charge. That negative charge then becomes unsatisfied on the surface of the clay. Of course, the hydroxyl ions become more available as the soil pH increases. Furthermore, aluminum hydroxyl ions can become strongly adsorbed between the layers of 2:1 type clays and through their tight binding reduce the exchangeability of the internal surface charge sites. This may be important in moderately acid soils (see the later section on soil acidity).

If the addition of more OH^- ions causes more of the clay surface charges to become negative, one would most naturally expect the opposite to happen if H^+ ions are added to the solution, which is to say as pH is decreased. This is precisely what happens as the exposed hydroxyl ions combine with the protons, thus leaving a new positive charge. That is, beginning with Al-OH, which is neutral, after adding protons, we get $Al-OH_2^+$, which adds a positive charge to the surface of the clay.

Thus, we see how the pH of the soil solution is able to change both the positive and negative charges in the soil. In fact, both positive and negative charges exist simultaneously on the surface of soil colloids. In most temperate zone soils, however, the negative charges far outweigh the positive ones. This is due to the high proportion of smectite, vermiculite, illite, and chlorite clays, whose negative charges are not as variable as the kaolinite, and hydrous aluminum and iron oxides of tropical soils. The fundamental structure of these various clays makes it obvious why such variability exists, as the exposed hydroxyl ions are the main reactive components that respond to the changing pH and as those ions are sandwiched between tetrahedral sheets in the less weathered clays. In Table 4.3, the charge characteristics of various clays are given, along with the distribution of the charges that are subject to change with changing pH (the percentage variable charge).

Origins and Weathering Sequence of Silicate Clays

There are two mechanisms whereby silicate clays develop in soils: 1) direct modification of clay crystals that formed with the formation of the parent material and 2) weathering.

TABLE 4-3   Charge Characteristics of Various Soil Colloids

Colloidal type	Negative	Charge		Positive Charge
	Total Charge (cmol/kg)	% Constant	% Variable	(cmol/kg)
Organic	200	10	90	0
Smectite	100	95	5	0
Vermiculite	150	95	5	0
Illite	30	80	20	0
Chlorite	30	80	20	0
Kaolinite	8	5	95	2
Al and Fe oxides	4	0	100	5

From Brady, 1990. All positive charges are variable.

Most parent material contains primary minerals with basic layered structures as indicated previously—mainly alternating tetrahedral and octahedral structures with aluminum and silicon as the basic core and oxygen and hydrogen forming the crystalline connections. Depending on the actual mineral content of the original parent material, these crystals may contain a variety of "impurities," that is, something other than aluminum, silicon, oxygen, and hydrogen. It is possible in many cases to categorize this parent material in the same way we categorized igneous rocks, those with high concentrations of silicon, aluminum, potassium, and sodium (silic) and those with high concentrations of magnesium, calcium, and iron (mafic). This parent material typically contains mineral crystals that formed as the material cooled from its original molten state, the exact mineral depending on what was contained in the original molten material. These minerals are primary minerals. As the process of soil formation proceeds, many of the primary minerals have their basic structure altered through weathering processes. This is the source of most illites, vermiculites, chlorites, allophane, and some of the smectites. Especially when weathering is severe (and therefore rapid), bases are removed rapidly from the primary minerals and, rather than leaching out of the system, their sheer abundance at any one point causes new crystals to be formed from the ions released from the primary clays. These are the secondary clays and include smectites, kaolinite, and the oxides of iron and aluminum. By recalling the basic layered crystal structure, it is easy to imagine the sequence beginning with, for example, illite, the weathering sequence removing some of the potassium ions that connect the basic 2:1 sheets together to form vermiculite. Further weathering might remove much of the magnesium that remains connecting the 2:1 sheets of the vermiculite, and montmorillonite (a smectite) is formed. Yet further weathering removes the silicon from the tetrahedral sheet, forming the 1:1 sheets connected with hydrogen bonds of kaolinite. An examination of Figure 4.10 makes this particular sequence obvious; however, this is only one scenario of how the weathering sequence might occur. In Figure 4.10, the general patterns of formation and weathering of all of the classes of layered silicate clays are illustrated.

Cation and Anion Exchange

As indicated earlier, all of the soil colloids have surface charges, some dramatically more than others. These charges have a great deal to do with the long-term fertility of soils. Unsatisfied surface charges tend to attract whatever ions of opposite charge are in the soil solution, and consequently they become adsorbed on the surface of the colloid particle. If they are thus adsorbed, they cannot be leached out of the soil solution and are effectively "stored" in the soil. Consider, for example, a soil dominated by a smectite clay versus one dominated by an aluminum hydrous oxide. The first will have about 100 cmol/kg negative charge on the surface of the colloids, whereas the second will have about 4 cmol/kg (refer to Table 4.2). Suppose that bananas are being produced on this soil. Bananas are heavy users of potassium and rapidly deplete whatever potassium ions are in the soil solution, necessitating (in industrial agriculture) the application of potassium in ionic form

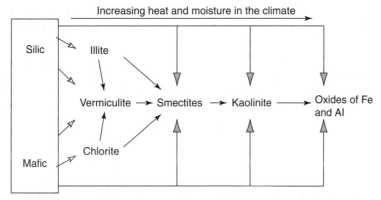

Figure 4-10 Formation and weathering of layered silicate clay minerals. The open arrowheads indicate modification of primary minerals. The hatched arrowheads indicate recrystalization, that is, formation of secondary clay minerals. Small black arrowheads indicate pathways of weathering.

to the banana plantation. In the first soil, the one dominated by the smectite clay, much of the applied potassium that is not immediately taken up by the plants is adsorbed on the colloidal surfaces because of the negative charge. On the other hand, the soil dominated by the aluminum oxide has very few negative charges, and most of the potassium not immediately absorbed by the crop will be leached out of the system as water percolates through the soil, taking dissolved ions with it. The soil dominated by the smectite clay thus acts as an ecological buffer, effectively storing the potassium for later use.

The fact that the ions are metaphorically "stored" on the surfaces of the colloids is a consequence of the exchangeability of the cations. For example, let us suppose that in a particular soil the cations Ca^{2+}, Al^{3+}, H^+, and K^+ are in the proportions 40, 20, 20, and 20, respectively. If we add an acid to the solution, we are effectively adding protons (H^+), and thus, the overall proportion of those four ions in the soil solution will be changed. During plant growth, for example, carbonic acid is formed from the release of carbon dioxide from the surface of the root hair, thus contributing protons to the general vicinity of that surface. Hydrogen from that carbonic acid will react with the clay surface as follows:

$$Clay–(Ca_{40}\ Al_{20}\ H_{20}\ K_{20}) + 5H_2CO_3 <->$$
$$Clay–(Ca_{38}\ Al_{20}\ H_{25}\ K_{19}) + 2Ca(HCO_3)_2 + K(HCO_3)$$

which can be more simply viewed as the addition of protons, and the substitution of those protons on the surface of the colloid in place of the cations that had been there. In this particular example, the carbonic acid provided 10 hydrogens, 5 of which were added to the clay surface (which is why hydrogen went from 20% to 25%), and to balance the charges, five positive charges in the form of two calciums and one potassium were removed from the clay surface and combined with the calcium and potassium carbonates. An exchange of cations was made, with hydrogens replacing other cations. The reaction is reversible, and thus, if potassium carbonate (for example) is added to the system, the reaction will move to the left, whereas the addition of more hydrogen ions will move the reaction to the right. This basic chemical reaction is extremely important in plant nutrition, as discussed in the next chapter.

The capacity of a soil to adsorb cations is referred to as its exchange capacity and is expressed in terms of moles of positive charge adsorbed per unit mass. In Table 4.2, we present it as centimoles of positive charge per kilogram of soil. Thus, for example, a soil with only smectite clay may be able to adsorb 100 cmoles/kg of hydrogen ion and could then exchange that 100 moles of hydrogen for 100 moles of potassium (another monovalent ion), or 50 moles of calcium (a divalent ion). In this sense, the cation exchange capacity is a measure of the ability of a soil to store cations, as only those cations adsorbed on colloidal surfaces will be protected from leaching out of the soil. Cation exchange capacities are typically low for tropical soils, the reason they are notoriously poor at retaining fertility.

A further useful measure associated with the adsorbed cations on the surface of soil colloids is the (perhaps inappropriately named) "percent base saturation." This refers to the percentage of the exchange sites that are occupied by cations other than H^+. Cations such as Ca^{2+} and K^+ are not, of course, bases, but their effect when substituted for H^+ on the exchange sites is to decrease the general acidity of the soil. Thus, it is of use in interpreting the pH (it is another measure of the acidity) in that high base saturation means a high pH and low acidity. Thus, for example, 50% base saturation means that of all the exchangeable sites in the soil, 50% of them are occupied by non H^+ cations. As is seen in the next chapter, percent base saturation is also critical for understanding plant nutrition.

Anion exchange may also be important in some soils. This principle is simply the opposite of cation exchange, in that positive sites on the surface of colloids may be the sites of adsorption of anions. The basic principles are precisely the same as in cation exchange. The reason anion exchange is not given as much attention as cation exchange can be seen in Table 4.2. Only Kaolinite, the hydrous oxides, and allophane have significant positive charges on their surfaces, and the abundance of those charges is much lower than the negative charges on organic colloids or many of the layered silicate clays. Nevertheless, anion exchange is occasionally important for plant nutrition in some soils.

The Acidity of the Soil

The pH of the soil is correlated with so many different aspects of soil fertility that it has become one of the most studied aspects of soils. The simplest reaction is as follows:

$$\text{Soil colloid} \underset{\text{Adsorbed}}{\overset{}{\bigg|_{\displaystyle H^+}}} \rightleftharpoons H^+$$

in which the hydrogen ions reach an equilibrium between number adsorbed and number in solution. Obviously, with low base saturation, most of the hydrogen ions will be adsorbed, which ironically tends to be the case in highly acid soils. This apparent contradiction (low base saturation should mean that most of the hydrogen ions are adsorbed and thus not contributing to soil acidity, yet low base saturation is in fact associated with highly acid soils) is resolved when we examine the role of aluminum and iron in highly acid soils. Here the role of aluminum is discussed, although the chemistry is basically the same in the case of iron.

With regard to aluminum in highly acid soils, the key reaction is as follows:

Thus, adsorbed aluminum is released into the soil solution in ionic form, forming an equilibrium. Aluminum ions then react with water molecules (hydrolysis) to form aluminum hydroxide and hydrogen ions, thus reducing the pH of the solution. This reaction is typical of highly acid soils.

In soils with lower acidity, there is usually higher base saturation, and aluminum can no longer exist in ionic form, being converted to aluminum hydroxyl ions, some of which are very complicated (e.g., $[Al_6(OH)_{12}]^{6+}$ or $[Al_{10}(OH)_{22}]^{8+}$). These ions can be adsorbed and act as exchangeable ions, coming to equilibrium with the soil solution. In the soil solution, they can engage in hydrolysis reactions, such as

$$Al(OH)^{2+} + H_2O \rightarrow Al(OH)_2^+ + H^+$$

$$Al(OH)_2^+ + H_2O \rightarrow Al(OH)_3 + H^+$$

thus adding hydrogen ions to the solution and lowering the pH.

In more neutral to alkaline soils, the aluminum hydroxyl ions react with hydroxyl ions to form the insoluble $Al(OH)_3$, but of course, without releasing a hydrogen ion. Excess hydrogen ions react with hydroxyl ions to form water. The exchange sites occupied by H^+ in more acidic soils are taken by other cations, especially Ca^{2+} and Mg^{2+}. It is a general rule that H^+, Al^{3+}, and Fe^{3+} promote lowered pH (by the mechanisms described previously), whereas Ca^{2+}, Mg^{2+}, K^+, and Na^+ promote higher pH (by occupying the sites otherwise occupied by H^+).

Various forms of acidity exist in soils. The active acidity is due to the actual concentration of H^+ in the soil solution and is what is usually measured when one measures the pH of the soil. The salt-replaceable acidity is associated with the Al^{3+}, and H^+ ions present in large concentrations in acid soils. Residual acidity is caused by the hydrogen and aluminum that is bound with silicate clays and organic matter in nonexchangeable form.

Although plant growth is largely responsive only to the active acidity, attempts to change the acidity by liming or other methods must deal with the salt-replaceable acidity and, more important, with the residual acidity. As calcium, for example, is added to a soil in an attempt to raise the pH, if aluminum and hydrogen are common and present in nonexchangeable form (and therefore do not enter into pH estimates from either measurement of direct acidity or measurement of both active and salt-replaceable acidity), the added Ca^{2+} can frequently have the effect of releasing the nonexchangeable aluminum and hydrogen from the clays and organic matter, thus subverting its intended role in raising pH.

Generally, it is not easy to change the pH of the soil. This is because soils are frequently buffered. To see how this occurs consider the following reaction:

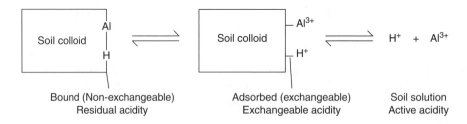

| Bound (Non-exchangeable) | Adsorbed (exchangeable) | Soil solution |
| Residual acidity | Exchangeable acidity | Active acidity |

If a liming agent is added to the soil, the active acidity is partly neutralized, but then the entire reaction moves to the right and the hydrogen ions are rapidly replaced. Similarly, if more hydrogen ions are added (e.g., by the breakdown of humic materials), the reaction moves to the left. Clearly, the soil colloids control the buffering capacity of the soil.

Soil Water

The physical movement of water in soil is largely determined by the structure of soil pores, which in turn is strongly influenced by soil biology, as discussed in the next chapter. Movement of water through pores is similar to movement of water through tubes, and a useful metaphor for soil structure and how it influences water movement is an unordered mass of flexible tubes, representing the complex interlocking bending and turning of soil pores through which water moves. It is useful to think of soils as on a continuum, ranging from those with a high proportion of large pores to those with a large proportion of small pores (see Figure 4.11).

Considering the problem of water availability, if the process is simply (1) rain or irrigation provides water at a given rate and (2) water percolates out of the soil at some other rate, then the fine textured soils (right of Figure 4.11) would be far better water-holding soils than the coarse textured soils (left of Figure 4.11). After a saturating rain, water tends to flow out of the pores through gravity and through evaporation. A soil that is completely saturated with water is in an unstable state because gravity and evaporation will happen no matter what the management options employed. After 2 or 3 days (it is not a fixed period), the majority of drainage and evaporation that is going to occur has happened and the soil

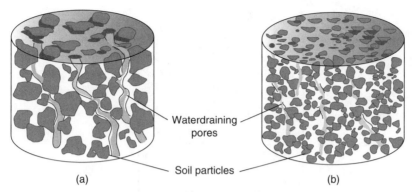

Waterdraining
pores

Soil particles

(a)   (b)

Figure 4-11   Two columns of soil illustrating the relationship between particle size
and pore size. On the left is a coarse-textured soil (dominated by sand) in which the
effective water-draining pores are relatively large and water will drain rapidly. On the
right is a fine-textured soil (dominated by silt and clay) in which the effective water-
draining pores are relatively small and water will drain slowly.

reaches a quasi-stable state of water content. Water will still be removed through these same
processes, but at a much lower rate. This point is referred to as "field capacity" of the soil.
Generally, field capacity is always larger with more finely textured soils, for obvious reasons.

Another force is operative and crucial for agriculture. To appreciate this force, consider
the diagram in Figure 4.12. The fine-textured soil of Figure 4.11b has been placed in a
cylinder and a piston inserted into that cylinder with the goal of sucking the water out of
the soil. The key question is: How much force does one have to apply to the piston to
remove the water from the pore spaces? That force is referred to as the suction capacity or
matric capacity of the soil, and it is intuitively obvious (with reference to Figure 4.12) that
a finely textured soil has a higher potential than a more coarsely textured soil.

The key concept of soil–water relationships lies in a contradiction between soil loss
to gravity and evaporation (as in Figure 4.11) and soil retention due to matric potential. On one hand, for proper functioning of the agroecosystem water storage capa-
bility is essential, a huge problem in the typically coarse sandy soils of desert regions.
Whatever rain falls is quickly lost from the soil through the very large pore spaces. On
the other hand, the plants need to remove the water from those soil pores, much as the
piston of Figure 4.12 removes the water from the cylinder of soil. A soil with a low
matric potential will allow plants to absorb water easily. Thus, we have a fundamental
contradiction in that water-holding capacity is both good and bad for agriculture—
good for its ability to conserve water and bad for its resistance to plant roots taking up that
water. This is largely the reason that a loam soil is generally regarded as the best texture
for agriculture.

However intuitive all of this may be, it is enormously complicated by the facts of soil
biology. The pore spaces so clearly delineated by the particle sizes have their properties

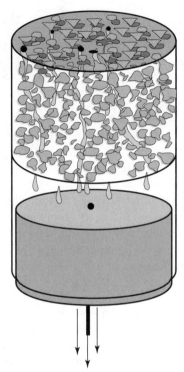

Figure 4-12 The column of fine texture soil from Figure 4.11 has been placed in a cylinder with a piston inserted at the left. The force necessary to apply to the piston in order to suck out the water from the small bore pores is the matric potential of the soil.

modified by the "skin" of bacteria that forms around each soil particle, which in turn causes the particles to stick together in a "ped" structure, which effectively changes the apparent particle distribution, at the same time that worms and insects burrow around, creating an almost infinite labyrinth of pore spaces. We treat this subject in more detail in the next chapter.

The Chemical Industry and Soil Science

As noted in the introductory notes to this chapter, early capitalization of agricultural chemicals was based on the production of superphosphate, yet phosphate was not the major limiting nutrient in agriculture in most areas of the world. As Liebig repeatedly pointed out, all nutrients would eventually become limiting if agriculture continued its development along the lines required by the developing capitalist system. He was particularly concerned, as were many analysts of his day, with the separation of town and countryside. He wrote extensively about the ecological irrationality of a system in which the nutrients of the countryside (agriculture) were transported to the city (in the form of food) and then dumped in the local waterways (in the form of excrement) only to create the foul conditions of 19th century European cities at the same time as it created an ecological crisis in agriculture—something we might consider in today's supposedly more sophisticated world as we continue pumping agricultural runoff into our waterways the world over, having by now created a crisis that makes the sewers of Victorian London seem mild.[5] Chief among the relevant nutrients was and is nitrogen.

Before the invention of the Haber-Bosch process (discussed later), the main source of nitrogen was guano, found on tropical islands in the form of massive bird droppings accumulated over years by breeding shore birds. The crisis of nitrogen in agriculture created a land rush for these tropical islands, contributing in a small way to the colonial structure of the world that we see yet today. In the "Guano Islands act," passed in 1856, U.S. interests managed to gain control of 94 islands around the world, eventually 66 of which were recognized as property of the United States by 1903. Consequently, by the turn of the century, the natural cycles of fertility, which still dominated the new husbandry, were beginning to be replaced by industrial appropriation. A piece of what might be thought of as the natural agricultural cycle had been taken over by industry. Ecosystem thinking had not yet evolved, and few questioned what might be the secondary or indirect consequences of imposing such a massive perturbation on natural ecological cycles. The idea that secondary negative effects might accompany the obvious positive benefits of increasing crop yields apparently never occurred to anyone.

The problems created by this basic philosophy are still not fully appreciated. Mineral cycles are very complicated, involving complex interactions among biological and chemical processes the details of which we are still only beginning to understand. For example, Figure 4.13a shows a simplified diagram of the nitrogen and phosphorous cycle, both of which are discussed in more detail in the next chapter. Although ecological cycles are certainly not fine tuned like Swiss watches, they nevertheless are very much like a complex set of interacting gears. The gears may not operate in a smooth fashion, may occasionally slip out of position, and at times even go temporarily haywire, but they all interact together and ultimately determine the overall functioning of the ecosystem. Figure 4.13a is only a minimal representation of the complexity of such cycles. Yet the vision the 19th century agricultural scientists had was more like that presented in Figure 4.13b. With this basic model, it is easy to visualize the solution to low crop production as simply an increase in the stores of nitrogen and phosphorous.

On the other hand, if the basic model is more like the modern version pictured in Figure 4.13a, the prediction of what would happen with an increase in mineral nitrogen (for example) would be far less direct. Indeed, recent increased understanding of these cycles really ought to invoke a bit of humility regarding our ability to predict what will happen. For example, could it be that directly increasing mineral nitrogen would increase the nitrifying microorganisms to the extent that they break down organic nitrogen too fast and thus reduce the humus content of the soil, which in turn decreases the soil's capacity to maintain nitrogen stores over the long term? Probably so, but maybe not.

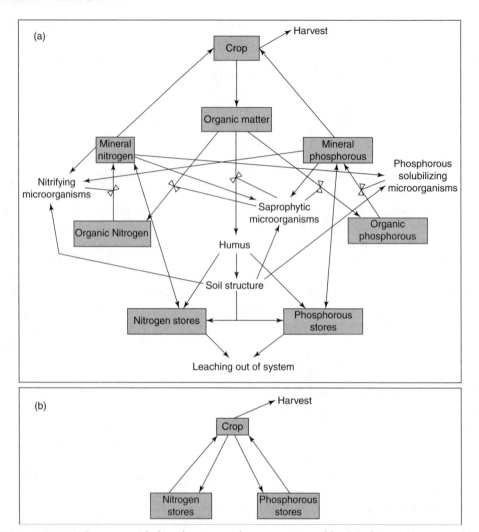

Figure 4-13 Nitrogen and phosphorous cycles as envisioned by (a) the modern agricultural scientist and (b) the 19th century protoindustrial farmer. In (a), the arrows refer to "effect on," but the arrows that point to the "valves" on the other arrows indicate an effect on the rate of the change of the original effect.

Or could it be that directly increasing mineral phosphorous increases the saprophytic microorganisms, which then increase the rate of decomposition, thus putting mineral nitrogen into the soil more rapidly than it can be consumed by the crops or maintained in the stores? Probably so, but maybe not. In short, it is not easy to predict with any degree of confidence what will happen, especially over the long term, through the provisioning of mineral nutrients into the system. If there is anything approaching a balance in the ecosystem, the brute force approach of supplying large quantities of mineral ions into the system could easily upset that balance. Yet such a conclusion would have been difficult to see if one's vision of the nutrient system was more like that of Figure 4.13b. This was more or less the vision of 19th century agricultural scientists, which led to the direct application of Liebig's law, which continually required ever-larger quantities of individual nutrient items and the industrial structures necessary to supply them.

Based on imported minerals, sulfuric acid (for the production of superphosphate), guano, and sulfate of ammonia (a byproduct of the gas works), the fertilizer industry in Britain was already big business by the turn of the century. The major breakthrough for the industry was discovering how to synthesize ammonium directly. The first commercial production occurred in the early years of the last century through an electrical process called the Frank-Caro Cyanamid process, which was the basis for the formation in 1907 of the American Cyanamid Company. The watershed breakthrough, however, came with the invention of the Haber-Bosch process, patented by the Badische Anilin- und Soda-Fabrik (BASF). It is difficult to exaggerate the importance of this process for the history of agriculture. As historian L. F. Haber (son of one of the inventors of the Haber-Bosch process) noted, the Haber-Bosch process "cemented the relationship between chemicals and agriculture and also between chemical companies and suppliers of process plant."[6]

With the new process of ammonia production, the fertilizer industry expanded rapidly. An especially important element of this expansion came, not from the needs of agriculture, but from the need for nitrogen-based explosives for the First World War. In fact, the world wars provided temporary solutions for companies like BASF and IG Farben as overproduction had begun to reduce their profits. The need for large quantities of explosives on all sides of the conflict was a godsend for the chemical industries of all countries involved. During the interwar years, improvements in technology again created a crisis of excess capacity that was largely solved by the Second World War. As might be expected, the end of WWII created, once again, an underconsumption crisis. This time the wartime production levels were sustained by the dramatic expansion of the industrial model of agriculture throughout Western Europe and North America in the postwar years, interrupted only intermittently until now.

As with so many other industrial activities, the fertilizer industry began to be questioned in the 1960s as part of the general awakening of environmental awareness. The most obvious problem was and still is contamination of ground and surface water.[7] The most serious pollutant is nitrate, both because of its ubiquitous occurrence and its negative health effects. In one study, 73% of wells in the United States were found to contain nitrate levels above the safe level.[8] Wells in shallow aquifers generally show higher nitrate levels than those in deep aquifers, largely because nitrate slowly leaches to lower levels. In an Illinois study, it was found that 23% of wells in shallow aquifers had nitrate levels exceeding safe levels, whereas only 1.4 % of the deep wells were so contaminated. The slow leaching means that nitrate can be expected in the future in the lower aquifers also. As Soule and colleagues note,

Even if all fertilizer application were suspended immediately in areas where deep aquifers are beginning to show nitrate pollution, the inputs of past years would still slowly percolate toward the aquifers, and nitrate input to the groundwater would continue for many years.

The application of commercial nitrate fertilizer is not the only culprit, however. Groundwater contamination associated with feedlots and alfalfa has been reported, and recently, organic agriculture has been implicated in groundwater nitrate contamination also.[8]

Although the bulk of public attention has been directed at nitrate contamination as the major problem, excessive phosphorous is also a major problem. Because many aquatic ecosystems are limited by phosphorous, runoff of phosphate from agricultural fields has contributed significantly to the problem of eutrophication.

In the end, the apportionment of low ionic forms of nutrients, especially nitrogen and phosphorous, although enabling dramatic increases in production with little effort at soil management, also had two negative consequences: disrupting the overall functioning of the soil nutrient cycling system and pollution of groundwater and waterways in general. Nothing in the physics or chemistry of soils suggests even approximately that the direct application of low ionic forms is somehow better than more complicated forms, such as is found in compost. Indeed, much analysis associated with the newer forms of more ecologically based agriculture begin with a challenge to this discourse, noting that from the plant's point of view, it matters little whether its source of nitrogen is ammonium directly applied or ammonium that is a consequence of natural mineralization, as described in great detail in the following chapter.

Appendix 4.A: Fundamental Ideas of Inorganic Chemistry

The Atom

Following the fairly obvious desire of the ancients to find the "ultimate" particle from which all matter is made, the idea of the atom is precisely that piece of matter that is indivisible. That holy grail, the ultimate particle, has been hijacked by particle physicists who have discovered that the standard model of the atom can be broken down further (quarks and such). We are not concerned with these "atoms of the atoms," as the chemical bonds and reactions used in understanding soil physics and chemistry are completely understandable with reference to the standard model.

The atom is composed of two parts, the nucleus, which constitutes almost all of its mass (about 99.9%), and electrons that orbit around it, which constitute almost all of its volume. The nucleus is composed of protons, which have a positive charge, and neutrons, which have a neutral charge. Electrons, which have almost no mass at all, have a negative charge, and an atom is balanced in its charge if it contains an equal number of electrons and protons.

It is important to have a feel for the mass and volume of an atom, as the fundamental structure is not particularly intuitive. If St. Peter's Basilica were the size of an atom, its nucleus would be the size of a pinhead in the middle of its space, the electrons whizzing around that nucleus somewhere within the walls of the Basilica. The space in which the electrons are orbiting the nucleus defines the volume of an atom, and the mass is determined almost exclusively by that nucleus. The speed of the electrons is effectively the speed of light so that if you could take a snapshot of an atom, no matter how fast the shutter speed, you would see nothing but a cloud of points; however, that cloud of points would not be a uniform cloud; that is, the probability of finding an electron at any particular point inside the Basilica (the volume of the atom) is not equal from point to point, but rather has a distinct structure. Thus, for example, what you would likely see if you could photograph a hydrogen atom would be a cloud of points, concentrated near the center, where the nucleus is, something like Figure 4.A1.

Each small point in Figure 4.A1 represents the position of a single electron, appearing over and over because it is moving so fast for the shutter speed of the camera that it appears billions of times in the photo, but it is always the same electron. In Figure 4.A1, it is obvious that there is not just a random distribution of the electron, but it appears to be more commonly found nearer the center of the atom, which is where the nucleus is located. There is another regularity that needs to be appreciated, however. Although hydrogen is characterized by having a single electron, it is the only element so sparsely endowed. The next largest element is helium, which has two electrons. If you could take a picture of helium, it would look more or less the same as hydrogen, except that its volume would be smaller. Looks are deceiving, however, because the electron cloud of

Figure 4.A1   Diagram of what a hydrogen atom might look like if you could take a photo of it.

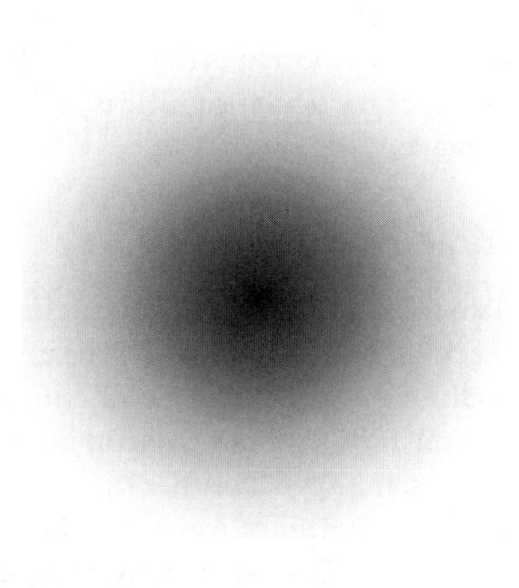

helium contains two electrons rather than one. The hydrogen nucleus contains a single proton. The helium nucleus contains two protons and two neutrons. Commonly, the elements are ordered according to the number of protons contained in their nucleus, with that number being their atomic number—hydrogen is atomic number 1, and helium is atomic number 2.

The dense regions where the probability of finding an electron is relatively high are energy levels, usually called orbitals because the electrons more or less fly around the nucleus, tracking these positions. Orbitals are classified based on their angular momentum, which specifies the general shape of the orbital. The important point here is that these orbitals have discrete positions with regard to the nucleus. Their particular orientations have to do with the characteristic angular momentum, and each orientation can contain two electrons (based on opposite "spins" in the quantum mechanical world). Why electrons indeed do behave this way, why they must track these very specific and discrete energy levels is referred to as quantum mechanics and need not concern us here.

Each orbital occupies a discrete region located at a particular distance from the nucleus, and its energy level is directly correlated with that distance. It is symbolized with a number that indicates its order in distance from the nucleus for that type of orbit, followed by a letter indicating its type (s, p, d, . . .) and usually followed by a number indicating how many electrons are in the orbital. For the s orbitals, there can be at most two electrons (of opposite spins). For the p orbitals, there can be at most six electrons (two of opposite spins for each of the orientations, as shown in Figure 4.A2), and for the d orbitals, there can be at most eight electrons.

An electron within a particular orbital type at a particular distance from the nucleus contains an amount of potential energy that is proportional to that distance, as illustrated in Figure 4.A2. As a general rule, the electrons in the s and p orbitals are the ones that are involved in chemical interactions (at least the ones we are concerned with in this text), and thus, special attention is given to them—they are referred to as the "valence electrons," and their number, in the s and p orbitals at a particular distance (referred to loosely as the "shell") from the nucleus, gives the "valence" of the atom. The atoms of particular significance to agriculture, with some facts about them, are listed in Table 4.A1.

Ionic Bonds

Two basic opposing tendencies exist in the atom. The s and p orbitals in each electron shell "seek" to be filled (two electrons in the s orbital and six in the p orbital). Thus, for example, sodium has the outer "shell" (in this case, the third energy level for both the s and p orbitals) with a single electron in the s orbital and none in the p orbital (see Table 4.A1). It can reach its "goal" either by losing that single electron in the s orbital (thus having the full complement of eight electrons in the second shell—two in 2s and six in 2p), or it can somehow gain seven electrons to fill not only the 3s orbital but also six extra electrons to

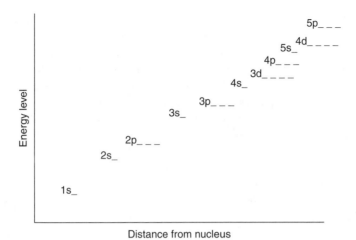

Figure 4.A2 Relative energy level of each orbital as a function of the distance from the nucleus of the atom.

TABLE 4-A1   The Major Elements of Significance to Agriculture with Some Critical Facts About Them

Element	Atomic Number	Electrons per Orbital	Symbol	Valence Electrons
Hydrogen	1	$1s^1$	H	1
Carbon	6	$1s^2\,2s^2$	C	4
Nitrogen	7	$1s^2\,2s^2\,2p^3$	N	5
Oxygen	8	$1s^2\,2s^2\,2p^4$	O	6
Sodium	11	$1s^2\,2s^2\,2p^6\,3s^1$	Na	1
Magnesium	12	$1s^2\,2s^2\,2p^6\,3s^2$	Mg	2
Aluminum	13	$1s^2\,2s^2\,2p^6\,3s^2\,3p^1$	Al	3
Silicon	14	$1s^2\,2s^2\,2p^6\,3s^2\,3p^2$	Si	4
Phosphorous	15	$1s^2\,2s^2\,2p^6\,3s^2\,3p^3$	P	5
Sulfur	16	$1s^2\,2s^2\,2p^6\,3s^2\,3p^4$	S	6
Chlorine	17	$1s^2\,2s^2\,2p^6\,3s^2\,3p^5$	Cl	7
Potassium	19	$1s^2\,2s^2\,2p^6\,3s^2\,3p^6\,4s^1$	K	1
Calcium	20	$1s^2\,2s^2\,2p^6\,3s^2\,3p^6\,4s^2$	Ca	2
Manganese	25	$1s^2\,2s^2\,2p^6\,3s^2\,3p^6\,4s^2\,3d^5$	Mn	2
Iron	26	$1s^2\,2s^2\,2p^6\,3s^2\,3p^6\,4s^2\,3d^6$	Fe	2

fill the 3*p* orbital. The other tendency, however, is for electrostatic equilibrium, which is to say an equal number of protons and electrons in the atom. Sodium has 11 protons, and thus, its neutral state is 11 electrons, which it attains with that single electron in the outer shell. When the sodium atom becomes "fulfilled" in its electron complements (i.e., when it loses that electron in the 3*s* orbital), it still has 11 protons, but now only 10 electrons, which is to say it has a net positive charge. As a general rule, atoms do not gain or lose more than three electrons, and thus, the "solution" to fill the outer shell for sodium is to lose that one electron.

Thus, we have a basic rule that an atom that loses electrons in its outer shell will have a positive charge, whereas one that gains electrons in its outer shell will have a negative charge. Chlorine, for example, with seven valence electrons will have to gain an electron to fill its total complement of eight in the outer shell, thus gaining a net negative charge of one. An atom that has its outer shell filled and is thus generally out of electrostatic equilibrium is called an ion and can be either positively charged (like sodium), in which case it is a "cation" or negatively charged (like chlorine) in which case it is an "anion." Basically, if you could throw a bunch of sodium atoms in a solution and a bunch of chlorine atoms in a solution, the tendency to fill the outer shell with eight electrons means that the sodium atoms tend to lose an electron and that electron tends to occupy the outer shell of the chlorine atoms with the result that the sodium atoms turn into cations (positive) and the chlorine atoms turn into anions (negative). Given this solution filled with cations and anions, they attract one another almost as if they were opposite ends of a magnet. That attraction of a cation to an anion is referred to as an ionic bond.

Covalent Bonds

The hydrogen atom, in its electrostatic neutral state, has a single electron in the first *s* orbital. Thus, it could attain its neutral valence state by borrowing an electron from somewhere else. In a solution of pure hydrogen atoms, it could be hypothesized that one hydrogen atom gains an electron and becomes a negatively charged hydrogen ion, whereas another atom loses that electron and becomes a positively charged hydrogen ion. The basic idea that an atom tends toward neutral valence and that an atom like sodium transfers an electron to an atom like chlorine would not be possible here. Sodium loses an electron to become valence neutral and chlorine gains an electron to become valence neutral—both atom types move toward that valence neutral position. In the case of hydrogen, however, losing an electron would push it further from valence neutrality (it needs two electrons in its *s* orbital), meaning there is no reason to expect it to lose that electron. How is it, then, that hydrogen, when not combined with other atoms, occurs in the diatomic form in nature (H_2—two atoms connected to one another)?

The solution to this problem happens because two atoms of hydrogen share their electrons in their *s* orbitals. Basically, the nucleus of one atom tends to attract not only the

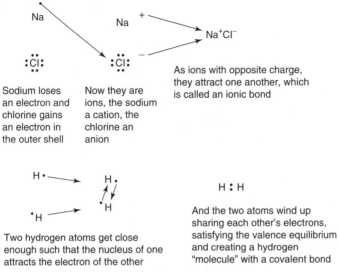

Figure 4.A3 Comparison of ionic and covalent bonding.

electron in its own *s* orbital, but also an electron in the *s* orbital of another atom of hydrogen, statistically filling its *s* orbital with two electrons. That is, both atoms share each other's electrons, creating a kind of super *s* orbital around the two of them. This type of bonding is referred to as a covalent bond. The difference between an ionic and covalent bond is illustrated in Figure 4.A3.

Hydrogen and Hydroxide Ions

Of particular importance are the ions formed from the basic constituents of water. A water molecule is a covalently bonded combination of one oxygen atom and two hydrogen atoms. Intuitively, if you were to dump a large number of oxygen and hydrogen nuclei into a bin and then add a large number of electrons into the same bin, the electrons would fill up the orbital shells around the nuclei, forming covalent bonds between pairs of hydrogen atoms, between pairs of oxygen atoms, and between the oxygen and the hydrogen atoms. The hydrogen nucleus is composed of a single proton, whereas the oxygen nucleus contains eight protons (sometimes hydrogen ions are simply referred to as protons as, when stripped of its only electron, the nucleus is all that is left and the nucleus of the hydrogen atom contains a single proton). If in this imaginary world we had an abundance of H, not so much O, and not so many electrons, we can imagine that the electrons would begin filling the orbital spaces around the atoms, forming covalent bonds among the atoms. Among those covalent bonds would be some that included oxygen and hydrogen. If all of the electrons are used to fill all the orbitals in the covalent bonds that now connect the oxygens with the hydrogens, but not enough left to fill the orbitals around the hydrogen atom, we will be left with a bunch of protons (hydrogen atoms without surrounding electrons), with no way to bond because there are no electrons left to fill the s shell. Thus, our bin will contain an excess of hydrogen cations (H^+).

We could imagine an alternative scenario in which there is an excess of oxygen nuclei. Some of the oxygen nuclei will form covalent bonds with two hydrogen atoms (i.e., water), but others will be able to form only a single such bond before running out of hydrogen atoms to pair with. Thus, there will be, in addition to normal water molecules, many oxygen–hydrogen covalently bonded pairs, which means a molecule that has an extra electron in its outer shell, or an anion. This is the hydroxide anion (OH^-).

Acidity and Alkalinity

The basic definition of the pH of a solution is the negative $\log_{10}$ of the concentration of hydrogen ions, which is to say the higher the pH, the less concentration of free protons in the system. An acid solution has a large concentration of H^+ and a low pH value. Accordingly, an acid is a substance that, when dissolved in water, yields hydrogen ions and a base is a substance that, when dissolved in water, yields hydroxide ions. According to the metaphoric bin in the previous section, if an abundance of hydrogen nuclei (protons) compared with oxygen nuclei is in the solution, the oxygen atoms will effectively scavenge up the hydrogen atoms, covalently bonding with them, until all the electrons are used up. Because oxygen was in abundance, there will be free protons (hydrogen nuclei) in the solution, and it will have a large concentration of H^+, which means an acid solution. The opposite, an alkaline solution, would occur if the metaphorical bin were to have contained an abundance of oxygen nuclei compared to the hydrogen nuclei.

Hydrogen Bonds

In addition to ionic and covalent bonds, another type of bond is important in soils, especially in the formation of clay crystals. To understand hydrogen bonding, it is first necessary to appreciate the idea of polar covalent bonding. In a covalently bonded molecule in which the atoms are different from one another, it is frequently the case that one atom attracts the electrons more strongly than the other does. Thus, for example, in water, the oxygen nucleus containing eight protons will obviously attract electrons more strongly than the hydrogen nucleus. Consequently, the part of the molecule closest to the hydrogen atom will have less of the shared electron cloud, and the part of the molecule closest to the oxygen atom will have more of the shared electron cloud, as represented in Figure 4.A4.

Because the covalently bonded molecule has a directionally oriented charge, it is referred to as a polar covalent bonded molecule. Generally, polar covalent bonds result in molecules that arrange themselves according to the charges, with the positive ends of some molecules attracting the negative ends of other molecules (i.e., molecules with the same identity). Water is just such a molecule. Thus, the negative ends of the water molecules tend to be attracted to the positive ends of other water molecules. When hydrogen is combined with oxygen (or some other atoms also, such as nitrogen or iron), it forms a polar covalent bond,

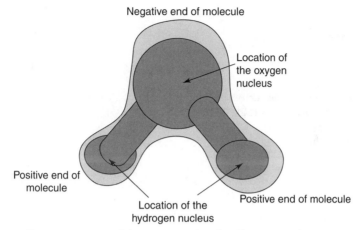

Figure 4.A4 Representation of the water molecule, illustrating the attractive oxygen atom compared with the less attractive hydrogen atoms and the resulting bias of the electron cloud surrounding the oxygen end of the molecule.

and thus, the molecules in solution tend to attract one another in a nonsymmetrical way. This attraction of the hydrogen ends of a molecule to the nonhydrogen ends is referred to as a hydrogen bond. Although it is very weak compared with a normal covalent bond (approximately 5%), it is nevertheless strong compared with other dipole/dipole bonds and is quite important not only in the way molecules interact with one another in solution, but also in the structure of many larger molecules. Hydrogen bonding in water is illustrated in Figure 4.A5.

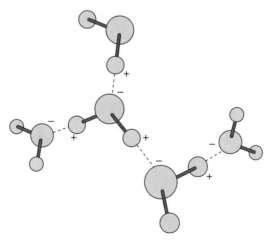

Figure 4.A5 Hydrogen bonding in water (dotted lines represent the hydrogen bonding).

Endnotes

[1]Goodman et al., 1987.
[2]Jenny, 1941.
[3]Darwin, 1881.
[4]U.S. Department of Agriculture, 1999.
[5]The hypoxic zones in coastal areas across the globe present a challenging problem at this point. For the Gulf of Mexico, see Nassaur et al., 2007.
[6]Haber, 1971.
[7]Soule et al., 1990; Nash, 2006.
[8]Singh and Sakhon, 1978/1979.
[9]Singh and Sakhon, 1978/1979.

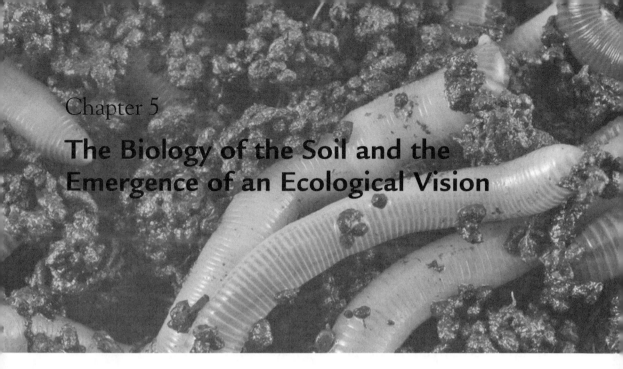

Chapter 5

The Biology of the Soil and the Emergence of an Ecological Vision

Overview

The historical sequence leading to an alternative vision for agriculture seems to come mainly from a coupling of fundamentally biological processes with a growing knowledge of how soil works as a seemingly chemico/physical factor. Thus, this chapter begins with a brief historical framework, beginning with the important insights of Albert Howard in India. Then we move on to a basic discussion of the way in which nutrients make their way to plants, drawing on material from the previous chapter on soil colloids and including the basic structure of the nitrogen and phosphorous cycles and the critical role of mycorrhizal fungi. This is followed with a description of one of the most important systems in agroecology, the decomposition subsystem. Finally, the chapter ends with an overview of the types of organisms involved in soil biology and the beginnings of treating the soil system as a dynamic system.

Changing the Soil Discourse: The Origin of an Alternative Discourse

Sir Albert Howard and the Early Pioneers

Perhaps it is not surprising that one of the first and most important moves toward a more ecological approach to agriculture occurred in one of Britain's main colonies, India. According to the Council for Agricultural Science and Technology, Sir Albert Howard, Director of the Institute of Plant Industry in India early in the 20th century, was the first to describe an "organic" system as something distinct from the pathway that world agriculture seemed to be taking. Howard perceptively noted that von Liebig's approach to the soil was strongly contrasted with that of Darwin, as reflected in the latter's treatise on earthworms. Whereas Liebig insisted on a reductionist approach, reducing the soil to its chemical constituents, Darwin correctly saw the soil as a complex biological system.

Howard made a point of promoting not only agricultural expansion in India, which had already been extensively underway when he started his directorship, but also was a keen observer of local customs and mores. In particular, he observed the techniques applied by traditional Indian farmers in all aspects of their production. He was particularly interested in the composting process commonly practiced on almost all traditional farms. From these observations, he developed a system of composting (called Indore composting) that effectively remains the underlying idea behind modern composting.[1] He regarded Indian traditional farmers as his main instructors in agriculture, noting: "by 1910 I had learnt how to grow healthy crops, practically free from disease, without the slightest help from . . . all the . . . expensive paraphernalia of the modern experiment station."[2]

He also noted that the Chinese traditional farmer was likely even more efficient than his Indian counterpart. His approach, which we can presume came from extensive observations of and conversations with Indian farmers, was focused on the health of the soil, arguing that a "healthy" soil, by which he meant one that contained a well-balanced mixture of worms, fungi, bacteria and other microorganisms, would produce healthy food, whereas a soil devoid of those healthy elements would not. Indeed, the connection between ecological health on the farm and the health-promoting qualities of food was a key element of the early organic movement more generally.

An explicit goal of this healthy food/healthy body approach is evident in the establishment in 1926 of the "Pioneer Health Centre" by two medical doctors, George Scott Williamson and his wife, Innes Pearse, in Peckham, South London. These two physicians, from vast experience in medical practice, concluded that much of a person's natural immunity to disease, as well as ability to fight infection when it occurs, derives from the diet. A healthy diet, composed of healthy food, they argued, is the key to good health. The Pioneer Health Centre was originally established as a family health clinic in the mid 1920s and moved to its ultimate location in Peckham in 1935, in a facility that included "clinics, a swimming-pool and various social and recreational facilities, which became internationally famous."

During its first year of operation at Peckham, the center bought a farm in Kent to supply the healthy food necessary for the health-promoting diets offered at the center. The developing philosophy of the new organic movement, inspired to no small extent by Albert Howard's writings, was to have a natural and obvious input to the goals of the new "Home Farm," as the new farm in Kent was labeled. Howard himself became actively and personally involved in the farm, and under his guidance, it adopted all elements of the organic system, as it was known then, especially the Indore composting process.

A variety of other tendencies began to emerge during the early 20th century, many of which emphasized the explicit study of how biological and ecological forces operate in agroecosystems generally, with the vast majority focused on the life of the soil. For example, France's book on the life of the soil (*Das Edaphon, eine neue Lebensgemeinschaft*, published in 1911), might be called the midwife of soil ecology and profoundly affected researchers and farmers alike in Germany and Austria.[3] Of perhaps even more importance were the writings of Richard St. Barbe Baker. Baker served the British Empire mainly in Africa and, partly as an outgrowth of his upbringing as an evangelist's son, felt a religious tie with forested lands. After obtaining his forestry degree from Cambridge in 1919, he was assigned to Kenya where he found, unlike the Edenic farming practices of India observed by Howard, a scarred landscape, the result of a century of irresponsible farming under British rule. He set out to reforest large sections of the country, a task for which he formed the "Men of the Trees" movement (which remains active today under the title International Tree Foundation). Much of interest is to be found in Baker's writings about forests, but for the purposes of this text, he saw in the forests lessons that would become extremely important for the emerging organic movement—the natural cycling of nutrients occurred in natural forests but not in many of the agricultural systems that displaced them—virtually identical to the ideology encompassed in the more modern "natural systems" agriculture. In 1938, the first Men of the Trees summer school was held, and one of the featured speakers was Albert Howard.

Some Prewar Entrants: Biodynamic and Other "Spiritual" Movements

In 1924, philosopher Rudolph Steiner gave his famous lectures on agriculture, thus giving birth to the biodynamic movement, a rather bizarre system that is, in my experience, not fully understood by even those who practice it. It presents us with a view of the soil that is at once mystical and ecological, with an unfortunate confusion between the two views. Biodynamic agriculture is at its base a form of the political movement called organic agriculture. But it must be understood that it began as, and remains a part of, a larger movement, the Anthroposophism Movement, especially important in Germany, Switzerland, and Austria, with thousands of adherents, most of whom demand certified biodynamic products on their tables. Consequently, Western Europe has seen an impressive growth of biodynamic agriculture, especially in the last 30 years. The idea of anthroposophism is to lead a balanced life in harmony with all of nature. It is also a semicultish form of antimaterialism, advocating a spirituality that is in some ways at odds with a scientific approach.

The sense of biodynamic agriculture can be gleaned from the idea of "preparations," compounds applied to compost or fields in very small amounts. Some invite ridicule if one looks into their origins. One preparation, for example, is made by putting cow manure in a cow's horn, burying it for some time, digging it up, removing the manure from the horn and mixing it with a large quantity of water in a wooden barrel, stirring the mixture so that vortexes form repeatedly in opposite directions. Where could such an idea come from? The basic idea comes from Steiner's particular notion of cosmic and life forces. For example, on the "nature" of the cow horn,

> The cow has horns in order to send into itself the formative astral and etheric forces, which, pressing inwards, are meant to penetrate right into the digestive organism. Precisely through the radiating forces from horns and hoofs, much work arises in the digestive organism itself.
>
> Thus in the horn you have something well adapted by its inherent nature to ray back the living and astral properties into the inner life. In the horn you have something radiating life—indeed, even radiating astrality . . . .
>
> We take manure such as we have available. We push it into the horn of a cow and bury the horn a certain depth into the earth. You see, by burying the horn with its filling of manure, we preserve in the horn the forces it was accustomed to exert within the cow itself, namely, the property of raying back whatever is life-giving and astral. Through the fact that it is outwardly surrounded by the earth, all the radiation that tends to etherealize and astralize is poured into the inner hollow of the horn. And the manure inside the horn is inwardly vitalized with these forces, which thus gather up and attract from the surrounding earth all that is etheric and life giving. . . .
>
> Thus in the content of the horn we get a highly concentrated, life giving manuring force. . . . [which we spray] over the tilled land so as to unite it with the earthly realm.[4]

The logic is clear if you accept the premises. The horn gives life force to the cow, which uses it up and excretes it in the manure. The manure is thus the repository of some sort of force, but somehow expended. By putting it back in the cow horn and closely affiliating it with the earth, which is the ultimate provider of the life force, the manure can be rein-vigorated and concentrated with that life force, if done properly with reference to the planets. Unfortunately, such ideas only invite ridicule in the context of modern knowl-edge. Even so, biodynamic agriculture attracted and still attracts many adherents. Most notable was Ehrenfried Pfeiffer who worked directly with Steiner on the experiments that formed the basis of the biodynamic prescription and became a vocal proponent of the techniques. His contribution was to bring the biodynamic movement to the United States when he settled there in 1938.

During these prewar years, two less-than-desirable trends must be acknowledged. As might be expected from the son of an evangelist who first went to Canada as a missionary and later found an epiphany and "rebirth" under the influence of the forests of his native home, St. Barbe Baker saw the organic movement as simply an outgrowth of the larger Christian movement. Williamson and Pearse, founders of the Pioneer Health Centre, noted that "through the process of biological development the Word of God becomes the deed of Man." In 1926 the Chandos Group (named after the restaurant of the same name where they regularly met) began meeting, eventually becoming the editorial committee of the "New English Weekly" that, although commonly associated with the organic agriculture movement, was much more generally focused on "Christian Sociology." Undoubtedly, despite the fact that many of its "founders" were scientifically trained, a strong Christian ideology was closely associated with many of the most influential of the early strands of the movement. This association would, similar to the biodynamic movement, erect a subtle but persistent barrier to a scientific approach to the alternative agriculture movement.

A second trend was a Fascist and semi-Fascist tendency, probably most strongly identi-fied with Jorian Jenks, an agricultural scientist and farmer who wrote regularly for the *New English Weekly* and who eventually became known more for his journalistic abilities than either his farming or agronomy. He is intimately associated with the origins of the organic movement and was also an open member of the British Union of Fascists. The association of Jenks with both the organic agriculture movement and British Fascism is not unusual. What is unusual to a modern reader is the interlocking of the political left, the political right, and the organic movement in a weird panoply of ideas, publications, and move-ments. Jenks was both a member of the British Union of Fascists and editorial secretary of the Soil Association (discussed later) and promoted such ideas as "agricultural land must cease to be regarded as a commodity," clearly a left-wing idea in contemporary discourse.

The Beginnings of the Modern Alternative Movement

For real institutional representation of an alternative to the industrial system, we must wait until the end of World War II, when the modern industrial system really began to dominate. In contrast to the postwar chemicalization propaganda (see Chapter 6), in the

1940s, Lady Eve Balfour began a series of long-term observations on her farm—she had decided that the farm would be wholly organic. These became quite famous in organic agriculture circles and are known collectively as the Haughley Experiment, described in her book *The Living Soil*.[5] This was the first time that detailed observations were made on organic production on a whole farm level, and many of Balfour's observations have become conventional wisdom in the organic agriculture movement. In 1946, inspired principally by Balfour's book, the Soil Association was formed in Great Britain, the first organization specifically devoted to the promotion of agriculture that would be self-sustaining, what we would call today organic or ecological agriculture. Its stated aim, at its founding, was "to create a great body of biological knowledge of the life of the soil and to distribute that knowledge far and wide to the consumer as it accrues to the cultivator." The society still exists and is the main agency for the establishment of standards for organic produce in the United Kingdom.

In 1921, Jerome Irving Cohen, son of a Jewish immigrant family settled in New York, changed his name to Jerome Rodale, largely for reasons of expected commercial opportunity from which being a Jew would have excluded him. His entrance into the organic movement marked a distinct deviation from the extreme Christian influence that had been so important in England, although he was strongly influenced by the biodynamic movement. Having made a substantial amount of money in publishing, by the late 1930s, he came under the influence of the organic movement, in part through the ideas of the biodynamic movement, especially the works of Ehrenfried Pfeiffer. He began publishing *Organic Farming and Gardening* in 1942, the first issue of which featured an article about biodynamic gardening by Pfeiffer as well as a description of the Indore process. Howard became an editor shortly after the magazine's first appearance. Rodale's *Pay Dirt*, published in 1945, was influential in launching the organic movement in the United States, but it was his publication of the magazine that had, and still has, such a continuing influence. The Rodale Institute does cutting-edge research on organic production techniques and remains an important source of research results for organic agriculture.

The Historical Legacy Through the Lens of the Soil

Throughout this short historical narrative, one can see a continual critique, even if it is muted, of the Liebig approach. The soil cannot be seen as a passive sponge through which chemicals pass to plant roots. Modern soil science has come to accept this view on how the soil actually works, even if its origin in Liebigism compels it to continue its support of the chemical bias in the practice of agriculture. That is slowly changing, and the importance of "organic" methods, writ large, are gradually penetrating the mainstream of soil science. The biological side of soils is now recognized as not only a legitimate component of soil science—it is acknowledged as essential.

On the other hand, a certain institutional inertia exists that has more to do with politicoeconomic assumptions than either chemistry or biology. The various soil management schemes that are normally used by either indigenous or organic farmers usually consist of a series of rules, sometimes cultish (e.g., manure in horns), yet sometimes surprisingly rational once properly understood (planting with the moon phase so as to coordinate crop phenology over a large area). Because the dominant social form is the MPM system (the money-product-money chain so characteristic of the Capitalist system—see the appendix to Chapter 7 if this terminology is unfamiliar), traditional practices do not fit into the fundamental scheme. Purchasing nitrate to satisfy the nitrogen demands of the crops means a product can be purchased by an entrepreneur and sold to a consumer, thus satisfying the underlying necessity of the system. If nitrate is supplied through biological nitrogen fixation, the entrepreneur is simply not needed at all. Converting the soil management system to one based on an integration of biological, physical, and chemical forces is thus not expected to be popular. A brute-force chemical approach is, indeed, more profitable according to standard accounting assumptions (including the standard assumptions about the "subsidy" structure of agriculture).

In a very general sense, von Liebig pretty much said it all. Plants require nutrients in particular proportions to one another, and the nutrient in shortest supply limits the growth of plants. However, knowledge of the intimate details of chemical and biological interactions within the soil remained fundamentally incomplete until Louis Pasteur who, in 1870, suggested that ammonium was converted to nitrate through the active mediation of bacteria. We can indeed understand much of the science of soil as a combination of these two key insights—the chemical nature of what plants need from the soil and the biological nature of how those chemicals are transformed in the soil. Beyond that, it is mainly details.

Although understanding the basics of soil physics and chemistry, as covered in the previous chapter, is certainly a prerequisite for understanding the ecology of agroecosystems, the real key for more ecological forms of agriculture comes with the biology of the soil. From the moles, to the ants and termites, to the earthworms, to the springtails, to the ciliates, the bacteria, the roots of the vascular plants, the algae and fungi, an enormous range of living organisms is involved in the formation, function, and maintenance of the soil. A great many of the physical and chemical changes that occur in soils are a direct consequence of living organisms. And there is general agreement that the extent to which this complex biological community is disrupted the "health" of the agroecosystem may become compromised.

Making sense of this enormously complicated subject is not easy. Indeed, attempts at presenting it in a systematic fashion are remarkably eclectic, as evidenced by the diverse presentations of the subject matter in soils texts. For purposes of this text, a convenient starting point is illustrated in **Figure 5.1** (many other frameworks are equally legitimate). Plant growth, the ultimate "goal," is followed by death of the plant, which provides dead organic material that is acted on by living organisms to 1) enter into the decomposition process 2) change the physical and chemical structure that 3) ultimately affects the supply of nutrients for plant nutrition. Because it is really all one continuous process, partitioning

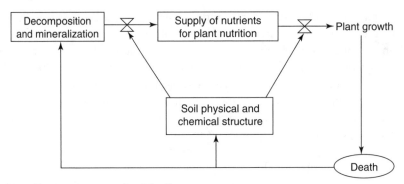

Figure 5-1 Basic processes of soil biology.

into such categories is always fraught with the errors of simplification and reductionism. Nevertheless, for pedagogical purposes, it is necessary to break such a complex subject into analyzable bits, and these three categories seem not only appropriate, but they are also fairly standard. Thus, these three compartments form three of the main headings of what follows in this chapter: supply of nutrients, decomposition, and effects on soil physical and chemical structure.

Supply of Nutrients for Plant Nutrition

Generalities

Nutrients arrive at the rhizosphere through three basic mechanisms: 1) root interception, 2) mass flow, and 3) diffusion. Root interception is caused by the growth of roots, the dynamics of which are a hotly debated topic. Root extension appears to be a function of nutrient availability such that when a limiting nutrient becomes less available, the plant tends to put more energy into root extension and less into shoot growth. At a finer scale, the response of the growing root tip (and possibly other more complicated mechanisms) results in effective root foraging, in which, through differential growth, root mass concentrates in local patches of nutrients.[6] Several authors[7] have suggested that plants can be roughly categorized into two basic types: those whose roots grow very rapidly and live in high nutrient environments, necessitating a large tissue turnover rate, as opposed to those that grow very slowly and live in a poor nutrient environment, with less tissue turnover and high investment in herbivore defense. This pattern could be important when considering strategies for designing future agricultural systems, especially in tropical areas where the second strategy is likely to be more important.

The second mechanism for getting nutrients to the rhizosphere is mass flow. Ions in solution are carried along with the flowing water. This is by far the most rapid means whereby nutrients arrive at the rhizoplane. The third mechanism is diffusion, which can be exceedingly slow. The relative importance of these three mechanisms is shown for a particular case in **Table 5.1**.

TABLE 5-1 Percentage Contribution of the Three Mechanisms of Approaching the Rhizoplane for Maize

Mechanism Nutrient	Mass Flow	Root Interception	Diffusion
N	99	1	0
P	6	3	91
K	20	2	78
Ca	71	29	0
S	95	5	0
Mb	95	5	0

Different nutrients are brought to the rhizoplane through different mechanisms. Furthermore, the pattern shown in Table 5.1 may be quite different for different soils and different crops. In this particular case, nitrogen was brought to the rhizoplane almost exclusively through mass flow, a very rapid process, whereas phosphorus was almost exclusively brought there through diffusion, a very slow process. Potassium was somewhat in between. Such figures illustrate how it is that some nutrients move quickly through the soil solution and others move slowly. Those whose movement is dominated by mass flow are generally called mobile nutrients, whereas those whose presence in the rhizosphere is mediated mainly by root interception or diffusion are known as immobile nutrients (other mechanisms may immobilize nutrients also, as discussed later). Particular mobility depends on the soil type as well as the vegetative material involved, but a general rule of thumb, for many soils and many plant species, is that nitrogen is a highly mobile nutrient, whereas phosphorus is highly immobile.

Ions arrive in the rhizosphere either in solution or adsorbed on the surfaces of soil colloids. Through the normal process of metabolism and sometimes through active physiological mechanisms, the root pumps H^+ ions into the rhizosphere. This means that the rhizosphere normally has a lower pH than the surrounding soil matrix. What might one expect to happen when a soil colloid with ions adsorbed on its surface comes into the rhizosphere? Suppose that there are just two types of cations in the soil solution (which is never the case, but for heuristic purposes assume), say K^+ and H^+. Away from the rhizosphere, the relative concentrations of H^+ and K^+ in the soil solution will determine the proportions of those two ions on a soil colloid. But as soon as the particle comes into the rhizosphere, it encounters a larger concentration of H^+ cations, which means some of the K^+ ions must be replaced from the surface of the colloid to maintain the proper balance between the H^+ and K^+ on the surface. This means that K^+ ions are suddenly released from the surface of the colloid, become free in the rhizosphere, and thus are available for absorption by the plant. This process is cation exchange, one of the most important characteristics of a soil with regard to plant nutrition as already discussed in the previous chapter.

The collection of cations adsorbed to the surface of the colloid is a function of the concentrations of those cations in the general soil solution, although the precise ratios of the various adsorbed ions will not be equal to the ratios in the soil solution, as the strength of adsorption differs for different cations. The relative strengths of adsorption for some key cations are as follows:

$$Al^{3+} > Ca^{2+} > Mg^{2+} > K^+ = NH_4^+ > Na^+$$

so that, for example, if the soil solution contained equal concentrations of calcium and potassium ions, there would be more calcium ions adsorbed on the colloidal surfaces than potassium ions. Consequently, the cations are not given up to the rhizosphere in the concentrations they are in the soil, but rather in the proportions in which they occur adsorbed on the colloidal surfaces.

A similar mechanism occurs in those soils having significant anion exchange capacity, except that the ion responsible for the exchange in the rhizosphere is $COOH^-$, another product of the normal metabolism of the plant. The entire process of releasing free ions to the rhizosphere is illustrated in **Figure 5.2**.

Thus, in the rhizosphere, free ions come to exist because of simple mass flow (those ions that are not adsorbed on colloidal surfaces or otherwise bound in the soil matrix), or because of ion exchange. The next question is, how do they enter the root from the rhizosphere? Absorption of the nutrients across the cell membrane of the root hair may be simple passive transport if the ionic concentration in the rhizosphere is larger than in the cytoplasm, or it may involve complexes of large protein molecules that actively pump the ions against large osmotic gradients, an energy-demanding process.[8] The ability to accumulate ions against a concentration gradient is not only restricted to the cell membrane, but within the cell, the vacuole is able to store very large quantities of ions at much larger concentrations than in the outer cytoplasm. This ability is crucial to the general regulatory metabolism of the plant cell. Furthermore, plants can selectively absorb some ions over others, and it is even the case that different positions along the root may be specialized for absorption of particular ions.[9]

Moving from the micro to the macro level, at a high level of abstraction, all nutrients have the same basic cycle: 1) brought to the soil as dead plant and animal matter, 2) acted on by decomposers to become organic matter, 3) mineralized and deposited in the soil solution from organic matter by the action of microorganisms and from primary minerals by weathering, 4) "immobilized" by microorganisms for use in their own metabolism, 5) adsorbed on the surface of soil colloids, or 6) leached out of the system, and eventually 7) absorbed by the plants in the system (**Figure 5.3**). Although this is always the basic sequence, each nutrient has certain characteristics that distinguish its cycle from that of the others. These differences can be critical in managing the nutrient in either conventional or alternative forms of agriculture. Despite the differences, all

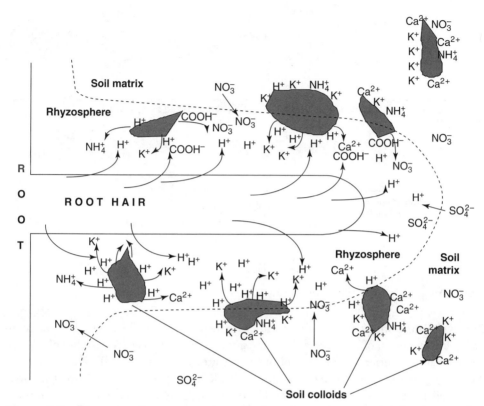

Figure 5-2 Diagrammatic representation of how cation exchange makes free ions available in the rhizosphere. Cations (such as K^+, Ca^{2+}, and NH^{4+}) are adsorbed on the surface of soil colloids according to their relative strength of adsorption and their concentration in the soil solution. As the soil colloids come into the region of the rhizosphere, they encounter large numbers of H^+ ions (i.e., a higher acidity). Many of the cations are released from the surface of the colloids to make way for the H^+ ions to maintain the proper relative concentrations on the colloidal surface (now more highly filled with H^+ ions because of the local high concentration of them in the rhizosphere). The usually less important anion exchange phenomenon is similar, with $COOH^-$ ions acting to exchange for the adsorbed anions, such as NO^{3-} and SO_4^{2-}. Most of the anions remain in the soil solution as free ions and are mass transported directly to the rhizosphere.

elements share some basic principles, the most basic of which are deducible from a careful examination of Figure 5.3.

For some elements (e.g., nitrogen), the cycle is so complicated by other factors that such a simple flow diagram is of little use. For other elements (e.g., potassium), the major features of Figure 5.3 are reasonably representative of what needs to be known about the element.

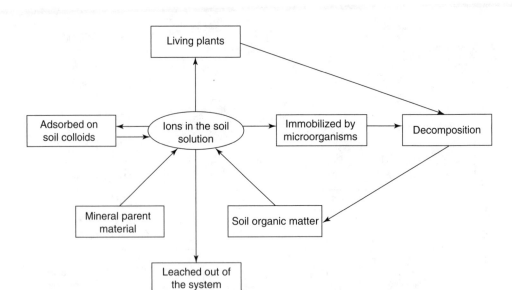

Figure 5-3 Basic processes in nutrient cycling.

One of the most important ideas that can be easily deduced from Figure 5.3 is the intensity/quantity notion of a nutrient element[10] and the buffering capacity of the soil (not in the sense of pH buffering, but metaphorically related). The intensity (I) of an element is the quantity of that element that is readily available to the plant for absorption (the "ions in the soil solution" oval in Figure 5.2). The quantity (Q) of the element in the soil is the amount adsorbed on soil colloids plus the amount that is readily mineralized from the soil organic matter and the mineral parent materials (by far the most important of this pool is that adsorbed on soil colloids because through a direct chemical reaction they can be made available to the ion pool in the soil solution). It is ultimately important to distinguish between two components of Q: 1) adsorbed on colloidal surfaces versus 2) available in organic matter, dissolvable from parent material and immobilized by microorganisms at any one point in time. The adsorbed ions are immediately available when the growing plant needs them for nutrition, whereas the other sources are stored more tightly for later use, an important consideration when concerned with questions of sustainability rather than just immediate nutrition needs for production. For now it is only necessary to note that a soil may have a high value of I, but because of low levels of soil colloids, have a low level of Q—that is, few colloidal surfaces on which the ions can be adsorbed, which results in their rapidly leaching out of the system. A measure of the potential buffering capacity (PBC) of the soil is then the ratio "change in Q" to "change in I." When a change in intensity is accompanied by the same change in quantity, the soil is said to be highly buffered (the perfectly buffered soil is PBC = 1.0, where every change in intensity is exactly balanced by an opposite change in Q, so the concentration of the ion in the soil solution remains constant even when there is a large uptake by the plant). The concept of PBC can be used for any nutrient element but is most commonly used in consideration of phosphorus and potassium.

As mentioned previously, although all elements follow the basic pattern shown in Figure 5.3, each has its own peculiarities. Perhaps the most complicated cycle is that of nitrogen.

The Nitrogen Cycle

Pasteur's speculation was prescient. Indeed, soil chemistry is intimately tied up with biological activity, and the nitrogen cycle is the canonical example. Although the cycle is complex, its elementary structure can be understood as the element nitrogen existing in six distinct forms and those forms transformed from one to another by seven distinct processes. The six forms are organic in plant (and animal) material, organic in soil microorganisms, organic in soil organic matter, ammonium (NH_4^+), nitrate (NO_3^-), and molecular nitrogen (N_2) in the air. The seven processes are fixation, death, decomposition, mineralization, nitrification, denitrification, and absorption. The states (forms) and processes are shown in **Figure 5.4**.

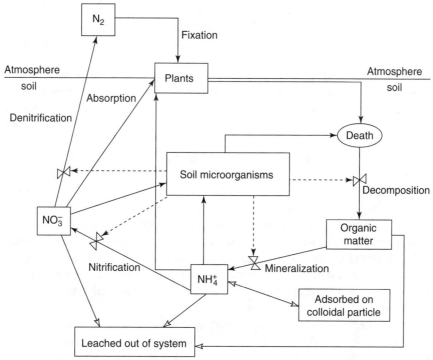

Figure 5-4 The basic nitrogen cycle.

A cursory glance at Figure 5.4 reveals the simple structure of the cycle. Atmospheric nitrogen (N_2) is fixed, mainly by bacteria symbiotically associated with vascular plants, thus converting it to organic nitrogen. When the plant dies, the organic form, under the influence of soil microorganisms, decomposes into smaller units, ultimately existing in the soil solution as large organic molecules, the soil organic matter. The soil organic matter is broken down, through the action of soil microorganisms, to form ammonium ions, in the process called mineralization. The ammonium ions, some of which become adsorbed on the surface of clay particles, are mainly transformed, through the action of other soil microorganisms, to form nitrate ions. The ammonium and nitrate ions may be absorbed by either plants or microorganisms. The nitrate ions may be denitrified, through the action of yet another set of microorganisms, to form atmospheric nitrogen once again.

Fixation is commonly thought to occur mainly through the symbiotic relationship between *Rhizobium* bacteria and nodulating leguminous plants. An intimate symbiotic relationship exists between the bacteria and the plant. After infection by the *Rhizobium*, a nodule is formed on the root of the plant, and the *Rhizobium* lives its life entirely within this nodule, converting atmospheric nitrogen to a form that the plant can use and receiving a supply of carbohydrate from the plant in return. Chemically, molecular nitrogen is reduced to ammonia, which is then combined with organic acids to form amino acids and finally proteins. The reaction can be simply represented as follows:

$$N_2 + 6H \xrightarrow[\text{Fe, Mo}]{\text{nitrogenase}} 2NH_3$$

$$NH_3 + COOH \longrightarrow \text{Amino acids}$$

The amino acids are then used in proteins, which are formed and used by both the plant and the bacterium. As nodules die and begin decaying, in the same way that the parent root material dies and decomposes, some of the fixed nitrogen is left in the soil as organic material. Although some nitrogen may leak into the soil from the living nodules and/or plant roots, probably the bulk of what is supplied to the soil comes from the decomposition of the plant after death. There is considerable variability in how much fixed nitrogen is actually retained in the plant doing the fixing. Soybeans, for example, retain almost all of the fixed nitrogen in the seed, whereas common bean appears to be less efficient and some nitrogen thus "leaks" out of its roots.

The enzyme nitrogenase, supplied by the *Rhizobium*, is a two-protein complex, the larger member of which contains molybdenum and the smaller of which contains iron. These two elements are thus critical to the process of nitrogen fixation. A shortage of either one of these elements can thus produce symptoms of nitrogen deficiency, bringing up the obvious conundrum of what indeed is in short supply, molybdenum and iron or nitrogen itself.

The legume/*Rhizobium* symbiosis is highly evolved in the sense that specific species of *Rhizobium* are associated with specific species of legumes, and even within a species, different varieties of *Rhizobium* may perform with different degrees of efficiency. It is thought that particular varieties or species perform better or worse under particular conditions, such that species A may be more efficient in dry years and species B in wetter years.

Although it has been common to assume that most nitrogen fixation is accomplished by the symbiosis of legumes and *Rhizobium*, recent work suggests that other sources of fixation may be important also, perhaps rivaling the legume/*Rhizobium* symbiosis in terms of world nitrogen fixation. These other sources include symbiotic associations of other microorganisms with plants, loose associations of microorganisms with plants and with free-living microorganisms.[11]

One of the "other" microorganisms that form nitrogen-fixing symbioses with plants are cyanobacteria. One of the most striking is with the tropical shrub *Gunnera* sp., a colonizer of roadsides and other nitrogen-poor habitats in the neotropics. Of particular significance to agroecosystems is the association between *Azolla*, an aquatic fern, and the cyanobacteria *Anabaena*, traditionally used in some areas of tropical Indochina in rice paddies. A typical management pattern is to flood the paddy, let the *Azolla* grow on the surface, drain the paddy and incorporate the *Azolla* mulch, fill and drain the paddy a couple of more times, and then plant the rice.

A very different kind of association is formed with bacteria in the genera *Azotobacter*, *Azospirillum*, and *Clostridium* and various grasses, where the fixation occurs in the rhizosphere. The bacteria use root exudates from the grasses as sources of energy for nitrogen fixation, and according to some reports, especially from the tropics, they can fix considerable quantities of nitrogen. Other authors minimize the importance of this form of nitrogen fixation, as most reports show relatively small quantities of nitrogen actually fixed.[12]

Nitrogen fixation is also accomplished by free-living bacteria and especially the cyanobacteria.[13] Several genera of aerobic bacteria are capable of nitrogen fixation (*Azotobacter* in temperate soils and *Beijerinckia* in tropical soils), and the genus *Clostridium* is an anaerobic bacteria that also has this habit. It is thought that normally there are always small oxygen-poor pockets in the soil, and the anaerobic and aerobic bacteria may work simultaneously in any soil.[14] It would also seem obvious that simultaneous maintenance of both anaerobic and aerobic nitrogen-fixing bacteria could constitute part of a risk aversive strategy since over time the proportion of area that is oxygen poor in the soil certainly must vary substantially.

Death is not usually thought of as a process, but in the context of agroecosystems, it seems heuristically valuable to include it as such. Normally, plants and animals die because of senescence, predators, parasites and diseases, and accidents. But in agroecosystems, death is a planned activity. When it is time for harvest, the crops die and, depending on the crop, leave some degree of organic material on and in the soil and/or are physically managed to enter into a directed decomposition, composting. In the context of agroecosystems then, understanding the process labeled death is closely associated with the farming practices employed by *Homo sapiens*. This issue is discussed more completely in a later section. In more complex agroecosystems, the harvest is not the main episode of death, but rather, leaf senescence and other processes similar to nonmanaged ecosystems supply the raw material to the decomposition process, which is discussed in detail later. In particular, in systems with animals, it is the cycling of plant materials by the herbivores that provides the raw material, manure, that initiates the process.

Decomposition is the process whereby soil organisms, both macro and micro, act on dead organic material to turn it into organic matter. Although it is a continuous process, it is convenient to separate organic material from organic matter. The latter is seen and analyzed as molecular structures, whereas the former is seen and analyzed as something that retains remnants of the anatomy of the parent material. Involved in this process are earthworms, small arthropods, fungi, and bacteria. The process is discussed in more detail later.

Mineralization is the process in which organic molecules containing nitrogen are broken down into ammonium. The nitrogen contained in organic molecules is thought to be largely unavailable to plants. It must be converted to ionic form first, a process requiring the intervention of specially adapted bacteria. The basic problem in mineraliza-tion is the conversion of rather large polymers first into monomers and then into ammo-nium. It is natural to break it down in this way because the microorganisms in question generally cannot ingest the polymers, and thus, the polymers must be broken down out-side the cell of the bacteria. But there is a basic problem. Breaking the bonds of the large polymers does not generally provide energy, and some of the stronger bonds actually require energy to break. Furthermore, the polymers in question are largely insoluble in water, yet the organisms in question require an aqueous medium. Thus, the breakdown of these polymers is not an easy task, and it is not surprising that it is accomplished by highly specialized organisms. The resulting monomers are the energy-yielding chemicals for the microorganisms involved.

The natural polymers are broken down outside the bodies of the bacteria. For this to happen, the bacteria must release special extracellular enzymes into the environment. The extracellular enzymes break the chemical bonds that hold the polymers together, thus producing monomers (or dimers or trimers—i.e., smaller molecules) that the bacteria can ingest. The monomers inside the bacteria's body are then metabolized by normal means to generate the energy for the bacteria.

The exact bacteria involved in these reactions depend on the nature of the organic matter. For example, when the organic matter is cellulose, bacteria capable of producing cellulase as an extracellular enzyme are necessary to break the molecule down, as cellulose is too large to be ingested. Only specialized bacteria are capable of producing cellulase, for example some members of the genera *Streptomyces*, *Cellulomonas*, *Pseudomonas*, *Chromobacterium*, *Bacillus*, *Clostridium*, and *Cytophaga*.[15] On the other hand, if the organic matter is protein, a wider array of bacteria (and fungi) is capable of producing peptidases as extracellular enzymes (nevertheless a restricted set). The monopeptides are then ingested and go through the normal metabolic processes, releasing the ammonium from the amino acids of the peptide.

Nitrification is the conversion of ammonium to nitrate. The natural conversion of ammonium, a cation, to nitrate, an anion, has never been observed within a single organism. Two highly specialized bacteria are required to carry out the process, *Nitrosomonas* for the conversion of ammonium (NH_4^+) to nitrite (NO_2^-) and *Nitrobacter* for the conversion of nitrite to nitrate (NO_3^-). The basic reaction can be illustrated as follows:

$$\overbrace{NH_4^+ \xrightarrow{\ [O]\ } HONH_2 \xrightarrow{\ -2H\ } \tfrac{1}{2} \ HONNOH \xrightarrow{\ [O]\ } NO_2^- + H^+ + Energy}^{\text{\textit{Nitrosomonas}}}$$

$$\overbrace{NO_2^- \xrightarrow{\ [O]\ } NO_3^- \ + \ Energy}^{\text{\textit{Nitrobacter}}}$$

Under normal conditions, both reactions occur almost simultaneously, thus usually preventing any buildup of nitrite. However, it obviously requires adequate population densities of both types of bacteria to function properly.

Several practical consequences of the process of nitrification are evident. Because ammonium is a cation and because in most soils negative charges dominate the colloidal surfaces (see previous chapter), ammonium is easily held as an exchangeable cation and can be absorbed by plants as a cation. Ammonium cannot be stored in great quantities in cells because of its toxicity. Nitrate, on the other hand, is an anion, and in most soils there is no appreciable positive surface charge on which it can be adsorbed (although this may not be true always, as discussed in the previous chapter). Thus, the nitrate remains as a free ion in the soil solution and can easily leach out of the system. Furthermore, nitrate may be subjected to denitrification and lost to the atmosphere (discussed later). Unlike ammonium, nitrate can be stored in great quantities in vacuoles but then must be reduced to ammonium before being processed into amino acids. Consequently, both plants and the soil environment face a fundamental contradiction dictated by physical laws. It would generally be better for both plant and environment to use ammonium as a source of nitrogen (it does not rapidly leach out of the soil so can be conveniently stored there, and the plant does not have to expend extra energy converting nitrate to ammonium before making amino acids). If ammonium were to be the source, however, the plant would not be able to store significant quantities of nitrogen and thus would have to rely on a noninterrupted supply from the rhizosphere. Consequently, most crop plants are more efficient at taking up nitrate than ammonium, and thus having a ready supply of it in the soil is important. The trick then is to have the nitrification balance the absorption. If nitrification goes too fast, some of the nitrate will not be absorbed by the crops and runs the risk of leaving the system through denitrification or leaching. If nitrification goes too slow, not enough nitrate will be available for the absorptive demands of the crops. Furthermore, the absorptive demands of the crop are constantly changing through the growing cycle, making the theoretical balance even more difficult.

Denitrification is another bacteria-driven process whereby nitrate is converted into volatilizable materials, eventually molecular nitrogen. The basic process can be represented as follows:

$$2NO_3^- \xrightarrow{-2[O]} 2NO_2^- \xrightarrow{-2[O]} 2NO \xrightarrow{-[O]} N_2O \xrightarrow{-[O]} N_2$$

Nitrate	Nitrite	Nitric	Nitrous	Molecular
ions	ions	oxide	oxide	nitrogen

Exactly how the bacteria drive this reaction sequence is not fully understood, but at each of the latter three steps, a volatilizable material is produced, and nitric oxide, nitrous oxide, and/or molecular nitrogen can be released into the atmosphere. The magnitude of these losses depends on the exact type of soil, the cropping system, soil cultivation, and so forth. Generally, the more humid the climate, the more important is denitrification. Also, nitrous oxide is an extremely potent greenhouse gas.

The organisms involved in this process are common anaerobic bacteria, found in all soils, but the process occurs only in the anaerobic part of the soil, either in an anaerobic layer in the subsoil, or in oxygen-deficient pockets that may exist in the general soil matrix. In some rice production systems, the loss through denitrification can be as much as 70% of applied ammonium, which has obvious implications not only for nitrogen use efficiency, but also for global climate change, given the importance of nitrous oxide as a greenhouse gas. A means of controlling this loss (sometimes reducing the loss by 50% or more) is to incorporate the ammonium into the deeper anaerobic layers, thus reducing the rate of nitrification.

Absorption is the process whereby the mineralized nutrients are taken up by the plant. The details of the general process of absorption have already been discussed. Nitrogen is a rather unusual case in that two ionic forms, one an anion the other a cation, are both taken up by plants. Because most nitrogen is used in the formation of proteins, it is only logical to expect evolutionary selection pressure for the uptake of ammonium ions, since they can be used directly in protein synthesis. Nitrate must be reduced, an extra cost the plant must assume if nitrate is the source of nitrogen.[16] Yet no such general evolutionary trend seems to exist. If there is any pattern at all, soils of low pH and low temperatures tend to favor the uptake of ammonium ions, whereas alkaline soils at higher temperatures tend to favor the uptake of nitrate ions, a simple function of the rate at which nitrification is occurring.[17]

From the point of view of the growing plant (or the farmer), all of this is a very dynamic process. Nitrogen is fixed biologically at a rate partially dependent on the biomass of the plant involved in the fixation process. It then is converted to organic dead matter by the rate of death of plant material and ultimately to organic matter by the process of decomposition. The ammonium ions are adsorbed on clay particle surfaces (organic matter too) in proportion to the other ions in the soil solution. At a rate dependent on the biomass of the proper microorganisms, the ammonium cations are transformed into nitrate ions and leached out of the system if they are not taken up by plants or denitrified to return to the atmosphere. In other words, it is sort of like filling a bucket with many leaky holes. The goal is to supply the nitrogen to the growing plant at exactly the correct rate, which implies controlling all the other leaks in the bucket, all the time, realizing that as the plant grows it changes its needs. It is indeed a tremendously challenging problem for the plant (and the farmer).

The Phosphorus Cycle

The importance of phosphorus in plant metabolism has long been recognized. Most recently, emphasis on tropical ecosystems has encouraged a greater focus on this element, as it seems to be the limiting nutrient in many tropical soils, unlike most temperate soils in which nitrogen is usually the major limiting nutrient. As a component in ATP and ADP, to say nothing of its central position in the structure of DNA and RNA, its importance to the most basic processes of biology is clear.

The basic problem with phosphorus is that 1) it is usually not very abundant (for a macronutrient) in the soil and 2) it frequently occurs in insoluble form, transferring from soluble to insoluble by everyday soil processes, especially in tropical soils. At a more fundamental level, there is no parallel to nitrogen fixation, so the concentration of phosphorus in the soil is all that is there, and the long-term consequence of agriculture on unfertilized soil is that phosphorus of necessity will become limiting. In principle, phosphorus makes a closed-cycle sustainable system impossible. Although this important truth obtains for almost all nutrients, the demand for phosphorus compared with its usually low concentration in the soil makes it particularly noticeable in this regard.

The basic cycle is illustrated in **Figure 5.5**. Soluble phosphorus is taken up by plant roots in either ionic or small molecule organic form and is incorporated into plant and animal tissue. After death, it becomes part of the soil organic matter, is taken up in the bodies of microorganisms, and eventually reverts back to the soluble pool. The problem is that frequently it becomes "fixed" in forms that are effectively as recalcitrant as the original primary minerals. (Note the difference in the use of the word "fixed" for phosphorus versus Nitrogen. Fixed P is that which is effectively unavailable in the soil, as explained here. Fixed N is organic N that was derived from atmospheric N_2). Phosphorus is the only nutrient element that effectively returns to the original mineral state, metaphorically speaking, and must be converted anew to its soluble form. This has been especially irksome in conventional chemical agriculture, as sometimes much of the phosphate added to the soil as fertilizer is converted to the fixed form before being absorbed by the crops.

Several mechanisms exist whereby phosphorus becomes fixed. To understand these various processes, we must first understand the basic ionic forms and how they are related to the pH of the soil. The following scheme summarizes this relationship:[18]

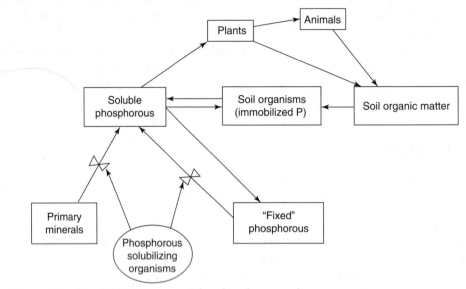

Figure 5-5   The basic processes of the phosphorus cycle.

$$H_2PO_4^- \xrightleftharpoons[H^+]{OH^-} H_2O + HPO_4^{2-} \xrightleftharpoons[H^+]{OH^-} H_2O + PO_4^{3-}$$

Very acid   Very alkaline
solutions   solutions

Thus, in very acidic soils, such as many of those in the tropics, you would normally expect $H_2PO_4^-$ ions to be available for plant absorption, whereas at the other extreme, in very alkaline soils, you would expect PO_4^{3-} to be the main form.

The $H_2PO_4^-$ ions in very acid soils could be moved very rapidly by mass flow, or they could be held on positive sites on the soil colloids, available for anion exchange, much like other anions. Frequently, however, high concentrations of aluminum and/or iron are also found in acid soils (especially in the tropics), and the following reaction may occur:

$$Al^{3+} + H_2PO_4^- + 2H_2O \leftrightarrow 2H^+ + Al(OH)_2H_2PO_4$$

Although the $H_2PO_4^-$ is soluble, the $Al(OH)_2H_2PO_4$ is not (basically the same reaction could occur with Fe or Mn ions rather than Al). In most cases in nature, acid soils contain far more aluminum and/or iron ions than $H_2PO_4^-$, meaning that the above reaction moves to the right, and the phosphorus moves invariably to its nonsoluble form.

An even more common reaction is with $H_2PO_4^-$ and hydrous oxides of aluminum and iron, the ultimate breakdown products of mineral clays under extreme weathering conditions, and the dominant clay colloid in many tropical soils (see previous chapter). Basically, an OH group of the hydrous oxide is replaced by the $H_2PO_4^-$, releasing an OH^- into the solution and consequently rendering the phosphate a part of the insoluble hydrous oxide. As with the case of binding with the aluminum and iron ions, here too we see that the very acidity that made the $H_2PO_4^-$ available for adsorption on the surface of the hydrous oxide clay particle insures that it will rapidly be converted to insoluble form. Worse still, after adsorption on the surface of the hydrous oxide, phosphorus tends to migrate to the interior of the crystal, becoming even less available than it was when bound to the surface of the clay particle.

In less acid soils with high concentrations of kaolinite, a similar reaction may occur. The $H_2PO_4^-$ becomes insolubly bound to the clay molecule. The exact mechanism whereby this happens is not known, but it is easy to imagine a similar reaction as that with hydrous oxides, as kaolinite is characterized by exposed OH groups on its surfaces. A similar effect can occur with other silicate minerals, but to a lesser extent than with kaolinite.

A final mechanism of phosphorus fixation is in less acid soils when calcium is available. Calcium phosphate and hydrogen phosphate, both insoluble, are thus formed.

The various mechanisms for phosphorus fixation are illustrated in **Figure 5.6**, illustrating the unfortunate fact that there is only a limited range of pH in which phosphorus is easily available and, even then, not a large percentage of the total contained in the soil. Phosphorus-solubilizing microorganisms can change this picture considerably and are thus probably quite important, especially in tropical soils. Various bacteria (e.g., members of the genera *Pseudomonas, Bradyrhizobium, Rhizobium, Bacillus*) and fungi (e.g., *Aspergillus niger*) are known to cause the solubilization of insoluble forms of phosphorus, although

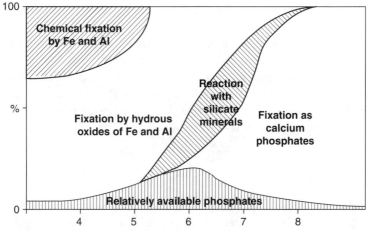

Figure 5-6 Summary of mechanisms for phosphorus fixation—the proportion of phosphorus in various categories as a function of soil pH.

the mechanisms whereby this occurs are not well known. Collectively known as phosphorus-solubilizing microorganisms (PSO—or PSB for phosphorus-solubilizing bacteria), they may have some special adaptation for dealing with phosphorus, or it may simply be due to local changes they force in the chemical environment.[19]

As with all other nutrients, phosphorus is also immobilized by microorganisms for their own nutrition. However, immobilized P is more readily available than fixed P because microorganisms eventually die. In a sense then, the problem of phosphorus nutrition is maintaining a balance between fixed and immobilized phosphorus, such that the maximum ionic form is continuously being supplied to the soil solution at quantities demanded by the growing plants (review the dynamics shown in Figure 5.5). Unlike nitrogen, in which the problem is mainly immobilization or leaching, in phosphorus, it is immobilization versus fixation. Furthermore, as noted previously, no parallel exists with nitrogen fixation so that the phosphorus that is available in the soil is the entire pool unless phosphorus from outside the system is added. As a theoretical principle, it makes fully sustainable agriculture impossible without external inputs. Indeed, in soils with a low absolute quantity of phosphorus (in remaining parent material, in fixed form, in immobilized form and ionic form), it is always a question of time horizon when planning on phosphorus use. Adding phosphorus-solubilizing bacteria, for example, to the soil may increase the rate of ion release into the soil solution but may accelerate the rate of "mining" of phosphorus from the nonionic forms that exist, closing the time gap between the present and that point when the absolute quantity of phosphorus in the soil will be nil. Recall the balance between quantity and intensity, as described previously, and how it relates to the "buffering capacity of the soil."

This long-term problem does not generate much concern in industrial agriculture since the paradigm is to apply phosphate as much as is needed at any point in time, even though the peculiarities of phosphorus fixation sometimes make the application of phosphate pointless. However, in the pursuit of more sustainable forms of agriculture, the long-term question cannot be ignored. One solution is the application of rock phosphate to fields. Conventional phosphorus fertilizer is manufactured from rock phosphate through a chemical process involving treatment with an acid to convert the phosphorus into an available form. In principle, the slow weathering process in the soil or the actions of phosphorus-solubilizing microorganisms could take the place of the acid treatment, making a single application of rock phosphate last for a very long period. This solution to the phosphate problems, especially in tropical soils, has received considerable empirical attention.[20]

Mycorrhizae

It has only been in the past 30 years that one of the most common symbioses in nature has become fully appreciated. Almost all terrestrial plants are symbiotically associated with fungi, which aid in nutrient absorption, a phenomenon first labeled mycorrhizae in 1885.[21] The tight symbiosis now so much appreciated is evidently extremely old, perhaps as old as the evolution of land plants themselves.[22] In a tight association with the plant's root system, the fungal mycelia spread throughout the soil matrix, effectively forming a massive extension of the root surface. The basic symbiosis is formed by the plant delivering carbohydrate to the fungus and the fungus delivering nutrients to the plant.

Currently, seven different types of mycorrhizal associations are recognized, two of which are especially significant for agroecosystems: the arbuscular mycorrhizae (AM) and the ectomycorrhizae. Almost all crop plants are characterized by the AM association, with ectomycorrhizae occurring mainly in the families Pinaceae, Fagaceae, Caesalpiniacea, Dipterocarpaceae, and Myrtaceae. Only a few families are nonmycorrhizal, most significantly for agroecosystems the Brassicaceae and the Cyperaceae.

AM are regarded as important components of the phosphorus cycle. Because so much phosphorus can be unavailable to plants, the ability to mine more extensive areas of the soil with mycelia, plus the possible active role of mycelia in solubilizing fixed phosphorus[23], makes AM especially important in soils where phosphorus availability is a critical factor. Recent work in Cuba has concentrated on the use of AM in pursuit of a solution to the phosphorus problem. It is possible that AM will be one of the main keys in the development of a future, more ecological agriculture because of the intractable nature of the problem of phosphorus nutrition, especially in tropical soils.

Decomposition

The amount of carbon flow involved in decomposition is approximately equal to that involved in primary production, yet the attention accorded the former is orders of magnitudes less than accorded the latter.[24] Decomposition is at least as important in ecosystem dynamics as primary and secondary production, forming one of the three main subsystems of general terrestrial ecosystem dynamics,[25] as pictured in **Figure 5.7**. It has one key dynamic feature that distinguishes it from the other subsystems, the internal recycling component, and is likely far more complicated than the other two subsystems.

Decomposition of any piece of plant or animal matter may take a period of hundreds or thousands of years, yet the bulk of decomposition occurs within one or a few years, depending on climatic conditions. The soil organic matter, the result of partial decomposition, consists of two fractions: 1) the cellular fraction, frequently referred to as organic "material," and 2) humus. The cellular fraction contains recognizable biological tissue, such as leaf fragments, dead microbial cells, and so forth. The humus content is amorphic, composed of polymeric molecules, and may persist in the soil for extremely long periods. It typically enters the pool of soil colloids and frequently forms complexes with clay minerals. As discussed previously, it is intimately associated with the chemical properties of the soil and is one of the main factors contributing to cation exchange capacity. In **Figure 5.8**, the rates of decomposition for six distinct fractions are shown for a typical temperate soil. Clearly, in this case, the sugars are broken down extremely rapidly, whereas the phenols are extremely recalcitrant. Waxes, lignins, cellulose, and hemicellulose are in between.

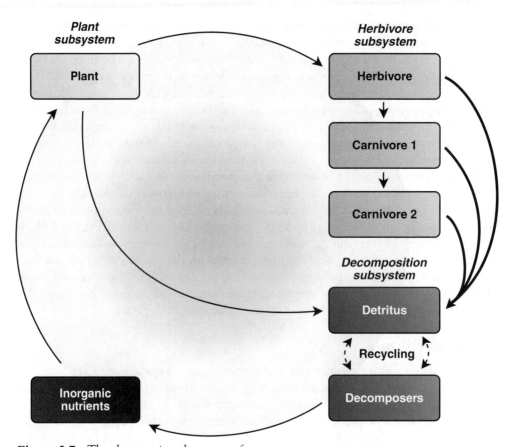

Figure 5-7   The three main subsystems of an ecosystem.
Adapted from: Swift, M. J., Heal, O. W., & Anderson, J. M. *Decomposition in terrestrial ecosystems. Studies in ecology 5.* Oxford, UK: Blackwell Scientific, 1979.

Rather than considering the range of "recalcitrants" suggested in Figure 5.8, normally the decomposition process is simplified by dividing it artificially into two components: the rapid cycle and the slow cycle. The rapid cycle is responsible for the mineralization of nutrients and the slow cycle for the production of humus, which in turn contributes to soil physical and chemical structure. Which materials go to which cycle depends on the chemical makeup of the constituents. Generally, the easily degraded molecular structures (e.g., sugars, carbohydrates, some proteins) are the materials that go into the fast cycle, whereas more structural kinds of materials (e.g., lignins, waxes) or storage and transport materials (e.g., fats, resins) go into the slowly decomposed material. The basic process is illustrated in **Figure 5.9**.

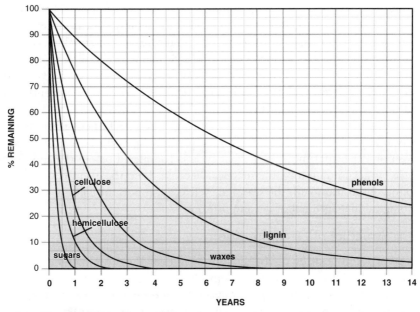

Figure 5-8 Decomposition curves for various organic matter components.

Adapted from: Swift, M. J., Heal, O. W., & Anderson, J. M. *Decomposition in terrestrial ecosystems. Studies in ecology 5.* Oxford, UK: Blackwell Scientific, 1979.

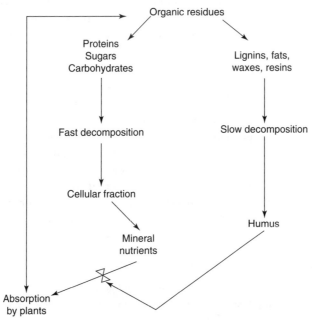

Figure 5-9 The elementary processes of decomposition.

In natural ecosystems, the process is initiated with the introduction of litter to the soil. This is almost always a seasonal phenomenon, even in seemingly aseasonal environments such as tropical rain forests. In most forms of agriculture, this seasonality is exaggerated because of planting and harvesting cycles, although many forms of permanent culture, especially in tropical regions, do not have such a strict cycle imposed by *Homo sapiens* and can be expected to correspond more or less to the patterns of natural systems. In a non-managed system, it is a reasonable expectation that the yearly seasonal cycle will result in something close to a complete cycling of nutrients through the fast cycle. But many agroecosystems are seemingly not at such an equilibrium state and have either very rapid decomposition, such that most of the fast cycle is completed before the end of the seasonal cycle, or accumulate organic material because decomposition is very slow. This timing of decomposition with the seasonal cycle is an important concept when it comes to the ecological management of decomposition.

The Basic Structure of Decomposition

The process of decomposition can be thought of as involving three main transformations, all occurring simultaneously at a given point, but more easily discussed if treated separately: leaching, catabolism, and comminution.[26] Leaching refers to the transport of soluble materials out of the system, presumably to somewhere else in the soil where they enter the decomposition process and continue to be decomposed. It is also convenient to include the loss of volatile products of decomposition in the same category as leaching. Catabolism is the chemical process whereby larger molecules are broken down into smaller ones. The chemical reactions are complex and immensely diverse, some producing mineral nutrients that will be subsequently absorbed by microorganisms and resynthesized into their tissues. Comminution is the physical reduction in size of the organic material, a physical rather than a chemical process. These three forces and their consequences are illustrated in **Figure 5.10**. A second aspect of the decomposition process is its recursive nature, such that organic material is sent down a cascading series of catabolism/comminution/leaching cycles until the original organic material has been completely decomposed, as illustrated in **Figure 5.11**.

The partitioning of organic material into the three processes, leaching, comminution, and catabolism (Figure 5.10), depends on a host of factors both physicochemical and biological, as does the rate at which the material cascades through the cascading sequence (Figure 5.11). Evidently, a scaling factor enters here, in that the time interval from one set of boxes (in Figure 5.11) to another is arbitrarily set yet in nature is a completely continuous process. Nevertheless, for heuristic purposes, this way of looking at the decomposition process is useful.

The Decomposition Subsystem in Larger Context

Although it is true that our present state of knowledge of the remarkably complex interactions that occur among all of these biological components is slim, the general pattern seems clear and well represented by this model of comminution, leaching, humus

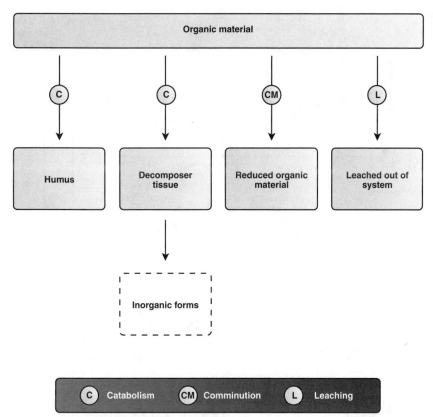

Figure 5-10 Basic processes in decomposition.
Adapted from: Swift, M. J., Heal, O. W., & Anderson, J. M. *Decomposition in terrestrial ecosystems. Studies in ecology 5.* Oxford, UK: Blackwell Scientific, 1979.

formation, and catabolism in a cascading framework, as presented previously (Figures 5.10 and 5.11). The first pass made by decomposing organisms at the organic material placed on or in the soil is dominated by comminution, and the products of that comminution are acted on mainly by organisms that promote catabolism. Note that the products of comminution are not only a more finely dissected detritus but also the microorganisms that have incorporated the carbon from that detritus. So the product of comminution might be most properly referred to as the soil organic "soup" because it includes both soil organic matter and soil microorganisms. It is thus not far off the track, as a heuristic device, to simply think of the detritus first being comminuted and then catabolized to form nutrients and humus. For the most part, the initial comminution is accomplished by fungi and macro ingesters, whereas the catabolism is accomplished mainly by fungi and bacteria. This view of the decomposition system is certainly too simplified to be useful in any active management plan, but it is a useful way to visualize the process, especially in context with the rest of the ecosystem.

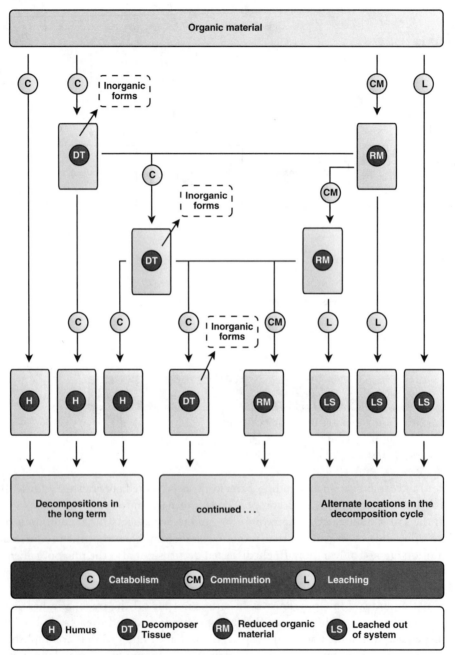

Figure 5-11   The decomposition cascade.

Adapted from: Swift, M. J., Heal, O. W., & Anderson, J. M. *Decomposition in terrestrial ecosystems.*
Studies in ecology 5. Oxford, UK: Blackwell Scientific, 1979.

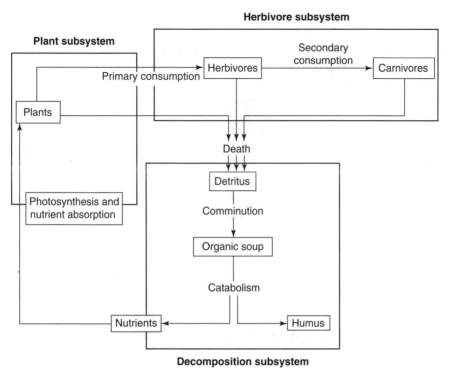

Figure 5-12 A model of the ecosystem.

The ecosystem, from this point of view, can be thought of as having seven compartments into which matter enters and exits and six processes whereby matter is transferred between compartments (**Figure 5.12**). The seven compartments are plants, herbivores, carnivores, detritus, organic soup, humus, and nutrients; the six processes are photosynthesis and nutrient absorption, primary consumption, secondary consumption, death, comminution, and catabolism (Figure 5.12). The fast and slow decomposition cycles are still represented in that some fraction of the organic matter flows to humus (the slow cycle), whereas the bulk flows to nutrients (the fast cycle).

The Organisms of the Decomposition Subsystem

The organisms involved in the decomposition process are remarkably diverse, ranging in size from bacteria to snails. Furthermore, these organisms do not fall into the same neat trophic categories as those in the herbivore subsystem. This is largely due to a curious accident in evolution. In the herbivore subsystem, as usually construed, animals ingest plants. But those animals generally do not have the ability to synthesize cellulase and thus cannot digest their food directly. This problem has been solved mainly by the symbiotic association with bacteria able to produce the necessary enzymes to break down plant cell walls. The specialization needed to become symbiotic with cellulase-synthesizing bacteria apparently made it quite difficult to evolve the habit of regularly eating both cellulose-rich material and animal matter (but not impossible, as is obvious from our own rather unusual eating habits). Thus, ingesters of plant material tend to be specialized on only plant material. Furthermore, again because of historical accident in evolution, most primary producers are absorptive organisms, with only highly specialized forms capable of ingestion (e.g., pitcher plants, Venus fly trap). This forms the basis of the classic notion of the producer trophic level. Plants absorb nutrients from the environment and herbivores ingest plants. Thus, the trophic levels of primary producer (plants) and primary consumer (herbivores) are reasonably well defined.

Even in the herbivore subsystem, however, the definition of trophic level becomes ambiguous when we reach the carnivore level. Normally we think of carnivores, those who eat herbivores, and top carnivores, those that eat carnivores. What, precisely, is the difference between a carnivore and a top carnivore? When a spider eats a carnivorous beetle, it is a top carnivore, but when the same spider eats an herbivorous beetle, it is a simple carnivore. Clearly, the trophic levels carnivore and top carnivore are vague.

With the absence of either primary producer or herbivore in the decomposition subsystem, it is not surprising that trophic levels are even more difficult to define—all organisms are carnivores and top carnivores in a sense. Several classification schemes based on manner of obtaining nutrients have been devised. One of the simplest is the three part carnivores (feeding on animals), microbivores (feeding on micro-organisms), and saprovores (feeding on the remains of dead organisms).[27] An alternative tripartite classification is necrotrophs (short-term exploitation of living organisms resulting in death), biotrophs (long-term exploitation of living organisms not resulting in death), and saprotrophs (exploitation of food already dead).[28] Perhaps a more useful approach would be a simple two-way crossed classification, with the mode of nutrient acquisition (absorptive versus ingestive) crossed with the size of the organism (micro versus macro). This classification in no way is meant to preclude the classic plant/herbivore/carnivore trophic system, but may be more useful when thinking of the decomposition subsystem. This simple classification is presented in **Table 5.2**. Specifically for decomposition, we are concerned with microabsorbers and both micro and macro ingesters.

The microabsorbers involved in decomposition are the bacteria (including the important actinobacters—or filamentous bacteria) and fungi. They all have a similar fundamental operation, although special mechanisms are important for various species and species groups. The organism excretes extracellular enzymes, which create a digestive zone around its body. The extracellular enzymes break down relatively large molecules outside of the body of the decomposing organism, and these smaller molecules then are absorbed and enter the normal intracellular metabolic process. Thus, part of digestion is external to the organism.

The bacteria are distinct from fungi in that the fungi, although normally engaging in external digestion, are also capable of penetrating tissue and digesting that tissue from

TABLE 5-2 Simple Two-Way Classification of Organisms Involved in Terrestrial Food Webs, Especially Aimed at the Analysis of Decomposition Systems, with Some of the Main Examples of Each Class

Method of Nutrient Acquisition	Size	
	Micro	Macro
Absorption	Algae	Higher plants
	Bacteria	Helminth parasites
	Fungi	
Ingestion	Protozoa	Arthropods
	Rotifers	Annelids
	Nematodes	Molluscs
		Chordates

within. Bacteria, for the most part, adhere to the surface of tissues or penetrate into spaces already there.

Absorptive organisms are also conveniently categorized according to their source of carbon and energy (**Figure 5.13**). They are the primary forces for decomposing highly commutated material and are effectively the last stage in the decomposition process, promoting the catabolism that releases nutrients. For any given input into the cascading decomposition process, the absorptive organisms become more important as the system

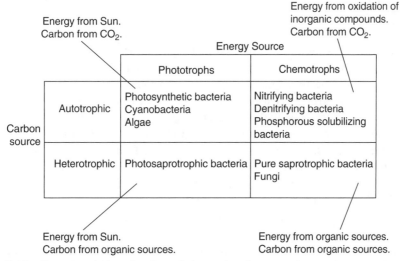

Figure 5-13 Functional classification of absorptive decomposition microorganisms.

cascades through its full long-term cycle. Nevertheless, there is a clear distinction between fungi and bacteria in that fungi are probably most important for breaking down primary resources and contributing significantly to comminution, whereas bacteria are more significant in breakdown of smaller secondary resources and more recalcitrant material. Furthermore, bacteria tend to dominate in anaerobic environments.

Absorptive organisms also share a characteristic the extent of which is not yet known very well, but the importance of which may be great. A large number of studies report on complex interactions among various absorptive microorganisms.[29] Furthermore, some bacteria concentrate near the surface of fungal hyphae, presumably to scavenge on the products of the exodigestion of the fungi.[30] In recognition of the basic processes involved in exodigestion, it would be surprising indeed if bacteria and perhaps even fungi did not scavenge on the products of the exodigestion of other absorptive organisms. This process, if indeed it occurs and is common, has important implications for the long-term sustainability of decomposer ecosystems and places great emphasis on the potential role of biodiversity therein. Suppose, for example, the decomposition environment is composed of *Cellulomonas* (which produces cellulase as an exoenzyme) and *Bacillus* (which produces peptidases as exoenzymes). It is possible that *Cellulomonas* can obtain a good fraction of its nitrogen nutrition from the small peptides produced near the body surface of *Bacillus*, and the latter can obtain a good fraction of its carbon from the carbohydrate products produced near the surface of *Cellulomonas* (from its digestion of cellulose). Indeed, could *Bacillus* survive better in a cellulose-rich environment that contained *Cellulomonas* than one that did not through its ability to scavenge the products of exodigestion of *Cellulomonas*? If such a process actually operates to a significant extent in nature, bacterial strains that have no obvious function in the actual decomposition process may in fact act as ecological buffers, providing exodigestive products that other bacteria can scavenge on when their preferred resource is temporarily in short supply. The significance for issues of microbial biodiversity is obvious.

Ingesters, both macro and micro, are typically involved at a higher level in the decomposition process, usually more associated with comminution and less with catabolism. That is, ingested resources are converted to constituent tissues, cells, and extracellular macromolecules and ejected as feces, returned to the soil organic material pool on death, and to a far lesser extent, released into the environment as metabolic products of catabolized molecules. The action of ingesters may be thought of as two discrete processes: ingestion and digestion. Ingestion may or may not contribute to comminution. Protozoa and rotifers, for example, are effectively filter feeders and ingest particles that have already been reduced to a small size or organisms that exist as small packets (bacteria, other protozoa, fungal spores, macromolecules). Their overall effect then is during the process of digestion and mainly as catabolizers.

Ingestion in the case of macroingesters contributes mainly to the process of comminution and to a lesser extent to catabolism. Consider, for example, the earthworm, the great comminuter of temperate soils. A stylized alimentary tract of an earthworm is presented in **Figure 5.14**. Food is actively ingested through the mouth with the active

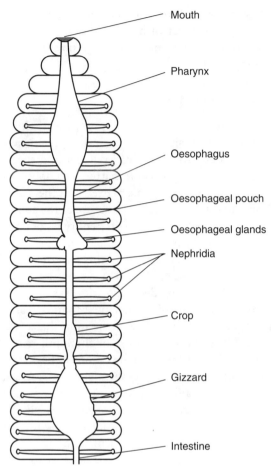

Figure 5-14 Diagram of earthworm anatomy.

participation of the pharynx, which operates as a pump. Food passes through the esophagus and arrives at the storage depot, the crop. From the crop, food is passed on to the gizzard, a muscular organ that grinds the food, using small mineral particles that are ingested along with the food. It is this physical grinding that has the main effect of comminution. The comminuted material is then passed to the intestine, where digestion and absorption occur. Although earthworms produce a variety of enzymes, much of the digestion is accomplished by symbiotic microorganisms or simply by the microorganisms ingested along with the organic matter. Other macro-organisms have similar digestive systems, most frequently with a microabsorptive symbiont involved, either within the gut or external to the gut (the so-called external rumen).

There is great geographic variability in the composition of the decomposition macroorganisms. Temperate systems are dominated by earthworms, collembola, mites, and so forth, whereas tropical systems tend to be dominated by termites and ants. The micro-fauna appears to be closely associated with the acidity of the soil, with bacteria dominating at higher pHs and fungi at lower ones, although no system is ever devoid of either of these major important groups.

Decomposition and Organic Matter

Two aspects of organic matter are of primary importance: 1) the basic partitioning of organic matter during the decomposition process and 2) the nutrient ratios that exist at any time, especially the nitrogen to carbon ratio. Regarding the basic partitioning of organic matter, as discussed previously, organic matter decomposes into two distinct fractions: the "fast" fraction, or the active organic matter, and the "slow" fraction. The slow fraction ultimately contributes to the physical structure of the soil, whereas the fast fraction is the main supplier of mineral nutrients. Different types of organic matter may contain different fractions of fast and slow raw materials and thus ultimately have different effects on the soil, although the details as to why this happens remain complicated and frequently elusive.[31]

The raw material that goes into the decomposition system is highly variable, both physically and chemically, and is one of the key determining features of what is made available to plants and how fast. The quality of the input has been studied intensively during the past decades.[32] The chemical nature of the input material has been conveniently characterized by nutrient ratios,[33] the most important one of which is frequently assumed to be the carbon:nitrogen ratio. If organic matter contains abundant carbon relative to nitrogen, the soil microorganisms are not able to grow fast enough to decompose the material efficiently and eventually must use nitrogen already in the soil. The application of organic matter may thus result in an initial decrease in nitrogen availability for the crops, caused by the needs of the microorganisms, the bane of the beginning organic gardener. C:N ratios vary enormously, from 250 for sawdust to 90 for cornstalks, to 20 for legumes, to 11 for humus, and to 4 for bacteria.[34] As a rule, a C:N ratio of less than 20 generates a net release of nitrogen for plant growth, whereas a C:N ratio of greater than 20 involves immobilization of nitrogen and therefore less available for plant growth. However, this is only a rule of thumb. It is well known that a great deal of variability exists in the decomposability of compounds with different structures, and it is not difficult to imagine an ideal C:N ratio made up of nitrogen-rich compounds that are relatively stable and carbon-rich compounds that degrade easily, resulting in a much higher "effective" C:N ratio from the plant's point of view. As a colleague recently remarked, "A plate of steak and potatoes is not the same as a plate of hair and sawdust, even if the C:N ratio is precisely the same." Furthermore, the C:N ratio changes over time as the carbon is used in respiration and the nitrogen is recycled—for plant residues. It has long been known that 20% to 40% of the carbon is actually incorporated into biological material, the rest being released as CO_2, raising some question about the role of soils and organic matter therein to the global CO_2 problem.[35] The rate of decomposition of various organic materials and thus the C:N ratio at a given time depends on O_2, moisture content, temperature, pH, substrate specificity, and available minerals and is also strongly influenced by management factors such as tillage and inorganic fertilizer applications.[36] For example, in one study, it was found that agriculture without supplementary fertilization was economical for 65 years on temperate prairie and for 6 years in a tropical semiarid thorn forest, whereas an extremely nutrient-poor Amazonian soil showed no potential for agriculture beyond the 3-year lifespan of the forest litter mat.[37] On the other hand, long-term experiments in England (the famous Rothamstead experiments, ongoing for over 150 years now[38]) and Denmark (100 years[39]) clearly show the benefits of organic matter for long-term soil structure and crop yields.

Mulching has long been a mainstay of organic and other forms of ecological agriculture. To decrease bulk for ease in transport and to reduce the C:N ratio of applied material, it is customary to compost the mulch before application. Most agroecologists regard compost as a central component of the ultimate sustainable agricultural system. The practical problem is that compost should be "biomature" when it is applied to the field, which is to say, the nutrients in the compost should be at least theoretically, in a state such that the demand from the plants is exactly met by mineralization in the compost. Because both organic matter breakdown and plant growth are dynamic forces, obtaining this synchrony is extremely difficult,[40] underlying the seduction of applying fertilizer in ionic form.

The problem of synchrony is illustrated with an extension of the cascade model of decomposition. Here the growth of a plant (B = biomass) is included along with the decomposition cascade, illustrating the basic problems of synchrony (**Figure 5.15**). This

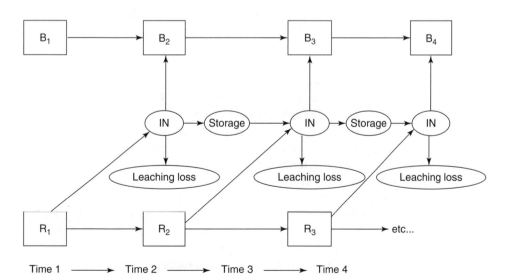

Figure 5-15 Illustration of the problem of synchrony using the cascade model.

model suggests that if storage is efficient there is a carryover of nutrients across time, thus allowing the available nutrients to remain in the available pool. Leaching losses mean, however, that all minerals that are not in storage or taken up by the plant will be lost to the system. The plant grows and demands nutrients, and if sufficient nutrients are not in the available pool, the growth rate of the plant is reduced. If the biomass of the plant is too small, however, the nutrient pool is subject to leaching. If the decomposition rates are exactly matched with the plant growth rates, leaching will not matter much since all mineralized nutrient will be absorbed by the plant. Thus we see the issue of synchrony as a complicated problem involving soil properties (for storage and/or leaching of nutrients), microbiology (for the rate of decomposition), and plant physiology (for the uptake of nutrients).

In the context of this model, a meeting of the Tropical Soils Biology and Fertility group in 1992 agreed on a set of 12 hypotheses that appear to emerge from the synchrony model. These are as follows:

S1. The maximum crop yield achievable by the use of inorganic (mineral fertilizer) inputs can be approached or exceeded by optimizing the time of application, placement and quality of organic nutrient sources.

S2. In environments where significant leaching or denitrification occurs, plant uptake of mineral N applied at planting can be increased by simultaneous application of a low N organic material which temporarily immobilizes N early in the crop growth cycle and remineralizes N later on.

S3. Stabilization of organic matter in the soil is enhanced by the addition of mineral nitrogen simultaneously with the addition of organic materials of high C-to-N ratio.

S4. Residues high in lignin will result in a low net mineralization and plant uptake in the first cropping season, but will produce a greater residual effect in subsequent seasons.

S5. Residues high in tannins exhibit delayed nutrient release, but will after a lag period release nutrients rapidly.

S6. Immobilization of P by microbes, or blocking of P sorption and fixation sites, following addition of organic material can prevent fixation of P, thereby improving medium-term availability of P. This phenomenon would be best exhibited in P-fixing soils that are poor in organic matter and high in Fe.

S7. Nutrient uptake efficiency increases with the longevity of the plant. Implicit here also is the notion that relays of short-lived plants may act in the same way as long-lived plants.

S8. The nutrient uptake efficiency of the system will be increased by plants that have more rapidly growing, deeper and more extensive root systems.

S9. Incorporation of organic inputs, as opposed to surface application, accelerates the release of nutrients, thereby providing another option for modifying nutrient use efficiency.

S10. Improvement of nutrient uptake efficiency due to the use of organic inputs is more likely when crop growth and soil processes are less constrained by water deficits.

S11. Quality and quantity of organic inputs can influence faunal composition and activity, and thus affect the synchrony of nutrient supply and crop demand.

S12. The need for exact synchrony and crop demand can be reduced by storage of nutrients within the crop in excess of the crop's immediate requirement for growth.[41]

These hypotheses derive from basic principles of soil science. The fact that they are presented as hypotheses to be tested is evidence of how much research is still needed on the most elementary ecological processes in soils.

Biology and the Physical Structure of the Soil

The Biological Elements

Recall the most elementary physical model of soil as presented in the previous chapter, a beaker filled with marbles. The obvious biological addition to this structure is the organic matter, which contains, on one hand, the gums and resins necessary to form the ped structure, which in turn determines the physical structure of the soil, and on the other hand, a key component of the overall chemical dynamics. Both of these subjects were covered in the last chapter.

The other biological forces operating on the physical structure of the soil are the digging and tunneling organisms, such as termites, ants, and earthworms, and the fungi whose mycelia act to bind particles together into peds as well as make physical connections among different peds, giving the soil a structural integrity beyond the simple ped structure. Termites are especially important in the tropical agroecosystems of Africa and South America. Earthworms are universally important, and ants are likely important but have not received the attention that they deserve. The activities of these organisms in effectively creating much of the physical structure of the soil have inspired many workers to refer to them as "ecosystem engineers."

Earthworms are ubiquitous in all the world's soils, although their distribution is heterogeneous.[42] From a functional point of view, earthworms can be categorized into three basic groups: epigeic, anecic, and endogeic.[43] Epigeic species are those that live in leaf litter. They are generally pigmented and mainly function to comminute the litter and aid in the fundamental processes of decomposition. They are generally of small size and do not make burrows and thus generally do not contribute significantly to soil physical structure.

Anecic earthworms are those that feed on litter but burrow in the ground, are usually large, and are frequently pigmented dorsally. This class is probably the most important from the point of view of physical structure because they not only aerate the soil by tunneling, but they bring organic matter down into their tunnels, thus acting like minute plows, incorporating organic matter into the soil. The familiar night crawler *Lumbricus terrestris* is an important, if controversial, example, being an important soil engineer on the one hand but an invasive European species in North America on the other. The burrows of anecic earthworms tend to be permanent and vertical, can be as much as 2 meters deep, and usually remain open at the top. They are typically lined with mucus, which, on abandonment, acts as an important binding agent for ped structure. The tunnels also act as important pathways for oxygen to enter the soil and CO_2 to exit, as well as the basic plumbing system allowing for water penetration to deeper layers. These tunnels are also frequently preferential sites for the growth of roots.

Endogeic earthworms vary in size, are rarely pigmented, and live in extensive ephemeral horizontal burrows. They feed on the soil itself, ingesting soil and digesting the organic matter contained therein. They ingest from five to 30 times their own weight of soil per day and deposit casts throughout their burrow systems.

All earthworms are comminuters of organic material, although the size of the material comminuted varies with the habit and habitat, epigeic and anecic species generally working on larger particles than endogeic earthworms. Anecic and endogeic species are responsible for the turnover of large amounts of soil (**Table 5.3**), but the amount varies enormously from site to site.

TABLE 5-3   **Amounts of Castings in Different Habitats and Areas (Edwards and Bohlen, 1996)**

Habitat	Locality	Casts Produced (Tons/ha)
Arable	Germany	91.6
Arable	India	1.4–5.0
Garden soil	Switzerland	17.8–81.2
Near White Nile	Egypt	268.3
Old grassland	England	18.8–40.4
Old grassland	England	27.7
Grassland	Switzerland	17.8–81.0
Grassland	Germany	91.4
Grassland	India	3.0–77.8
Heathland	Germany	5.2
Oakwood	Germany	5.8
Beechwood	Germany	6.8

In addition to surface casts, large numbers of casts are deposited below the surface of the soil, and thus, the amount of soil moved around is probably much larger than the data in Table 5.3 suggest.

One of the most obvious effects on the physical structure of the soil induced by earthworms is the creation of more pore space by tunneling activity. Sometimes this effect can be spectacular. It has been reported that an increase of from 75% to 100% porosity in orchard soils was due to earthworms.[44] Published results, however, are highly variable with some authors finding insignificant effects of earthworms, and most others reporting figures on the order of 5% to 25% of total soil pore space being earthworm burrows. It undoubtedly depends on the soil type, vegetation, and climate.

An important effect of increasing porosity is on water infiltration. Earthworm burrows contribute substantially to this process, especially those that are open to the surface, that is, those constructed by enecic species. It is also well documented that endogeic species contribute to the water penetration capacity of soils.[45] Burrows must be connected to one another to be effective at water infiltration, and tillage can significantly disrupt the network of burrows, thus reducing their function as water conduits.[46] Other agricultural activities, such as pesticide applications, have been shown to reduce water infiltration by as much as 93% because of increased earthworm mortality.

On the other hand, earthworms can contribute to soil erosion, a fact first noted by Darwin in 1881. Surface deposited casts are susceptible to being carried away by water, and the bare spots created by large anecic species on the surface of the soil near their burrow entrances make patches of bare soil that then can be eroded. In most studies thus far reported, however, the beneficial effect of infiltration due to earthworm burrows outweighs the effects of water erosion.

The role of termites and ants would seem obvious. Unfortunately, ants have been largely ignored as agents affecting soil physical structure, although their effect, by casual observation, could be enormous. For example, it has been estimated that the infamous leaf cutting ant, *Atta cephalotes*, could be responsible for complete soil turnover in a lowland rain forest in Costa Rica and that complete turnover could occur in as short a period as 200 years.[47]

It has long been part of conventional wisdom that termites' effect on physical structure is enormous, given the evident size of their nests in African and South American savannas. Termites have been linked with increasing aggregate stability, with increasing water penetration into the soil (i.e., porosity), with increasing hydraulic conductivity, and with pedogenesis itself.[48] However, many of these positive attributes have been questioned by other authors, probably because of different functions of different functional groups of termites. Termites have been divided into five different functional groups (grass feeders, grass/litter feeders, polyphagous, soil feeders, and wood feeders),[49] and the effect they will have on the soil depends to some extent on which functional group is being considered, perhaps accounting for the diversity of literature accounts. On the other hand, it is evident that the above-surface structures of termites present significant challenges to some forms of agriculture, just because of their physical presence (**Figure 5.16**).

(a)

(b)

Figure 5-16 Termite nests in South America. (a) Cattle pasture with obvious physical problems caused by termite nests. (b) Typical termite mound on an oxisol.

Consequences for Soil Physical Properties

Although a soil's "texture" is a property of the physical nature of the soil, its "structure" is a consequence of the physical nature interacting with the biota, as discussed in Chapter 7. The physical nature combines with various gums and resins of biological origin to produce an aggregate structure (recall that a natural aggregate is a ped), which dictates a certain natural pore space distribution, which in turn is modified by living macro biota such as arthropods, earthworms, and growing roots. Altering the biota can have dramatic effects on this structure, and it is generally acknowledged that, for a variety of reasons, one might speak of good versus bad soil structure with respect to a specific crop (e.g., enough pore space must be available for adequate infiltration of water into the soil, to say nothing of retaining proper hydrostatic pressure to allow the crops to take up what water is available, yet too much pore space may result in high levels of evaporation and general loss of moisture). Furthermore, pore space is dialectically related to root growth in the sense that growing roots not only create new pore spaces, but many roots grow through existing voids, some of which were created by formerly growing roots.[50]

That industrial farming has had an important effect on physical structure seems well documented,[51] and the assumption is naturally made that pulling back from the industrial system will improve things. In particular, a variety of studies have found greater soil aggregate stability when alternative methods are used,[52] possibly because of greater amounts of soil organic matter, and certainly due, in part, to changes in soil preparation technology.

Soil macroporosity is a major determinant of the structural quality of the soil, contributing not only to crop growth but also to general environmental quality. In an experimental "integrated farm," soil macroporosity and the proportion of pores influenced by soil biota were significantly higher under organic production than in a conventional farm, but both the details of the physical structure itself and the role of soil biota in forming that structure depended strongly on the details of the cropping system.[53] In another experimental system, it was found that the macroporosity was higher under an integrated system, but that total root density was higher in the industrial system.[54]

Soil organic matter is a major determinant of soil structure through its action of creating soil ped structure, it is sometimes the principal force determining cation exchange capacity, it obviously derives from the soil biota as well as partially determines the nature of the latter, and its breakdown supplies the soil with its mineral nutrients. The formation of soil organic matter at the most general level is nothing more than the consequences of decomposition as discussed previously. This process is immensely complicated as evidenced by the many published flow charts illustrating just the "simplest" processes involved in both managed and unmanaged systems. The model of Swift et al.[55] remains an important conceptual tool, presented previously as the major model for the decomposition process.

Ecological Theory and Soil Biology

Dynamic Consequences of Soil Communities

As articulated in the previous sections on decomposition, some of the traditional organizational modes of theoretical ecology are completely lost when we move to the soil subsection of the ecosystem. In particular, the cycling of materials within the soil subsystem has only vague counterparts in the above-ground section of the ecosystem. It is not unusual to suggest that ecological theory ought to contend with this issue.[56] One approach that has been suggested, but not yet completely explored, is the use of interaction networks, including a specific acknowledgment of indirect effects, but with an understanding that the key processes are fundamentally different than in the aboveground subsection. An initial attempt to formulate soil biology dynamics in this way is presented in **Figure 5.17**.

The general dynamics of this system have not been worked out in great detail. However, the existence of multiple and oscillating subsystems strongly suggests that it will be a complicated dynamic picture, even at this very simple level of abstraction. A glance at Figure 5.16 combined with a knowledge of what goes on in the soil will convince any reader that that figure is woefully incomplete, leading to the oft-repeated irony of ecological systems that even the simplest structures may lead to obscure and enigmatic dynamics.

An Integrated Model of Soil Dynamics

An integrated ecological approach to how general soil processes affect plant production is provided through an ingenious model proposed by van Noordwijk and de Willigen.[58] These authors consider inputs into the system as being transformed by four general functions. An input could conceivably be a variety of things, from rock phosphate to compost, from manure to nitrate. The input has some nutrient associated with it, and our desire is to determine how the crop responds to that input. The classic agronomic approach is to treat the soil as a black box and empirically create a graph relating response of crop to input of nutrient material. At the other extreme, modern agronomists may reduce this problem to a large number of micromechanisms,[59] providing a detailed picture of physiological and chemical processes. This latter approach in its most modern form typically results in large computer models purporting to predict the transformation of nutrient resources into plant productivity. The popular approach of developing large computer models has been criticized from the point of view that a large computer model may provide good quantitative predictions of what will happen, but it rarely provides understanding of why. A computer model that we do not understand is not much better than a natural phenomenon we do not understand.

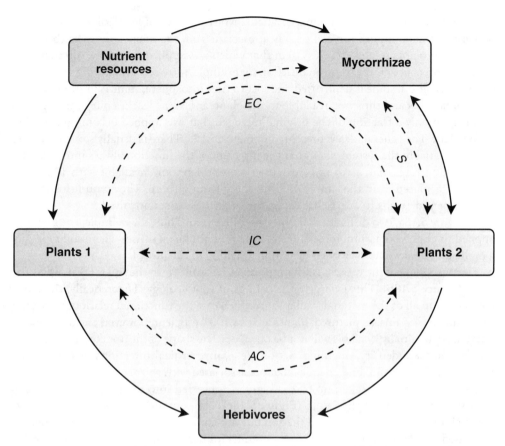

Figure 5-17 Minimal dynamic network of a soil subsystem with the key components (herbivores, plants (two types), mycorrhizae, and nutrient resources) with the key processes (AC = apparent competition; IC = interference competition; EC = exploitation competition; S = symbiosis). The solid arrows indicate energy transfer, and the dotted arrows indicate various forms of indirect interactions.[57]
Adapted from: Ohtonen, R., Aikio, S., & Vare, H. "Ecological Theories in Soil Biology," *Soil Biology & Biochemistry, 29*(1997):1613–1619.

The van Noordwijk–de Willigen model is something of a compromise between the black box model previously so common in conventional agriculture and the hopelessly grandiose computer systems models. It involves three transformations. First, the material applied must somehow become available to the plants in the soil (**Figure 5.18a**).

For example, application of ammonia to a field involves considerable volatile loss, as anyone passing a zone of application can readily appreciate from the odor. Thus, what becomes available for plants to absorb is less than that which was applied originally. Alternatively, applied compost contains some quantity of, for example, nitrogen, and during the process of decomposition, some is denitrified and lost to the atmosphere, some is immobilized by microorganisms and thus not available to the plant, and some leaches out of the system, resulting in only a fraction of the original nitrogen that was applied becoming available to the plant. The general idea is presented in Figure 5.18a. There is usually some base level of the material in the system, such that the intercept of the function always intersects the ordinate at greater than zero, and thus, the theoretical amount available with 100% efficiency is displaced from the simple 45° line going through zero. The amount lost to the environment depends to a large degree on the nature of the soil involved.

In Figure 5.18a, two different soil types are illustrated. Thus, for example, the material applied might be potassium ions, and soil type B might be an ultisol where soil type A is a mollisol. There would be a great deal more leaching in the ultisol.

The next phase of the process is when the material available in the soil is taken up by the plant (**Figure 5.18b**). This is largely the subject of root ecology. Theoretically, the plant could take up all of the material available, in which case the functional relationship would simply be the 45° line, as pictured. Plants are not 100% efficient, however, and some material that is potentially available will not be taken up. The shape of the uptake curve depends on root biomass density, microbial actions (e.g., mycorrhizal infections or pathogenic infections), and physical and chemical factors associated with rhizosphere dynamics.

Finally, the material taken up by the plant is converted into dry matter of the plant, largely a subject of plant physiology (**Figure 5.18c**). The point on the production function in which material is taken up from the soil but not converted into dry matter production (e.g., used to manufacture secondary compounds for defense against insect herbivores) is normally referred to as "luxury consumption."

The strength of looking at the system from this point of view is that each output (each ordinate) is an input (abscissa) to the next level. Thus, the three functions can be composed, as shown graphically in **Figure 5.19**. In quadrant I is the function illustrated in Figure 5.18a, the transformation of applied material to that available in the soil. In quadrant II is the function illustrated in Figure 5.18b, the transformation of material available in soil to the plant uptake. In quadrant III is the function illustrated in Figure 5.18c, dry matter production from the material taken up by the plant. In quadrant IV is the composed function, which is the black box frequently viewed in the classic agronomic literature. With this composition approach, one can easily visualize the component functions that lead to the basic relationship between dry matter production and the application of an input. Furthermore, various classic sciences can be more or less associated with each quadrant. In quadrant I is located processes commonly thought of in basic soil ecology. Quadrant II illustrates the process of nutrient uptake by plants and involves factors usually thought of as root ecology. Quadrant III is largely associated with plant physiology.

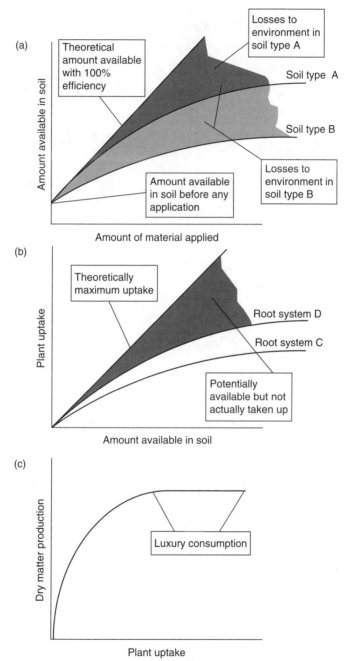

Figure 5-18 Basic transformations for the van Noordwijk–de Willigen model.

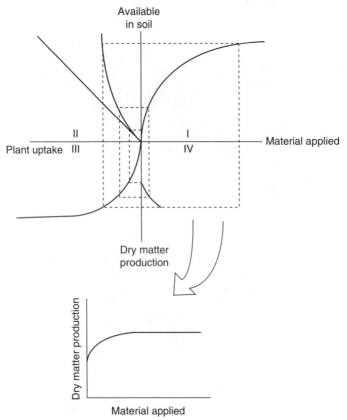

Figure 5-19 The van Noordwijk–de Willigen model (the shaded quadrants indicate identical quadrants oriented differently).

A General Qualitative Approach

The previously mentioned process-level models are certainly useful, even necessary, when focusing on a particular input, particularly nutrient inputs. An alternative, perhaps more general, way of viewing soil ecology is from the point of view of ecosystem health (a complex issue to which we return in Chapter 8), rather than from the point of view of analyzing inputs. There are four features of the soil that most soil scientists agree are crucial to good soil health: acidity, physical structure, organic matter content, and biodiversity. These four factors are obviously interconnected in complicated ways, and understanding soils requires an understanding not of each one of the subjects but rather of the way all four are interrelated. Some illustrative topics associated with various intersections of these topics are depicted in **Figure 5.20**. Indeed, a detailed knowledge of all four plus their interrelations with each other is needed for good soil management. Nevertheless, these four subjects are all measurable, and all provide indices that most soil scientists would agree are correlated with the "health" of the soils: low acidity, high organic matter, high biodiversity, and good

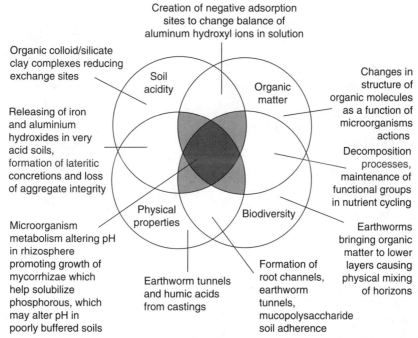

Creation of negative adsorption sites to change balance of aluminum hydroxyl ions in solution

Organic colloid/silicate clay complexes reducing exchange sites

Soil acidity

Organic matter

Changes in structure of organic molecules as a function of microorganisms actions

Releasing of iron and aluminium hydroxides in very acid soils, formation of lateritic concretions and loss of aggregate integrity

Decomposition processes, maintenance of functional groups in nutrient cycling

Physical properties

Biodiversity

Microorganism metabolism altering pH in rhizosphere promoting growth of mycorrhizae which help solubilize phosphorous, which may alter pH in poorly buffered soils

Earthworm tunnels and humic acids from castings

Formation of root channels, earthworm tunnels, mucopolysaccharide soil adherence

Earthworms bringing organic matter to lower layers causing physical mixing of horizons

Figure 5-20 Illustrative topics associated with various aspects of soil health.

physical structure. Perhaps such characterizations are oversimplified, but they seem to represent a first step in judging the health of the soil. Further refinements, such as cation exchange capacity, soil bulk density, earthworm density, the C:N ratio of the organic matter, and many others are obviously desirable pieces of knowledge. But as a first approximation, it seems perfectly reasonable to say that high organic matter, high biodiversity, low acidity, and good physical structure are the four pillars of soil ecosystem health.

Endnotes

[1]Niggli and Lockeretz, 1996; Howard, 1940.
[2]The historical notes in this chapter are largely taken from Conford, 2001, unless otherwise indicated.
[3]Niggli and Lockeretz, 1996.
[4]From Rudolf Steiner, *Agriculture: An Introductory Reader*, edited by Richard Smith Thornton (Forest Row: Sophia Books, 2004), courtesy of Rudolf Steiner Press.
[5]Balfour, 1943.
[6]Sattelmacher and Thoms, 1989; Cui and Caldwell, 1996.
[7]Loveless, 1961; Grime, 1977, 1991; Coley et al., 1985.
[8]Larcher, 1995.
[9]Marschner, 1991.
[10]Brady and Weil, 2007; Cadish and Giller, 1997.
[11]Boddey and Dobereiner, 1988.
[12]Mishustin, 1970; Shabaev et al., 1991.

[13]Havelka et al., 1982.
[14]Brady, 1990.
[15]Paul and Clark, 1980.
[16]Raven, 1985.
[17]Robertson, 1982.
[18]Details can be found in any standard soil text, for example, Brady and Weil, 2007.
[19]Mikanova and Kubat, 1994.
[20]Casanova, 1995; Dann et al., 1996.
[21]Munyanziza et al., 1997; Brundrett, 1991.
[22]Pirozynski and Dalpé, 1989.
[23]Smith, 1980.
[24]van Noordwijk (personal communication).
[25]Swift et al., 1979.
[26]Swift et al., 1979.
[27]Heal and MacLean, 1975.
[28]Swift et al., 1979; Lewis, 1973.
[29]Forlani et al., 1995; Prathibha et al., 1995.
[30]Swift et al., 1979.
[31]Cadish and Giller, 1997.
[32]Cadish and Giller, 1997.
[33]Swift et al., 1979.
[34]Donahue et al., 1977.
[35]Alexander, 1977.
[36]Rasmussen and Collis, 1991; Simard et al., 1994.
[37]Tiessen et al., 1994.
[38]AFRC, 1991.
[39]Schjønning et al., 1994.
[40]Meyers et al., 1997.
[41]Meyers et al., 1997, pp. 216–217.
[42]Edwards and Bohlen, 1996.
[43]Fragoso et al., 1997.
[44]Hoeksema and Jongerius, 1959.
[45]Joschko et al., 1992.
[46]Chan and Heenan, 1993.
[47]Perfecto and Vandermeer, 1993.
[48]Black and Okwakol, 1997; also see Lavelle et al., 1994.
[49]Black and Okwakol, 1997.
[50]van Noordwijk et al., 1993
[51]Logsdon et al., 1993; Reganold et al., 1987.
[52]Jordahl and Karlen, 1993; Schjønning et al., 1994.
[53]Brussard, 1994.
[54]Schoonderbeek and Schoute, 1994.
[55]Swift et al., 1979.
[56]Wardel and Giller, 1997; Ohtonen et al., 1997.
[57]From Ohtonen et al., 1997.
[58]van Noordwijk and de Willigen, 1986.
[59]Loomis and Connor, 1992.

Chapter 6

The Problem of Pests—Herbivory, Disease Ecology, and Biological Control

Overview

This chapter treats the issue that is perhaps most critical to an alternative approach: the way in which pests and diseases are dealt with in agroecosystems. The history of this topic is rich with specific examples that begin to set the stage for the way in which the issue was dealt with until the "chemical revolution." After presenting this historical outline, the chapter treats some of the basic ecological issues, casting the issue of herbivory as conceptually similar to that of pathogens. A section on the principles of biological control follows, placing an emphasis on the important concept of the paradoxical nature of seeking to lower pest populations through natural enemies. Then the conceptual basis of integrated pest management is discussed in its technical form along with the political ramifications that inevitably arise. Finally, through an extensive example of the control of two major pests in coffee production, the idea of autonomous biological control is discussed.

Historical Precedents

This issue is perhaps the most volatile of the controversial issues associated with agroecosystems, as it includes first some of the more spectacular problems faced by agriculturists throughout history and second some of the more spectacular environmental and human health effects that we have ever witnessed. The proverbial locusts that the God of Israel sent to destroy the crops of the Egyptians, to the coffee rust that completely wiped out coffee production in major European colonial outposts, to the threat the Asian rust poses for today's soybean producers in Brazil, agricultural production has always been and always will be vulnerable to the "pests" that predictably take advantage of the inevitable concentrations of relatively uniform photosynthate that agriculture inevitably produces. Paradoxically, dealing with this basic issue in the industrial system frequently generates more problems than it solves, from the Bhopal disaster in which a Union Carbide pesticide manufacturing plant accidentally released a volatile byproduct into the air and killed thousands of local people, to the hundreds of farm workers poisoned each year from misuse of pesticides, to the byproducts that continue to plague our food system, to the environmental health effects that range from hormone-mimicry to cancer, to the potentially massive effects on world biodiversity. Modern chemical pesticides, almost the symbol of the contemporary industrial system, have been vilified and defended, but by now, a consensus seems to have grown that their use should be minimized. But it was not always that way.

The perceived need for pest control began when agriculture began. Early methods were comical by today's standards. For example, a Roman text dated 50 CE has the following remedy for caterpillar pests: "A woman ungirded and with flying hair must run barefoot around the garden, or a crayfish must be nailed up in different places in the garden." Despite such blundering, earliest records suggest relatively sophisticated methods were also in common use. For example, before 2500 BCE, the Sumerians used sulfur compounds to kill insects and mites, and China was using plant extracts as well as mercury and arsenic compounds for seed treatment and fumigation by 1,400 years ago. Most important, Chinese farmers recognized the role of natural enemies at least 2,200 years ago. Indeed, the Chinese apparently had an appreciation of natural processes long ago, as indicated by a text from 300 CE:

> A factor which increases the abundance of a certain bird will indirectly benefit a population of aphids because of the thinning effect which it will have on the coccinellid beetles which eat the aphids but are themselves eaten by the bird.

Indeed, by 300, the ant *Oecophila* was being used as a biological control agent in citrus production, usually cited as the first recorded case in history of biological control. However, while China was increasing in sophistication, Medieval Europe seems to have been mired in superstition. For example, in 666, St. Magnus, Abbot of Füssen, repulsed locusts and other pests with the staff of St. Columba, and in 1467 Switzerland cutworms were taken to court, pronounced guilty, excommunicated by the archbishop, and banished from the land.

During the later phases of the evolution of the new husbandry in Europe and its colonies and ex colonies, the scientific revolution began penetrating agricultural practice, and more sophisticated understanding of pest problems was evident by the 18th century. The general relationship between parasite and its host and its significance for pest control was described in some detail early in that century. Linnaeus also suggested the use of a variety of predacious and parasitic insects for biological control. In general, the qualitative notions of pest control that today inform much of our nonpesticide thinking were established, in a relatively sophisticated manner, by these early prescient scientists.

Other corollaries of pest control in the new husbandry, during its later phases, were not so salubrious. In particular, five major events signaled the emergence of the same sort of recurrent pest problems that would plague agriculture throughout its evolution toward the industrial system. First, as discussed in Chapter 1, the Irish potato blight was the largest human tragedy ever caused by agriculture to that point. Potatoes were part of the intercontinental exchange associated with European conquest, and they rapidly became the main staple of several countries, most important Ireland. The later phases of the new husbandry system promoted the use of the "best" variety to maximize per land unit production (anticipating one of the fetishes that remain with us today), and soon all of Ireland was producing the same variety of potato as the number one starch product for its population. When blight struck in the 1840s, it was a disaster. No resistant varieties were immediately available because of the "rational" economic calculations of the incipient capitalist farmers, and massive starvation resulted. However, it would be a mistake to attribute the entire event to the monocultural philosophy—the substantial quantity of grain produced in Ireland was almost all exported to England. These grains were withheld mainly because of the then-popular Malthusian notion that the real problem in Ireland was overpopulation, and feeding the poor would just exacerbate the problem.[1]

Second, during the 1850s, the grape-growing areas of Europe were devastated by powdery mildew. This problem was solved with the accidental invention of the famous "Bordeaux mixture," which is hydrated lime mixed with sulfate. According to legend, in an attempt to reduce the theft of his grapes along the roadside, a farmer applied a mixture of copper and lime, which gave the grapes a poisonous, or at least unappetizing, appearance. Later he noticed that the grapes receiving this treatment were free of the powdery mildew, leading to the generalized use of Bordeaux mixture and the solution to the powdery mildew problem.

Third, an outbreak of coffee rust disease on Ceylon (today's Sri Lanka) was the first devastation of luxury export crops in the European colonies. So devastating was this disease that Ceylon was forced to abandon coffee production entirely, switching the island's economy from coffee to tea production. The same disease devastated Java and Sumatra, arguably creating difficulties, both political and economic, for both English and Dutch colonial powers. Most significant, however, was the fact that all three island nations effectively ceased producing coffee because of this uncontrollable disease—the island that inspired the popular phrase "a cup of java" was rendered unsuitable for the production of this crop (something that would change a century later).

Fourth, in another example of the intercontinental exchange, albeit a less direct and planned one, a small insect, the grape phylloxera, was introduced from America around 1850 and eventually became a major problem in the vineyards of Europe, especially France. The approach to this problem was remarkably sophisticated. The logical method of importing a natural enemy from the origin of the pest was tried in 1873 when a predacious mite was imported. Although the program was largely ineffective, it did signal the first coordinated attempt at the introduction of a natural enemy from within the pest's original range, the basic idea of classic biological control. Successful control of this pest was through the combination of a new, resistant variety, discovered in America around 1870 and the technology of grafting, which enabled the new variety to be grafted to the root stalks of the European varieties already in the vineyards. The new plants, a combination of popular European varieties and the resistant variety from the New World, were sufficiently resistant to the insect.

The fifth event was the accidental introduction of the cottony cushion scale in California in the late 1860s. It was a serious pest that threatened to destroy the highly successful citrus industry of that state. After determining that the insect's home was Australia, the U.S. government dispatched entomologists down under to seek the parasite or predator that controls the pest in its native habitat. They encountered two, a parasitic fly and a beetle, the vedalia beetle (Coccinelidae). The beetle proved to be extremely effective, and within a couple of years, the cottony cushion scale was no longer a problem.

These five events set the stage for pest control for the next half century, until the advent of petrochemically derived organic pesticides: 1) spraying of simple poisons (as the Bordeaux mixture), 2) resistant varieties (as in the potato and phylloxera-resistant grapes), and 3) introduction of natural enemies (as in the mite predator of phylloxera and the vedalia beetle), coupled with 4) a host of cultural techniques traditionally used by farmers the world over. The one final method, one not frequently cited, but nevertheless important as historical fact, is 5) the abandonment of a particular crop (as in the case of the coffee rust disease). Thus, by the dawn of industrialization, the technologies used for control of pests and diseases were diverse and eclectic but usually incorporated some underlying ecological understanding.[2]

The Ecology of Herbivory and Disease
Elementary Population Dynamic Assumptions

In most standard ecological texts, herbivory is subsumed under the general category of positive/negative interactions, or predator/prey interactions, broadly conceived. However, in the case of agroecosystems, the critical issue of herbivory is rarely associated with the long-term consequences normally studied in the predator/prey context. Instead, it is associated with the short-term question: "Will this animal become a pest?" It is similar to the question, "Will this disease become epidemic?" which is why it is convenient to treat herbivory and disease in the same section—for practical purposes, they are similar. This is not to say that all herbivores represent a "disease" or even a problem. It is evident that the constructed elements of the agroecosystem always have some herbivory, some subcritical pathogens—in short, a host of associated organisms. Sometimes these organisms have a crucial role in the implied ecological balance of the system. For example, the hemipterans that so frequently form mutualistic relationships with ants do suck photosynthate out of the plant and thus could be considered as an enemy of the plant. However, by maintaining the ants in the system, they provide protection for the plant against other herbivorous insects.[3]

Nevertheless, whether dealing with small traditional farms or large industrial farms, there are times when an herbivore is clearly recognized as a potential pest, and the question then becomes, "When will this organism become a pest?" which is, as stated previously, structurally similar to the question, "Will this disease become epidemic?" This critical issue, the timing of an epidemic (or epizootic), has long been recognized explicitly with respect to diseases, but it is my particular (perhaps peculiar) idea to include herbivory as if it were a disease also. Yet, in the more classic literature, it is not all that unusual, if viewed with the correct framing. Some 50 years ago, Vanderplank established the foundations on which much of plant pathology has since been analyzed,[4] within which the critical issue of herbivory quite obviously sits snugly.

If the principal concern is when an herbivore actually becomes a pest or, more formally speaking, when the herbivore passes some threshold defined as pest level, the most basic concern is population growth of the herbivore itself. Thus, in principle, the formal framework for study has been some simple form of exponential growth, usually one of the two forms,

$$\frac{dN}{dt} = rN \qquad (6.1)$$

or

$$N_{t+1} = rN_t \qquad (6.2)$$

which are simply two ways of representing the process of exponential growth of a population (see Appendix 6.A) and, as noted by Vanderplank many years ago, of disease also. The focus then is on determining the value of r and the impact that various forces, such as temperature and rainfall, have on that parameter. Knowing r and having some threshold defined by socioeconomic facts, it becomes a trivial exercise to compute when the population will reach that threshold (to be convinced of this, take the current population to be 4 and the threshold to be 400, and with an r of 2 per day, compute the number of days it would take to pass the threshold, using the second of these equations—not a completely trivial problem if you have not already done such a thing, but worthwhile thinking it through rather than having someone tell you from the start).

To the untrained, it is always somewhat surprising how exponential processes operate. If the lily pads double their population every week and it took 50 weeks to cover half the lake, how long will it take to cover the rest of the lake? Ask your nonscientist fishing enthusiast that question, and he or she will likely first think "well, something like another 50 weeks." The answer of just 1 more week is somewhat shocking if you do not think about it too deeply. That is the thrill of exponential processes.

The vast literature concerned with predicting herbivore growth rates is not worth reviewing here in detail. In truth, most herbivores of interest to agriculture have populations that have complicated internal structures (larvae, pupae, adult; fawn, adult; etc.) that must be taken into account for effective estimation of r. For example, a population of 1,000 tomato hornworms (the larvae of *Manduca sexta*) will have a dramatically different effect than a population of 1,000 adult moths of the same species. Thus, rather than a straight application of the exponential model, most workers would employ a structured model, with the structure including different age and/or stage categories.

The details of such models are rather complicated and of interest to only those seeking to actually estimate some value of r for some population, in which case one of the many other more advanced ecology texts must be consulted at any rate, and thus, I leave this point without further comment. However, once we admit that structure is necessary for even rough predictions of population trajectories of most herbivores of interest, two other factors come into play: nonlinearities and stochasticity. We have recently come to the knowledge that neither of these factors can be ignored if we wish to be realistic about these herbivore predictions.

Consider Equation 6.2 as an example of what can happen. If we change Equation 6.2 in a perfectly reasonable fashion, something seemingly unreasonable happens. We know that the parameter r from that equation cannot really be a constant because no population can grow without limits forever. A very large population is likely reaching the limits of its environment's carrying capacity and thus experiences a relatively low rate of growth, whereas a very small population knows no such limits and probably experiences as high a rate of growth as possible. In other words, the parameter r is not independent of the population size. If r is dependent on population size, Equation 6.2 must look something like:

$$N_{t+1} = f(N_t)N_t \tag{6.3}$$

where f is a function that depends on N (a review of the appendix to Chapter 3 may be appropriate at this point). Because f depends on N, whatever its form, it will add an N to the ultimate form of the equation, and that added N will have to be multiplied by the original N, which means that the equation now becomes nonlinear. One of the simplest of all possible forms of f, assuming that the rate of population increase decreases as the population becomes larger, would be $f = r(1 - N)$. If we simply substitute that function into Equation 6.3, we arrive at:

$$N_{t+1} = r(1-N_t)N_t = rN_t - N_t^2 \tag{6.4}$$

which, as you can see, is nonlinear (the variable of concern, N, is multiplied by itself). This most simple of all nonlinear forms results in some of the most complicated population dynamics that could be imagined, indeed more complicated than had ever been imagined. Populations obeying Equation 6.4, with certain values of r, are chaotic. In fact, Equation 6.4 is the precise equation that was used in the most celebrated elaboration of the facts of chaos in a series of classic papers by ecologist Robert May.

Of course, no one would suggest that Equation 6.4 actually represents any real population. Real populations are too complex to be rendered meaningfully with such a simple approach. However, the ideas of chaos, and even more complicated forms of population behavior, apply when we add nonlinearities to the various components of any structured model. If, for instance, the pupation rate is a function of larval density, that fact alone is sufficient to generate hopelessly complicated long-term population dynamics. If both the pupation rate and adult survival rate are also density dependent, that just complicates the issue further. To underscore the importance of nonlinearities, a team of ecologists (known to their friends as "the beetles") have been modeling the population dynamics of *Tribolium confusum*, the common flour beetle, and comparing their model predictions with the actual behavior of flour beetles in bottles of flour. Using a combination of mathematical modeling and data from the actual beetle populations, they have discovered that even in this remarkably simple environment (a small bottle of flour) the beetles exhibit unusually complicated population dynamics, including chaos and other complicated forms.[5] If even such a simple environment produces such complicated behavior, what might we expect from real situations that are not as homogeneous as a bottle of flour?

One of the most influential insights of the beetle team has to do with stochasticity (random forces). Many ecologists tend to think of stochasticity as forming a cloud of points around a deterministic kernel. That is, if a population grows according to the rules of Equation 6.2, we expect a regular exponential curve, but with a certain amount of fuzziness around a smooth curve. That turns out not to be true, even in this simple example.

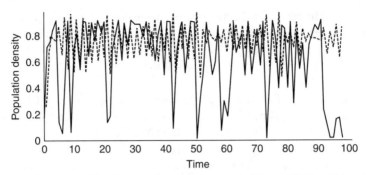

Figure 6-1   Realization of the discrete logistic equation, $N_{t+1} = rN_t(1 - N_t)$, with a stochastic effect, and a small immigration term added. Solid lines show when a stochastic effect is added to N at each time iteration. Dashed lines show when the same stochastic effect is added to r at each time iteration. Note the dramatic and qualitative difference between the two stochastic models.

One could imagine that the stochasticity enters with the parameter r, for example, or one could imagine that the stochasticity enters with the variable N. The outcomes of each of these two assumptions are different, as illustrated in **Figure 6.1**. For more complicated and realistic models, the situation is even worse. The underlying qualitative predictions made by a model are sometimes dramatically affected by the way in which stochasticity enters. This fact is now universally recognized in ecology and is one that needs to be taken into account when trying to understand the dynamics of potential pests.

Although most attention to herbivory in agroecosystems is concerned with the within-season buildup of potential pests and thus articulates with simple ideas of population growth, another focus is on the longer term consequences of pests. For example, the cotton boll weevil, while clearly a problem for individual cotton farmers within a given season, is a more general problem over a regional level and over a longer time. Thus, agricultural planners tend to conceive of the problem as multigenerational, in which the overwintering stage of the beetle is as important as its growth during the cropping season. When the within-season buildup of an herbivore is thought of within the framework of Equation 6.3, the time frame of t to $t + 1$ is on the order of days and/or weeks. However, when attacking the longer term problem, the time frame of t to $t + 1$ is 1 year. The population during the growing season this year is thought to be a function of the population during the growing season of last year. Naturally, the previous comments about nonlinearity and stochasticity apply to this longer time as well as to the within-season time.

In practice, herbivory is frequently expressed in terms of damage done to the crop rather than in terms of the number of herbivores in the population. This seemingly slight change in focus is philosophically critical. Indeed, with such a focus, the problems of herbivory begin to look more like the problems of disease dynamics. Concern with "the fraction of seedlings" attacked by the caterpillars of a moth changes the variable on which we focus our attention—from "number of caterpillars" to "fraction of seedlings attacked." That new variable is precisely the one normally examined when dealing with disease dynamics—the fraction of the host population infected. Rarely, if ever, is the concern with the actual population density of the bacteria or fungus or virus that is causing the disease, but rather with the extent of the infection. In this way, concerns with herbivory and disease can be viewed through a similar lens.

The study of disease dynamics usually begins with the idea that an infected host "transmits" the disease to a susceptible host. Although this is an obvious framework for a disease, it seems at first unusual to think of herbivory in this way, yet that is precisely the way in which some cases of herbivory are treated in practice. The population density of bollworms is actually measured not by the number of bollworms, but rather by the fraction of the crop that is attacked. Even though it is the bollworm that makes more bollworms, it is convenient to think of plants infected with bollworms to be the source of new bollworms, effectively the same as a host transmitting a disease to another host. Thus, understanding some of the basics of disease ecology is relevant not only to diseases in agroecosystems, but also to herbivory.

Basic Principles of Disease Ecology

The fundamental process in the ecology of disease is the so-called mass action assumption. If the number of infected individuals is symbolized as I and the number of susceptible individuals as S, the basic idea of disease transmission is that the rate of increase of the infected individuals must be proportional to the rate of contact of the infecteds with the susceptibles. Thus, the most elementary dynamics must be nothing more than

$$\frac{dI}{dt} = aIS \tag{6.5}$$

which is to say that the rate of increase in the number of infected individuals in the population is proportional to the probability that infecteds come into contact with the susceptibles. But we also know that the overall population of hosts is the sum of infecteds and susceptibles, or $N = I + S$, and we let $N = 1.0$ (i.e., we think of the proportion rather than the numbers of I and S). We can substitute $S = 1 - I$ into Equation 6.5 to obtain

$$\frac{dI}{dt} = aI(1-I) \tag{6.6}$$

which can be seen as identical to the classic logistic equation, where the carrying capacity is 1.0 (review the appendix to Chapter 3 if necessary). Thus, the expectation, from simple basic considerations, is that the buildup of disease in a crop will follow a logistic pattern, up to the point that 100% of the crop is infested.

It is a well-known pattern, however, that frequently a crop does not become completely infected, even if nothing is done to stop it. For example, the fungal disease Koleroga

(*Corticium koleroga*) is always present in coffee plantations in southern Mexico at a very low level (approximately 1%). Nothing is done to try to eliminate it, yet it rarely becomes a real economic problem. Certainly, disease organisms themselves have their natural enemies, as is likely the case with Koleroga, but diseases in vertebrates, some invertebrates and, depending on how you define it, in plants have another characteristic that needs to be taken into consideration. Individuals infected with the disease frequently recover from the disease and contain partial or complete immunity to the disease afterward. Thus, in addition to susceptibles and infecteds, we must also consider "resistant" individuals. With this addition, we arrive at the SIR (susceptible, infected, recovered/resistant) framework of basic disease ecology. Although almost all of disease ecology uses the SIR framework as a springboard, the issues with plant pathogens in particular are somewhat different.

First, the host is sessile, which means that the pest must move to find it, placing a focus on dispersion and the ability of the pathogen or herbivore to locate the host as much as on the direct effect, or virulence, of the disease/herbivore. Second, especially associated with many types of agroecosystems, the hosts are in a correlated growth phase, frequently one large cohort in the case of industrial-like production, but even with more ecological production, phenologies tend to be quite coordinated compared to most natural vegetation. This pattern in time creates a discrete-time flavor over long-term analysis and a restricted time and probably transient-like dynamics over the short term. Third, herbivore and disease are often intimately associated with one another since frequently disease organisms are vectored by herbivorous insects. These special issues suggest that plant diseases be treated with a slightly different framework than animal diseases,[6] although the approach taken in this text is to treat herbivory as if it were simply another form of disease, in keeping with the special place of herbivores as the socially constructed idea of "pest."

Two fundamental concepts enter into the dynamics of disease/herbivory, treated in a variety of different ways by different authors, but fundamentally the same idea. First, a disease (herbivore) must arrive to infect a host plant, which is to say it must be dispersed. Second, the disease must act locally to extract energy from the host. Relatively speaking, the more energy extracted, the more virulent the disease. Although the concept of virulence is not normally associated with herbivores, the parallel biological process is clearly operative, and I see nothing wrong with speaking of the "virulence" of an herbivore. A disease that infects a large number of plants in a given area has probably been dispersed there frequently and has a high "incidence" in the population, whether or not it causes the plant any problems. But a disease that increases its energy-gathering activities within the plant it has attacked has a high virulence whether or not its incidence is high in the local population. Equations 6.5 and 6.6 refer only to the incidence side of the disease/herbivore.

From a practical perspective, plant diseases have long been divided into monocyclic versus polycyclic.[7] Monocyclic diseases are spread through a plant population from inoculum that is present in the environment at the beginning of the season but do not reproduce and spread from plant to plant during the course of the growing season. For example, tomato wilt is caused by a fungus that survives in the soil and infects plants through their roots during the early phases of the cropping cycle, but does not spread significantly from plant to plant during the season, releasing spores that survive in the soil at the end of the season only to reinfect the field during the next year. Polycyclic diseases are those that reproduce and spread from plant to plant during the course of the growing season. For example, leaf blotch disease in barley spreads rapidly through a field of barley from separate foci that are established early in the cropping season but continue producing spores throughout the season. Vanderplank, considered by some to be the father of plant pathology, was the first to employ the simple equations of disease ecology to polycyclic diseases.[8]

Recently, it has been suggested that the disease cycle of a plant pathogen (and, I add, for any herbivore) includes not two but four fundamental issues: dormancy, reproduction, dispersal, and pathogenesis.[9] Dormancy exists when the pathogen has reduced physiological activity due to some environmental condition but, of course, can resume its activities if the conditions become appropriate. Reproduction is the process whereby a new generation of pathogens is produced, either sexually or asexually. Dispersal can be through a variety of mechanisms (wind, water, soil, insect vectors, etc.) and is nothing more than movement of the disease from one place to another. Classically, dispersal is thought of in two distinct categories: the dispersal of the disease organism into the agroecosystem and the local spread, the latter frequently in the form of identifiable "foci" that originated with the initial invasion. Pathogenesis is the interaction between the disease organism and its host.

The Chemicalization of Pest Management

As reviewed in the opening section of this chapter, by the turn of the century, five methods of control were well established, either by tradition (e.g., all the cultural methods listed by Harris[10]) or from some watershed example (e.g., the vedalia beetle): 1) biological control, 2) mechanical and physical control, 3) cultural control, 4) chemical control, and 5) resistant varieties. There was no particular emphasis on any one of these techniques, the discourse of pest management being one of using all methods available, frequently in combination with one another, to get the job done. Farmers generally regarded the farm as either healthy or not, almost as if the farm were the patient, the farmer the doctor, and pests and diseases as the metaphorical germs that made the farm sick. Maintaining the farm (the patient) in a healthy state was the principle goal of the farmer. The emergence of petroleum-based organic biocides, and especially the sociopolitical and economic realities of World War II, would change all of that.

The Emergence of Chemical Control

In 1932, a French patent was taken out on the chemical dinitro-ortho-cresal (DNOC). It was a complex molecule, far more complex than any of the simple chemicals used previously in agriculture (except the biologicals, of course, whose structure was simply not known) and was a product of the impressive gains in knowledge of chemistry. Its synthesis was a product of scientific knowledge of chemistry, whereas its suggested use resulted from the simple observation that it killed insects, with little concern over what other effects it may have had. One might say, perhaps cynically, that its synthesis came from the existence of scientific knowledge about chemistry, whereas its use came from a lack of scientific knowledge about ecology.

With the outbreak of war in 1939, governments generally became interested in DNOC and similar chemicals as possible biological warfare agents. This spurred a great deal of government-sponsored research into chemical poisons. The United States was particularly active in biological and chemical warfare research and as a result came up with a product that would see its application not in war but in the war against weeds, 2,4-D (although later, as a component of the infamous agent orange, it would see war service in Vietnam). Its development and use were described in Chapter 3.

The real revolution was the synthesis of the chlorinated hydrocarbon dichloro-diphenyl-trichloro-ethane (DDT) in the late 1800s and the discovery of its insecticidal properties in 1939, from work done in the J. R. Geigy laboratories in Switzerland. It was quickly adopted by the armed forces in Britain and the United States. Tropical diseases, many of which were vectored by arthropods, were more important killers than the enemy in the practice of warfare. Finding effective solutions was a priority of war research and DDT seemed to provide the final solution.

By 1945, other chlorinated hydrocarbons had been developed, and the arsenal of insect-fighting weapons was well established as an essential part of war preparedness. Germany had made the same preparations, emphasizing organophosphates (e.g., parathion) rather than chlorinated hydrocarbons. World War II had thus created a high capacity for the production of biocides, with the general class of carbamates joining the chlorinated hydrocarbons and organophosphates to make the three major classes of insecticides we know today. Combine that with the herbicides that originally were developed as a byproduct of biological warfare research, and it could be legitimately charged that World War II was the seed that germinated the agrochemical revolution.

Although the war was a watershed for the chemical industry, allowing it to build up immense productive capacity, peace turned out to be a problem. The capacity suddenly turned into an overproduction crisis, and the industry had to scramble for new outlets. Agriculture was the obvious target, although household use was not insignificant. The industry developed some ingenious marketing strategies in post-war U.S. and Europe. War fever having reached a pitch, the public was especially susceptible to wartime rhetoric. The original argument that we needed the chemicals to defeat the enemy in war was easily translated into the need for these chemicals to defeat the new enemy in agriculture.[12] For example, the Industrial Management Corporation noted, in an internal memo in 1946:

> It's a sales story that's simple, effective, and true. It clears up the confusion in the average person's mind when you tell him—Yes, INSECTO-O-BLITZ is still exactly the same as supplied to the U.S. Armed Forces.

The importance of this advertising blitz cannot be overemphasized. Throughout the times of the new husbandry, pest management had taken on the generalized goal of maintaining a healthy ecosystem through the five major categories of activities (biological control, mechanical control, cultural control, chemical control, and resistant varieties). Pests in agriculture were almost like germs in health, and the goal was to maintain a "disease-free" system. The dominant metaphor thus seems to have been a medical one. With wartime and especially postwar propaganda, the pests came to be seen as enemies to be vanquished rather than germs to be controlled. The "war metaphor" replaced the medical metaphor.

The armaments in this new war were, of course, the new organic chemicals produced by the same corporations that produced them for the war effort. The wartime-induced productive capacity of the chemical industry was thus saved by the appropriation of pest control in agriculture, and the attitude toward pests was altered, seemingly subtly, but in the end in such a way as to effect a dramatic transformation of agriculture. The new metaphor meant that farmers changed from stewards who maintained the health of their farms to warriors who vanquished their enemies. The consequences were massive spraying of biocides in the years after World War II and, effectively, the maturation of what we now usually refer to as the "industrial" agricultural system. It is worth contemplating the proposition that this industrial system owes its origin not to the need for increased production or production efficiency, but rather to the needs of a distorted economic system, the crisis of overproduction in the postwar world.

The attitude faced a major challenge with the publication of Rachael Carson's *Silent Spring* in 1962. Carson suggested that the massive use of pesticides was having a dramatic negative effect on the environment. Previously, there had been much popular commentary about the human health effects of pesticides, a concern shared even by the pesticide manufacturers, but *Silent Spring* was the first popular account of the environmental consequences of pesticides, contributing not only to concern about environmental poisons, but perhaps providing the main springboard for the entire subsequent environmental movement.

Carson's words are now well known. Pesticides kill not only the targets, but also many species that are not targeted. Pesticides may concentrate in the higher trophic levels, thus making nonlethal doses at lower trophic levels quite dangerous at higher levels. Pests develop resistance to pesticides. The poisonous effects of pesticides and their residues may persist for a long time in the environment. These were the basic themes of her book.

Silent Spring was an extremely well-documented book. Despite a Herculean effort at finding errors, mainly by representatives of pesticide manufacturers, only the most trivial errors were eventually encountered. Nevertheless, an immense and coordinated attack against the book was orchestrated by the pesticide industry, including an attempt by Velsicol to pressure Houghton-Mifflin not to publish the book. Book reviews were generally harsh and, as later discovered, frequently written by scientists receiving monetary rewards from the chemical industry. As carefully documented in subsequent works, most independent scientists received *Silent Spring* positively, and rereading the book even today suggests that understatement was its actual problem.

Silent Spring, despite its harsh yet substantiated warnings, was generally ignored, except for the growing numbers of people in the environmental movement. Carson's predictions have come true in many areas of the world, and further problems have emerged. Industrial accidents in pesticide manufacturing plants, plus worker exposure to toxics in such plants, have added to her concerns. The environment is now filled with chemicals that mimic estrogen, causing not only feminization in reptiles, but lowered sperm counts in humans, as documented in the remarkable book *Our Stolen Future* by Theo Colborn and colleagues. Unfortunately, the final chapter on pesticides has yet to be written. What began with the need to curb an underconsumption crisis ends with an environmental crisis of unprecedented proportions.

The Pesticide Treadmill

Although Carson's book was a watershed, a less acknowledged, but perhaps even more important book, was *The Pesticide Conspiracy*, by Robert Van Den Bosch (1978). This book accomplished three things simultaneously. First, it reinforced all of the analysis of Rachael Carson, sometimes updating information and pinpointing problems caused by pesticides that Carson had not foreseen. In this it was far more critical of pesticides than Carson had been, partly because 16 more years of experience had made more and better data available. Second, it made explicit what Carson had only left for the reader to conclude, that pesticides were being used largely for the purpose of making profits for those who manufacture them and that the "conspiracy" involved government and university scientists receiving direct and indirect financial rewards from this industry. Third, it elaborated a generalized scheme as to how, ecologically, pesticides came to be such a problem, strongly implying that the problem would continue to become worse. This scheme was called the "pesticide treadmill" and has since become part of the general discourse of the environmental movement.

Nowhere is the pesticide treadmill more evident than in the history of pesticide use in cotton in Nicaragua. Nicaraguan cotton farmers began the cotton years in the 1950s facing a single pest, the boll weevil. Ten years later the main pest was the bollworm, and in fact, a group of about nine other pests had also become important. Twenty years later, there were so many pests that it did not make any sense to tell the farmer which one the new magic formula was intended to treat. Furthermore, farmers went from spraying DDT once or twice during the growing season in 1955 to spraying almost every second day with whatever the newest product happened to be in 1985. Shortly thereafter, cotton was essentially abandoned in Nicaragua. When you start spraying for a single pest, more pests come, and you have to spray more, which leads to more pests, meaning that you have to spray yet more—a treadmill.

The pesticide treadmill includes three forces operating simultaneously: 1) pest resurgence, 2) pesticide resistance, and 3) secondary pest outbreak. Each of these forces had been well known even before *Silent Spring*, but they were first put together as a coherent theory of pesticide use in Van Den Bosch's book.

Pest resurgence is illustrated by the response of the cotton bollworm to the application of monocrotophos, an insecticide registered for cotton, in an experiment performed in 1965. Some cotton fields were treated with the insecticide, and others were left alone. Shortly after the first treatment, all of the fields were sampled, and it was discovered, as expected, that there were about 50% fewer bollworms present in the treated fields. Then, 7 days later, more samples were taken. This time, not only were the treated fields newly infested with bollworms, they had over 400% more bollworms than the untreated fields! After two more treatments with the insecticides, the dramatic 400% disparity was reduced, but in all future samplings, the insecticide-treated fields had more bollworms than the others.

Unfortunately, the scientific testing of insecticides focuses on their ability to kill insects, seemingly the obvious thing to do. However, as so frequently happens in ecology, one does not necessarily do what one sets out to do in a complicated ecosystem. Pesticide engineers saw the problem in an overly simplistic way—bollworms eat cotton. The negative effect of bollworms on cotton would thus be canceled by spraying monocrotophos. Unfortunately, the pesticide engineer's vision is not complete, and the complications necessary to make it more realistic include the natural enemies of the bollworm, a whole army of spiders, wasps, and other predatory insects. In consideration of this slightly more complex reading of the environment, one might ask the pesticide engineer how he or she knows that the ultimate effect of applying the insecticide will damage the bollworm more than the natural enemies, for any child knows that if an action designed to hurt an enemy hurts a friend more, that action is counterproductive.

Despite the obviousness of it all, failure to examine the problem ecologically rather than economically led to the resurgence of the bollworm. Application of the pesticide indeed did kill the target pest, but it also killed the nontarget natural enemies. The result was that the bollworm, now free of its normal predatory control, was able to "resurge" after the pesticide use. This phenomenon has been observed repeatedly in pesticide applications.

In addition to pest resurgence, pesticide resistance was likewise foreseen by entomologists, at least those familiar with the principles of evolution. Pesticide resistance is one of the most well-documented cases that we have of the operation of evolution through natural selection, and its occurrence is far from surprising. Most insect pests are herbivores. Because most plants have evolved the ability to produce toxins to protect themselves from herbivores, it is not surprising that insect herbivores have a variety of methods of detoxifying poisons. If a detoxifying enzyme is under genetic control and genetic variability is associated with it, we normally expect evolution to result from the application of selective pressure. Indeed, it should have been forseen by anyone willing to think about the system as a whole. In fact, the first case of resistance was reported as early as 1946 for the housefly. By 1966, 224 species of arthropods had evolved resistance to one or more pesticides. Today, it is recognized as a fact, and pesticide companies regularly talk of "managing" resistance.

The third component of the pesticide treadmill is secondary pest outbreak. It is illustrated in an experiment examining control of the lygus bug in cotton. Some cotton fields were treated with a mixture of toxaphene and DDT, and others were left as controls. After spraying, the treated fields had no lygus bugs, whereas the control fields had a significant number, as expected. However, the interesting part of the experiment is not the lygus bug, but rather the armyworm, another insect pest of cotton. When the experiment started, there were a very small number of armyworms in both control and experimental fields. A week after spraying, a few were encountered in the control fields but none in the experimentals. Yet another week passed, and the number of armyworms in the experimental plots was 10 times the number in the control plots. The armyworm had become a very significant pest. A month later, after another treatment with the insecticide mixture, the armyworm population in the controls was the same as before but had more than quadrupled in the experimental fields, making it not only a problem but a disaster.

The spraying of an insecticide for one pest, the Lygus bug, had caused another insect that had not previously been a pest (because it was too rare to do any damage) to become one—a secondary pest. How such a phenomenon occurs is no mystery and was anticipated even before the new hydrocarbon-based insecticides were ever invented. The pesticide was aimed at the Lygus bug, but actually killed predators, releasing the armyworms from their natural predatory control.

Putting these three forces together (resurgence, resistance, and secondary outbreaks), we have a pattern that is devastating in the long run for farmers. A resurgent pest suggests to the farmer that more pesticide is needed since the problem is worse. Because of the resistance, the pesticide is no longer as potent as it once was, suggesting to the farmer that more pesticide or stronger pesticide is needed, and the new pests that mysteriously appear suggest once again that more pesticide is needed. It seems much like a drug addiction story. The more pesticide you use, the more you need—the pesticide treadmill.

Biological Control—The Practical Side of Predator/Prey Interactions

Classic Biocontrol and Its Alternatives

Before World War II generated the dramatic change from the medical metaphor to the war metaphor, most control of pests occurred through some combination of biological and cultural control, as discussed previously. One of the traditional methods of biological control is "mass release," effectively an acknowledgment of a fundamental paradox, which is described in detail below. Mass release, the method normally thought of when the words biological control are uttered, involves rearing predators in the laboratory such that they can be released en mass at the appropriate time, that is, when the pests have reached some particular threshold. This procedure indeed works well under some circumstances. It has come to be known as "classic biological control," which actually involves two components: identification of a natural enemy and mass production and release of that natural enemy. As in the case of the vedalia beetle, the standard pattern is that once a pest is identified precisely, a taxonomist is marshaled off to somewhere within the original geographic range of the pest in search of a predator or parasite that exists in the natural range. The underlying assumption is obvious. Having been introduced from somewhere else, the new pest lacks whatever natural enemies it would have normally encountered in its native habitat. The goal is to determine what those natural enemies are and to bring some of them to be introduced, usually in mass release, in the new situation.

This procedure fits quite well under the assumptions of the modern industrial economy—jobs are created for people who work in those laboratories who must produce the biological control agents, which then can be commodified. In the Cuban system, for example, the centers of biocontrol production are an excellent example of this philosophy.[13] It may be the case that mass release will always be necessary under some situations, and it is not insignificant that the procedure fits well with the post-World War II industrial model—spraying a biological control agent with a backpack sprayer is not much different, from a labor requirement perspective, than spraying a biocide. Recently, there has been a great deal written about the dangers of biological control when the agent of control is an exotic species—the potential for the spread of a noxious invasive is enormous.[14]

However obvious the solution of classic biological control may be, another tendency, as old as organic agriculture itself, exists—the permanent maintenance of natural enemies in the system, setting up a sort of paradox between continual active manipulation of the control system and the "autonomous" control effected by the complex interactions among various control agents. Usually the resolution of the paradox is not even recognized for the resolution it is. Farmers simply understand that many insects and other organisms are potential pests and thus need to be kept under control. One of the controlling forces is the diverse community of natural enemies. At this level, the whole idea is easy to understand. And it is important to take into account that the idea of ecological "balance" was usually part of the general discourse, preceding the modern environmental movement by decades. Central to this idea of balance, then and now, is one very simple idea, encapsulated in the famous parable offered by Darwin and extended to tongue-in-cheek extreme by Huxley. Bumblebees nest in abandoned mouse nests, which are made available when domestic cats prey on those mice. The pollinating service the bees offer to the clover is thus ultimately provided by the cats, thus providing an indirect facilitation of clover by cats. Huxley's further elaboration was that cows eat the clover and provide bully beef to the British navy, which guaranteed the power of the empire, but cats were famously tended by elderly unmarried women. The conclusion then is that the power of the British Empire lay with what Huxley referred to as its "spinsters" (the common term for unmarried women in Victorian Britain). This sort of indirect interaction cascade has become a major focus of ecological research in modern times and is really not much more than a continuation of that ancient Chinese quote cited at the beginning of this chapter.

These simple ideas, when cast in a formal analytical framework, turn out to be not so simple as discussed later. Nevertheless, through what is increasingly referred to as "ecological complexity," the ecosystem can be self-regulating. It is an old idea, intuitively embraced by farmers in every corner of the globe, but sometimes thought to be romantic and naïve. Modern ecology is showing that the naïveté is perhaps among those who have embraced the industrial model, seeking to cancel politically what is recognized as immense and frequently confusing complexity.

Perhaps the most convenient place to enter a discussion of this morass of complexity is in the fundamental process of consumption, which from its inception includes something more complicated than the simple ideas of pest increase described in the previous sections of this chapter. Any increase in a population of organisms, pest or otherwise, is invariably accompanied by an increase in one or more of its natural enemies. Thus, the basic rule that N at time $t + 1$ is equal to some rate of reproduction times N at time t ultimately must be related to the simple fact that in the real world, as N increases, so do the things that consume it. Therein lies the origin of much enigmatic complexity.

The Foundational Principles

The standard way of approaching this subject is to start with the dynamics of each individual population, food and consumer (prey and predator). If we have two populations and we describe one of them as the food item and the other as the consumer, we ask the simple question: What connects the two of them together? The answer is obviously that the death rate of the food item is proportional to the size of the consumer population, whereas the birth rate of the consumer population is proportional to the size of the food population. This is the approach taken by Lotka and Volterra in 1926 in continuous time and by Nicholson and Bailey in 1932 in discrete time. Here, we assume that the reader is familiar with the elementary forms of these approaches, at least to the point of knowing what predator/prey (consumer/food) isoclines look like and how they generate the dynamics of the interaction. For those readers who may have forgotten these ideas, the basic formulations are described in detail in the appendix to this chapter.

Three ideas have come to be central to the theory of predator/prey relationships, as they evolved from the simple ideas of Lotka and Volterra: 1) they occur in cycles, 2) the cycles are controlled by intraspecific competition, and 3) predators get full. Almost all modern ecology that deals with predator/prey cycles expands from the base of these three principles.

From the large to the small, nature is filled with cycles, from the stars that cycle around the black holes in galaxies to the electrons that cycle in a radio transmitter—even the most recent version of particle theory postulates that ultrasmall pieces of string are vibrating in 10 dimensions. Predator/prey relationships are no exception. They always have four states that naturally give rise to one another ("rare predator/abundant prey" gives rise to "abundant predator/abundant prey" gives rise to "abundant predator/rare prey" gives rise to "rare predator/rare prey" gives rise to "rare predator/abundant prey," etc.). That such a relationship gives rise to cycles is completely unsurprising. In the early formulation of the theory, these oscillations were of a very strange form, a form that no ecologist could accept as reasonable. This strange form said that no matter how many predators you start with, eventually you would cycle back to that number sometime in the future. Thus, for example, if you begin with 100 lions on a game ranch in Kenya, they might increase and decrease in the future, but eventually, they would come back to 100 lions. On the other hand, if you stock your game ranch with 10 million lions, actually too many to fit in your game ranch, they would decrease in numbers and then increase again and inevitably come back to 10 million. No matter where you start, the population will oscillate back to that point. Ecologists regarded this as an absurdity and immediately discarded it, correctly. The basic qualitative idea of cycling, however, was an important insight and an important step in the development of a full theory of predator prey relationships.

The next major development was the realization that these cycles would be controlled if there were some form of density dependence, either of the prey or predator population. This idea is identical to the argument that gives rise to the logistic equation. If any population grows exponentially, eventually, it will result in a bulk of biomass extending into space at the speed of light. Thus, population growth rates must decline when populations become very large, which is the logic that gives rise to the logistic equation. If we apply the same idea to the cycles of a predator/prey system and suppose that the prey item is, as all populations, not able to expand beyond its energy source, the cycles become controlled. The 10 million lions crash because they overeat all of the zebras—their food, and as they recuperate when the zebras come back, the zebras themselves are only able to reach their own carrying capacity, which is not nearly enough to support 10 million lions. Essentially, the permanent oscillations of the most elementary theory are replaced by damped oscillations when you add density dependence. If the original permanent cycles are like a pendulum oscillating permanently in the absence of friction, the density dependence represents the friction that causes the arcs of the pendulum to become ever smaller as it reaches its resting point.

The third, and in many ways most important, idea is that of predator satiation. If you take the basic theory that gave us cycles and add to it the basic theory of density dependence, you get a theory that looks much like what we think happens in nature—predator/prey systems oscillate, but they do not explode to unreasonable levels. However, there is one thing about the theory that was just as wrong as presuming that the zebras would have no limit on their populations—permanent cycles were impossible. That is, all predator/prey systems would have to come to some stable point where they would no longer oscillate. Because it had been concluded from a variety of observations in nature that many populations do in fact seem to oscillate permanently, this was an unacceptable conclusion. The key insight that resolved the contradiction was about predator consumption. The basic theory (cycles plus a carrying capacity of the food) assumed a linear response of the predation rate to the amount of food available. Thus, a lion might have a consumption rate of one zebra a week when there are 10 zebras on the game ranch, which would be a consumption rate of 10%. Then if there were 100 zebras on the ranch, the lion would eat 10 of them. And if there were 100,000 zebras on the ranch, the lion would eat 10,000 a week. Obviously, there has to be a point where the lion is simply not physically capable of eating yet another zebra, which is to say that the lion gets satiated. The addition of this idea of predator satiation results in the complete modern theory as we now understand it. Predator/prey systems oscillate, but in a controlled fashion, sometimes reaching a permanent balance between predator and prey but sometimes oscillating permanently in a perfectly predictable cycle (i.e., no millions of lions in someone's small game ranch, even if she puts them there to start with). These three phases in the development of modern predator/prey theory are illustrated graphically in **Figure 6.2**, and a more complete development is presented in the appendix to this chapter.

The Paradox of Biological Control[15]

It is generally assumed that the goal of biological control is to reduce the pest population as much as possible, whether speaking of classic biological control or the more sophisticated ecological forms that require a management scheme that promotes a "balance" of the system over the long haul. In light of the fundamental principles of predator/prey interactions as described previously, the solitary goal of decreasing the pest species as

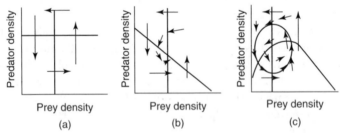

Figure 6-2 Development of classic predator/prey theory. In all three cases, the predator isocline is the vertical line and separates the arrows (vectors) going up (increasing predator population) from those going down (decreasing predator population). If this confuses you, go to the appendix to this chapter. The other line (horizontal, angled downward line, or hump shaped) is the prey isocline and separates the arrows going to the right (increasing prey density) from those going left (decreasing prey density). (a) The simple density-independent isoclines of Lotka and Volterra, which will result in permanent, uncontrolled oscillations. (b) Adding density dependence to the prey results in the prey isocline declining to its value of K, the carrying capacity, and the overall pattern of damped oscillations. (c) Adding predator satiation to the model, which results in the prey isocline having a hump and permanent controlled oscillations if the predator isocline crosses to the left of the peak of the hump (with simple damped oscillations if the predator isocline crosses to the right of the peak of the hump).

much as possible would mean moving the predator isocline to the left (as it defines the position of the equilibrium value of the prey density). Using the Lotka-Volterra formalisms (see Appendix 6.A), this means either increasing the value of the predation rate and/or decreasing the value of the predator's death rate. In either case, if the predation rate becomes too large, the prey is rapidly eliminated from the system, leading to the predator's collapse (because of a lack of food). Thus, the rational goal of biological control, driving the pest population to as low a level as possible, may lead to the collapse of the biological control system. The basic idea as it follows from elementary ecological theory is illustrated in **Figure 6.3**. Unfortunately, for most situations, the isoclines are unknown and there is tremendous economic pressure to reduce the pest to as low a value as possible, meaning that the unexpected consequence of elimination of the control is possible, perhaps even probable.

In trying to promote a permanent biological control, a planner must seek the "stable" situation of a predator with an intermediate rate of consumption, which means that the pest must remain in the system. This then implies that to "vanquish the enemy" is an ecologically inappropriate metaphor that could ultimately lead to the elimination of whatever biological control had been there in the first place.

The extensive work of Murdoch and his colleagues has contributed in an important way to this issue.[16] For example, with very detailed studies on the red olive scale and its parasites, this team found that the system appeared to be unstable. Yet the system remained extant for years in the California orchards they studied. Upon closer examination, they found that scales congregated on the trunks of the trees, where the parasites did not forage. Thus, the scales on the trunk lived in a sort of refuge, out of contact with the parasites. The system on the leaves and branches indeed was unstable, but by adding the refuge, the system, writ large, was stable. It is a simple exercise to add a refuge to the elementary Lotka-Volterra approach and show how that can stabilize an inherently unstable system (effectively place a new isocline vertically, which allows prey to increase at any predator density at all, as long as the prey is below that "isocline").

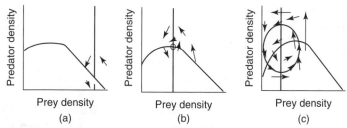

Figure 6-3 Paradox of biological control. The predation rate increases from (a) to (b) to (c). Concomitantly, the equilibrium value of the prey density declines; however, with that decline comes a tendency to oscillate ever more extremely, possibly leading to the elimination of the system (as in c), which implies the elimination of the autonomous control offered by the permanent presence of the predator.

Integrated Pest Management

Prophylaxis Versus Responsive

The key idea for pest control is predicting when the pest problem becomes too bad to ignore. Thus, simple qualitative arguments or exact quantitative methods have always been somehow at the foundation of control technologies. Applying, for example, Equation 6.2 to an insect population, one can easily project the population into the future to decide when the insect will reach an abundance that will cause problems and plan control procedures accordingly. Two concepts are involved in that procedure, both of which are more about economics than about ecology. First, pest control workers talk of the economic threshold, which is the critical density of the pest that begins to cause significant damage. The significance of the damage is always assessed with respect to the cost of whatever control procedure is thought to be necessary. Second, some workers distinguish the economic threshold from the action threshold—the critical density of the pest that indicates actions for controlling it must be planned. This second idea implicitly assumes knowledge about the population ecology of the pest because its current status must be projected into the future so as to decide whether action should be taken.

Precisely when the action threshold will be reached depends largely on management decisions that were made in advance of the time a decision has to be made. On the one hand, a rational control procedure will include continual field surveys to determine when the pest becomes so problematical that action must be taken. On the other hand, earlier decisions about what and how to plant will obviously affect the approach to the action threshold, and a host of "prophylactic" measures can be taken explicitly to minimize the probability that the action threshold will be reached in the first place. Considering pest-control strategy as a complicated management problem, one must balance the cost of prophylaxis against the cost of responsive control (taking action only after the action threshold has been reached), in light of the population ecology of the pest/disease in question.[17]

Control of pests through a variety of prophylactic measures has always been diverse and universal. These prophylactic measures are eclectic and usually specific to the pest situation involved. The diverse array of prophylactic techniques can be conveniently divided into those that directly employ a natural enemy of the pest—biological control—and those that manipulate the planning of the farm in some way so as to encourage natural enemies or discourage the pests in the first place—cultural control. General principles of biological control were discussed in some detail previously. Here we consider several forms of cultural control.

Rotating crops is perhaps the oldest means of pest and disease control. From the time of European new husbandry and even before, farmers understood that pests could build up in an area that was cultivated continually. Combined with the need for nutrient supplements, rotations have long been a standard form of husbandry in agriculture. Only recently, with the advent of agrochemicals, both pesticides and fertilizers, have rotations seen less use.

Increasing plant diversity either through planting multiple strains of the same crop species or through intercropping is often cited as a major prophylactic technique for controlling pests and diseases. Here we see a close parallel with certain principles of disease ecology more generally. Frequently, vaccination programs are based on the idea of herd immunity—as the number of resistant hosts in the population increases, the transmission of the disease vector decreases, to the point that the disease is eliminated from the host population. When the disease can no longer survive because the probability of encountering a susceptible individual has become so low due to the large number of resistant individuals, the situation is referred to as herd immunity. A similar situation happens in plant pathology.[18] By planting a mixture of cultivars, some of which are resistant others susceptible, one can plan a situation of effective herd immunity. Obviously, the trick is to balance the reduction in yield caused by the resistant varieties (presuming resistance implies a reduction in yield) with the reduction in probability of disease infection. Usually this will require some detailed knowledge not only of the elementary population ecology implied in Equations 6.5 and 6.6 (and their more complicated brethren) but also the spatial ecology of dispersal.

One of the classic prophylactic techniques is the encouragement of natural enemies in the system, which is where the paradox of biological control enters. As techniques for promoting natural enemies in the system progress, it is presumably the case that the need for a response to a pest outbreak will be lowered. In contrast, the entire concept of mass release is a responsive technique. Nevertheless, as a prophylactic technique biological control, of the autonomous sort, holds great promise as part of an integrated pest management (IPM) program that has potential for eliminating the need for biocides altogether. Yet, and this cannot be emphasized too strongly, the sorts of problems suggested by the paradox of biological control (and undoubtedly there are other, yet-to-be-recognized problems) are not insignificant. As noted in the description of the paradox itself, a producer or planner will not generally know the nature of the isoclines and thus will not be able to anticipate the point at which encouragement of yet more natural enemies might kick in a change in the ecosystem that will unbalance the relationship between pest and natural enemies, thus disrupting whatever autonomous control may have existed. In principle, it would be good to know where those isoclines are, metaphorically speaking. The many population-based approaches to biological control need to be continued and elaborated much more vigorously.[19]

Thus, elementary population ecology has been at the base of pest control technology from the beginning and should remain a critical component of future planning for agroecosystem development. Despite this historical reality, at the end of World War II, something changed dramatically, as has been noted repeatedly in previous chapters. The relatively sophisticated idea of predicting when a pest would become a pest and using a degree of foresight to apply a variety of control procedures was replaced by a new paradigm. The events leading up to this new paradigm had less to do with ecology than with economic and political facts, as described previously. As a consequence, the expansion of the chemical industry and its pesticide sector came to dominate pest control technology from World War II to the present day, with the challenge provided by Carson's *Silent Spring* echoing the same challenges today as it did when Velsicol tried to ban its original publication. Yet today's procedures are far more elaborate, with the chemical industry having been forced into a more ecological framework, and sophisticated prophylactic/responsive approaches to pest management going under the name "integrated pest management."

The Politics of IPM

The fundamental ideas of IPM, as formulated first by van den Bosch (discussed previously), provide a general framework. Dividing various techniques into prophylactic versus responsive could be a useful device for planners. The idea is to have the tools available to respond to a pest situation as it emerges, but also continually develop other tools that will provide a prophylaxis against that situation emerging so frequently. Generally, the responsive techniques are likely to involve petrochemicals for the near future, although the environmental consequences of such use should be fully assessed as part of the planning process. The general idea, however, remains that as prophylactic techniques are perfected, there will be ever less need for the responsive techniques. Eventually, the tools of the responsive techniques will naturally become very rare, and the situation will naturally arise in which farmers respond to the question "what kinds of pests do you have?" with the answer "none," fully understanding that a program of autonomous ecosystem management is underway. This is a hopeful scenario.

Other, more critical, analyses of IPM are also on the table. Altieri, for example, views IPM as a cynical attempt by the chemical industry to prolong the life of their "chemical agriculture," as it was called by them in their initial propaganda campaigns. According to this critical viewpoint, IPM is nothing more than a co-optation of ecological forms of agriculture to extend, and sometimes even promote, the use of dangerous pesticides. Nevertheless, historically, IPM must be seen as a recognition by that very pesticide industry that there are problems associated with their underlying philosophy, and furthermore, as a transitional strategy it contains all of the necessary ingredients for a major transformation of the industrial agricultural system.

Autonomous Biological Control

A calling card for much of the modern environmental movement has been "ecosystem services," the not so subtle suggestion that ecosystems in all their complexity serve humanity and that disturbing them, or implicitly, simplifying them, may result in the loss of those services. Nowhere is this principle more evident than when the complexity of ecosystems is thought to constrain certain organisms that would otherwise be classified as pests. The entire idea that the internal functioning of an ecosystem could result in regulation of potential pests is referred to here as autonomous biological control. A major goal of the sustainable agriculture approach is to move all recognized pests into this category. It is, in a sense, taking the idea of classic biological control and integrated pest management up a notch.

The classic studies of Morales, discussed extensively in Chapter 1, represent a sort of canonical example, finding that many insects known to be pests of maize did not reach pest status in traditional Guatemalan farming systems. The reason seems to be that the overall planning of the system is such that the ecological interactions in traditional management restrain the populations of those potential pests. The ecosystem as constructed by the farmer indirectly and autonomously keeps the problems at bay.

Although this idea is commonplace among practitioners and promoters of sustainable forms of agriculture, only rarely are the details of the ecosystem known well enough to demonstrate exactly how the autonomous control happens. One example is recent work in Mexican coffee production that has elaborated a complex ecological web that appears to contribute to an autonomous regulation of two potential pests of coffee.[20] The two pests are well known as occasionally devastating pests of coffee: the green coffee scale insect (**Figure 6.4**) and the coffee rust disease (**Figure 6.5**), the latter of which is infamous for its total destruction of coffee production in Ceylon as discussed in one of the opening scenarios of this text.

The green coffee scale is associated with an arboreal ant, *Azteca instabilis*, in a classic hemipteran/ant mutualism. The scale insect secretes a sweet honeydew that attracts the ants. The honeydew is the major source of carbon for the ants. As the ants frantically crawl around the scale insects, they scare away the parasites (small wasps that attack the scale insects) that continually try to attack. Thus, the scale insects benefit from the activity of the ants scaring away their natural enemies and the ants benefit from the scales providing them with their main carbon source. It is a classic mutualism.

The ant builds very large nests in the shade trees in traditional shade coffee systems and tends the scale insects in the surrounding coffee bushes. Even though the shade trees are uniformly distributed in the system, the trees with ant nests occur in clumps, sometimes

Figure 6-4 The green coffee scale insect, *Coccus viridis*, adults, young nymph (on the leaf veins) and crawlers.

Figure 6-5 The coffee rust disease on the underside of coffee leaves. The light spots are yellow in nature, and the darker areas near the bottom of the upper leaf are necrotic areas caused by the rust.

with a single tree having a nest in it, but most often with two or more neighboring trees containing nests. Consequently, the ant itself occurs in a very patchy fashion around the coffee farm, some places with high concentrations of ants and others with no ants at all (of this particular species). This spatial disparity, with clumps of trees with ants and other clumps of trees (much larger clumps) with no ants, is extremely important to the functioning of the system—because of a beetle.

The first complexity in the system arises from the question of why the ant nests occur in clumps or clusters in the first place. The trees in which they nest are uniformly distributed in the coffee plantation, yet the trees with nests are clustered. The ants themselves occur in nests that have multiple queens. When a nest gets too large, or perhaps for other reasons, one of the queens takes off with a group of workers and establishes a satellite nest in a nearby tree. Much as a plant would send out seeds to establish seedlings nearby, an ant nest sends out propagules in the form of queens with workers to establish a new nest nearby. Through this process, we would expect to see first the clustering of nests in neighboring trees, then the expansion of the ants placing nests in all the trees.

In another coffee farm a large study plot contained 11,000 trees, any of which could have contained a nest, yet only about 400 of those trees contained nests. If the formation of clusters continued as described previously, we would expect that the ant would eventually take over the entire plantation with nests in every single shade tree. But that does not happen. Why? The reason may be because of another parasite—this time a fly—that attacks the ant. The fly is in the family Phoridae and is known as a decapitating fly. It lays an egg on the back of the head of the ant, and when the larva hatches, it crawls into the head capsule of the ant and eats the contents. When the new fly is ready to emerge, the ant's head falls off (thus, the fly is called a decapitating fly), and the fly emerges, ready to mate and look for other ants (only of this species) to parasitize.

The fly parasite is attracted to the larger clumps of ant nests, more than to single trees with ant nests, that is, it is a density-dependent natural enemy of the ant. Thus, the clumps of ant nests seem to be formed by the ants sending out satellite colonies to nearby trees, but with the phorid flies limiting the growth of clusters of ant nests.[21]

In addition to the parasites that are continually trying to attack the scale insects, there is a beetle that is a voracious predator on the scale insect. However, the beetle has a complex life cycle that includes a larva that is protected from the ants by waxy filaments on the surface of its body. The waxy filaments act to gum up the ant's mandibles, thus allowing the beetle larvae to eat its favorite prey items, the green coffee scale, with minimal interference from the ant. Furthermore, because the ant is running around scaring all of the parasitoids in the area, it also inadvertently scares away some particularly devastating parasitoids that normally attack the larvae when the ants are not around. So, ironically, the ants, because they cannot get at the beetle larvae because of the waxy filaments, actually protect the larvae from its natural enemies, even though the larvae are voracious predators of the ant's favorite food, the green coffee scale.

The adult beetles, however, do not have any protection from the ants, and whenever they land on a coffee bush with ants on it, they are vigorously attacked by the ants and must quickly retreat, or be killed. When there are no ants, the beetles fly around and are able to efficiently find the scale insects that occur sporadically throughout the entire plantation. The larvae are not so lucky when the ants are absent, as the parasitoids that are constantly searching for them do not have to contend with the ants and the larvae experience very high rates of parasitism. Thus, the overall story about the beetles is that on the one hand the larvae do fine when ants are present (lots of food and enemy-free space), but the adults cannot eat because the ants keep them away from their food, the scale insects. On the other hand, the larvae do very poorly when the ants are not present (low density of food and surrounded by parasites), but the adults are able to fly around finding the rare individuals of the scale insects that are not protected by the ants. Basically, through the clustering of the ant nests, the beetle population is able to survive and effectively control the scale insect population over the entire plantation. The scale insects are effective pests of the coffee only when they are under the protection of the ants, which is in approximately 4% of the farm, but through the spatial dynamics of the beetle population, effective control over the scale insects is established over the other 96% of the farm.[22]

What about the historical devastation of the famous coffee rust disease? Here we must focus on yet another component of the spatial system not evidently related to the rust, a disease of the scale insect. The disease is caused by the fungus *Lecanicillium lecanii* and is called the white halo fungus disease, since a white halo of mycelium is formed around the edge of the dead and dying scale insects when they are infected. The disease seems incapable of sustaining itself unless the scale insects are at a relatively high local density, but they never reach that density unless the ant is around. So the white halo fungus *L. lecanii* may be endemic throughout the farm but reaches epizootic levels only when the scale is abundant, which only occurs when the ant is present.

Finally, the white halo fungus, in addition to attacking the scale insect, also attacks the coffee rust disease. Although it does not completely control the coffee rust disease, there is evidence that it at least contributes to the control of the rust.[23] Thus, the ant occurs in clusters, meaning that the scales do also, and the beetle can survive only because the ant nests occur in clusters, which they do because of the phorid fly parasite. The scales are attacked by the white halo fungus, but only when the ant is around. The white halo fungus is an antagonist to the coffee rust disease. Thus, both scale insect and fungus are kept partially under control, and the cause of that control is the complex set of interactions in the functioning ecosystem (**Figure 6.6**). It is an example of autonomous biological control, happening because of a complex network of ecological interactions.

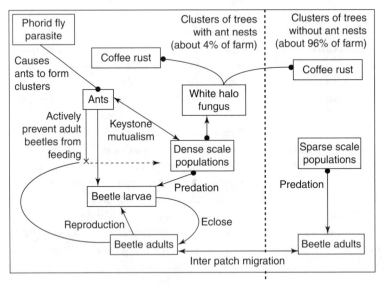

Figure 6-6 Diagrammatic representation of autonomous biological control of the green coffee scale and the coffee rust. The tendency of the ant to expand its colonies is restrained by the phorid fly parasite, which results in clusters of trees with ant nests and clusters of trees without ant nests. The ant forms a keystone mutualism with the green coffee scale, which allows the latter to reach excessively large population sizes, which in turn allow the disease the white halo fungus to become epizootic. The increased local density of white halo fungus generates spores that are carried by the wind to other areas of the farm and attack the coffee rust. The beetle larvae survive well in patches with ants because their food (green coffee scale) is abundant and they are in enemy-free space (the ants scare away their parasites). The adult beetles, however, are prevented from feeding by the ants and thus can feed only when they fly away to clusters of trees without ants, thus helping to regulate the green coffee scale over the entire farm.

Appendix 6.A: The Three Stages in the Evolution of Predator/Prey Theory

Some of the earliest attempts at understanding ecology involved, at least indirectly, the notion that eating food caused population increase of the consumer and population decrease of the food source—the ultimate "atomic theory" of predator/prey relationships. It is a three-part story, understandable completely in graphical fashion or through employment of dynamical equations. Here I show both approaches. The reader should understand that the graphical and equations-based approaches are in fact the same biological argument.

We start with a graph of predator versus prey, using lions and zebras as a running example. Graph the number of lions on the y axis and the number of zebras on the *x* axis (**Figure 6.A1a**).

Suppose you are managing a game ranch and you want it to be ecologically sustainable so that tourists will come and pay money to watch the wildlife on your ranch. Thus, you really want there to be a solid population of lions on your ranch. The key factor to maintaining a population of lions is their food source, so you want to make sure that there are sufficient numbers of zebras (I know that lions eat other things too, but for heuristic purposes, let's assume they only eat zebras). In this way, we define a particular threshold value of zebra numbers, above which the lion population can increase in numbers, but below which the lion population will decline. We indicate the threshold number of zebras as a vertical line and the dynamics of the lion population as arrows. The ones to the right of the vertical line indicate an increasing lion population. The ones to the left of the vertical line indicate a decreasing lion population, as in **Figure 6.A1b**.

Similarly, from the zebra's point of view, there is a critical number of lions, above which the zebra population must decline and below which the zebra population will increase (as shown by the arrows parallel to the zebra axis), as in **Figure 6.A1c**.

Thus, in Figures 6.A1b and 6.A1c, we see the dynamics of both lion (Figure 6.A1b) and zebra (Figure 6.A1c) populations. Now the point is to ask what will happen if we put the two dynamic pictures together, as is done in **Figure 6.A2**.

This basic arrangement results in an oscillatory system as is obvious if you just follow the arrows in Figure 6.A2d, going from "many lions/few zebras" to "few lions/few zebras" to "many zebras/few lions" to "many zebras/many lions" and over and over. It is obvious

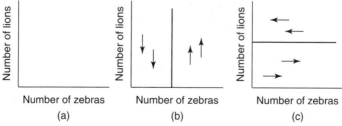

Figure 6-A1 Basic dynamics of a predator prey situation (see the text for an explanation).

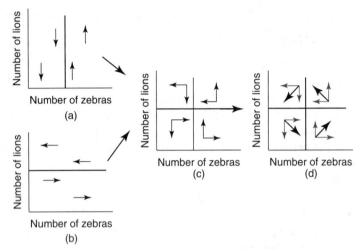

Figure 6-A2 (a and b) Dynamics of each of two species. (a) Dynamics of lion population. (b) Dynamics of zebra population. (c) Combining the vectors (arrows) of the two populations in (a) and (b). (d) Summing the vectors to indicate the overall dynamics.

that the most elementary population rules of food and consumption (prey and predator) result in population oscillations. But a further aspect is that those permanent cycles always repeat themselves, no matter where you start, returning to that precise point, what has been referred to as "neutral stability." So, if you put a combination of 10 lions and 10 zebras, for example, on your game ranch, both populations may increase dramatically, and then the zebras decrease, and then the lions decrease, but eventually the two populations would return to *exactly* 10 lions and 10 zebras. But if you started out with 50 lions and 50 zebras, again the populations would increase and then decrease, but ultimately, both would return to 50 and 50 again. No matter where you start, you will ultimately return again, as illustrated in **Figure 6.A3**.

Although most ecologists regard the notion of neutral oscillations as completely unrealistic, the beauty of this simple analysis is that without stipulating hardly anything other

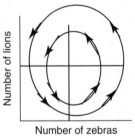

Figure 6-A3 Classic Lotka-Volterra formulation of predator/prey system, with "neutral" oscillations.

than food and eating (prey and predator) we come to the conclusion, a qualitative conclusion actually, that predator and prey ought to be oscillatory with respect to one another. This is the first, and perhaps most important, conclusion to be drawn from the elementary theory of predator and prey.

We can approach the exact same problem slightly more formally, although the basic analysis is precisely the same as the purely graphical approach. If we begin with the assumption that we have two generalized populations both obeying the standard exponential growth model (see appendix to Chapter 3 for a review) and then modify one of them to be the prey and the other to be the predator, we come up with precisely the same result. Thus, let the two populations be symbolized as P (e.g., the number of lions, predators) and V (e.g., the number of zebras, victims). Then apply the exponential equation to each of them to obtain

$$\frac{dP}{dt} = (b_1 - m_1) \tag{6.A1a}$$

$$\frac{dV}{dt} = (b_2 - m_2)V \tag{6.A1b}$$

We need two very elementary assumptions to make these equations correspond, qualitatively to the process of predator and prey (or host and parasite, or plant and herbivore, or consumer and resource, etc.). First, the birth rate of the predator should be proportional to the amount of food available (i.e., the b_1 in Equation 6.A1a should be b_1V), and second, the death rate of the prey should be proportional to the number of predators around (i.e., the m_2 in Equation 6.A1b should be m_2P). Putting these two assumptions into Equations 6.A1, we arrive at the classic Lotka-Volterra equations:

$$\frac{dP}{dt} = (b_1 V - m_1)P \tag{6.A2a}$$

$$\frac{dV}{dt} = (b_2 - m_2 P)V \tag{6.A2b}$$

We obtain the isoclines of these equations by setting the derivatives equal to zero and rearranging to obtain

$$V = \frac{m_1}{b_1} \qquad P = \frac{b_2}{m_2}$$

Thus, there is a constant value on each of the axes that corresponds to the isoclines, which are precisely the isoclines originally described in Figure 6.A2a and 6.A2b.

Density Dependence

Consider what happens when we relax the assumption of density independence of the prey (i.e., when we presume the zebras have some carrying capacity). Now the point that separates the increasing zebra population from the decreasing one will change depending on the density of the zebra population itself. Thus, for example, if the zebra population is already at its carrying capacity, there is no point in the space for which the zebra population can increase. The result is that the prey isocline is no longer parallel to the abscissa, but rather is a downward sloping line, intersecting the abscissa at the prey's carrying capacity. The predator isocline stays the same, and the equilibrium point, where the two isoclines cross, attracts all trajectories—that is, the system is one of damped oscillations, as shown in **Figure 6.A4**.

We can see this graphical result (as shown in Figure 6.A4) in the basic equations. Recall the development of the logistic equation. We began with the process of exponential

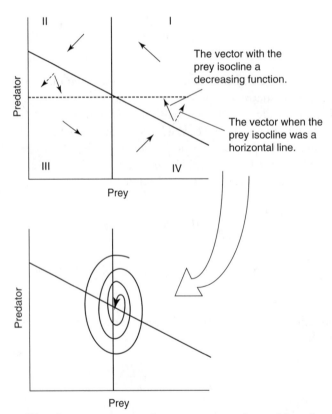

Figure 6-A4 The change in the population trajectory from adding density dependence to V (see the text).

growth and proceeded by adding density dependence. The assumption that the per capita growth rate remained constant was replaced with the assumption that it varied negatively with the population density. It would seem obvious to do the same with the predator prey situation. Thus, we begin with the basic predator prey equations:

$$\frac{dP}{dt} = bVP - mP \qquad (6.A3a)$$

$$\frac{dV}{dt} = rV - aPV \qquad (6.A3b)$$

(the parameters have been changed and subscripts eliminated for clarity; i.e., we have replaced b_1 with r and m_2 with a). Using the same reasoning that we used in deriving the logistic equation, we cannot reasonably assume that the per capita birth rate of the prey (r) is constant. This would be the same as presuming that the prey population could grow without limit if the predator was absent. This part of the equation can be made more biologically realistic by simply making the prey population obey the logistic equation, when the predator is absent. Thus, Equation 6.A3b becomes

$$\frac{dV}{dt} = rV \left(\frac{K-V}{K} \right) - aPV \qquad (6.A4)$$

where K is the carrying capacity of the prey population. The isocline for the prey equation (Equation 6.A4) is

$$P = \frac{r}{a} - \frac{rK}{a} V \qquad (6.A5)$$

which is a linear equation in the P, V space. The qualitative dynamics that result from adding density dependence to the equation for V are deducible from an examination of the isoclines and how they change when the density dependence is added, as noted earlier and illustrated in Figure 6.A4. In the upper graph of Figure 6.A4, the original isocline is indicated by a dashed line, and the original vectors are also dotted.

The new isocline (the isocline that exists after density dependence is added to the prey) changes a piece of the region that used to be part of quadrant II or quadrant IV, into part of what is effectively quadrant III or quadrant I, respectively. The change in vectors in this changed part of those quadrants is shown, whence it can be seen that a piece of the space now has vectors that point more toward the equilibrium point than they did with the original isocline. This causes the oscillations to move toward that equilibrium point. Thus, adding density dependence to the prey population causes the neutral oscillations to change into stable oscillations, which is to say the system is necessarily an oscillatory point attractor. This is not a trivial conclusion, for it restricts the behavior of predator/prey systems considerably, probably unrealistically so. It suggests that predator/prey systems are always stable! This is an unwarranted conclusion, as discussed in the following section.

Functional Response

The density independent assumption for the prey species seems quite foolish, biologically. However, the parallel assumption for the predator is not unreasonable. It simply says that the predator will have a positive birth rate as long as there is food to eat, which under the restrictive set of assumptions used in developing these models is sensible. But there is another assumption about the predator behavior that indeed is unreasonable. We have presumed that the ability of the predator to eat its prey is completely independent of the density of the prey (b is a constant), and similarly, the effect of the predator on the prey population is independent of the density of the prey population (a is a constant).

It is actually well documented (e.g., Hassell, 1978) that prey are not eaten independently of the prey density. Indeed, for most predators, if you plot the rate of prey consumption against the population density of the prey, you do not get a straight line, as stipulated by Equation 6.A5 (i.e., if the predation rate is [aVP], this means that the predation rate is a linear function of the population density of the prey). The original Lotka-Volterra equations presumed that predators will eat a certain fraction of prey, no matter how much is available. Thus, if a lion eats 1% of the zebra population each month, that means that it will eat 1 of a population of 100, 10 of a population of 1,000, and 10,000 of a population of 1,000,000. That is, it will never become satiated. This is obviously an unreasonable assumption for most predators. Nevertheless, that is exactly the assumption made by the classic form of the Lotka-Volterra equations. In terms of actual data, the assumption is obviously ridiculous.

The predation rate is always a function of the population density of the prey, and that is why it is usually referred to as the "functional response." However, the exact form it takes can have a dramatic effect on the qualitative outcome of the predator/prey interaction. As we have already seen, if the functional response is linear, the system is either of neutral stability (if the prey are not density dependent) or exhibits damped oscillations (if the prey are density dependent). What if the functional response is nonlinear?

This form of the function is most frequently modeled with the so-called Holling disc equation, which states[24]

$$\text{Predation rate} = \frac{cVP}{g+V}$$

an example of which is shown in **Figure 6.A5**, labeled II. The meanings of the parameters in the Holling disc equation are illustrated in the figure. The value cP is the asymptotic value of the predation rate (assuming, for now, that P is unvarying), and g is the value that V assumes when the predation rate is at half its maximum value. Thus, c is proportional to the asymptotic value, whereas g is a measure of the degree of curvature of the curve. Clearly, if $g = 0$, the functional response reverts to its classic linear form, and as g increases, the curvature of the functional response slowly increases. Changes in the form of the functional response are illustrated in Figure 6.A5b.

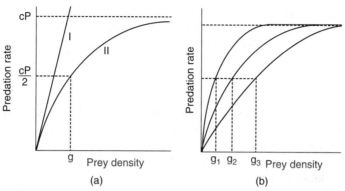

Figure 6-A5 The functional response. (a) Linear form (I) and asymptotic form (II). (b) Three versions of a type II functional response with different values of the parameter g.

Going back to the original Lotka-Volterra equations, recall the basic form:

$$\frac{dP}{dt} = bVP - mP$$

$$\frac{dV}{dt} = rV - aPV$$

The prey equation can be thought of as

$$dV/dt = \text{rate of growth in absence of predator} - \text{predation rate}$$

and "rate of growth in absence of predator" is rV, whereas predation rate is aPV. But this is assuming that the prey population is density independent (see previous section) and that the predation rate is linear. If the predation rate is nonlinear, we can employ the disc equation of Holling to obtain

$$\text{Predation rate} = \frac{cVP}{g+V}$$

which makes the prey equation

$$\frac{dV}{dt} = rV - \frac{cVP}{g+V}$$

which has the isocline

$$P = \frac{b}{c}(g+V)$$

Following this argument, we must also modify the predator equation. That is, if the predation rate is an inverse function of the prey density, that fact must be represented in the predation rate of the predator equation also. That is, if the predator equation is

$$dP/dt = \text{birth rate} - \text{death rate}$$

and, as argued before, the birth rate is a function of the prey density, that is, the birth rate becomes satiated. That is, the birth rate cannot be a linear function of V, for the linear assumption implies that the predator can eat, effectively, an infinite amount of food. So the birth rate of the predator must be modified in the same way as the prey equation. Thus, if the original equation was

$$\frac{dP}{dt} = rVP - mP$$

where rVP is the birth rate, we must modify that birth rate with the functional response. Thus, the predator equation becomes

$$\frac{dP}{dt} = \frac{rVP}{g+V} - mP$$

which has the isocline

$$V = \frac{mg}{r-m}$$

which is qualitatively the same as without the functional response term, that is, a simple vertical line.

The two isoclines are plotted in **Figure 6.A6**, along with a qualitative interpretation of the vector field. Note that here we have an unstable equilibrium point, of necessity. That is, according to the simple addition of functional response to the classic Lotka-Volterra equations, we conclude that all predator/prey systems are unstable and thus cannot persist!

So we see that from two separate simple and biologically sensible modifications of the classic LV predator/prey equations, we have first, all predator/prey systems are stable (have damped oscillations) and second, all predator/prey systems are unstable (have ever-expanding oscillations).

Functional Response and Density Dependence Together

The two nonlinear forces of functional response and density dependence seem to balance one another. That is, the tendency of predator/prey systems to be unstable (resulting from functional response) is counterbalanced by their tendency to be stable (resulting from density dependence). This can be seen if we simultaneously add both nonlinear components to the original equations. Thus, for the prey equation we have

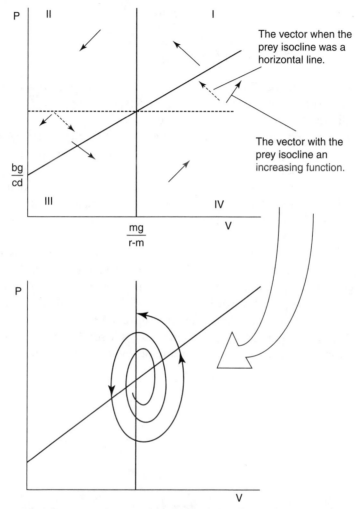

Figure 6-A6 The change in the population trajectory from adding type II functional response.

$$\frac{dV}{dt} = rV\left(\frac{K-V}{K}\right) - \frac{cVP}{g+V}$$

The isocline for this equation is a bit more complicated than before and is given as follows:

$$0 = rV\left(\frac{K-V}{K}\right) - \frac{cVP}{g+V}$$

which, after some algebraic manipulation, becomes

$$P = \frac{r}{c}\left[g + \left(1 - \frac{g}{K}\right)V - \frac{1}{K}V^2 \right]$$ (6)

It is important to note that Equation 6.6 is quadratic in the space of P, V, which means it is shaped like a parabola, as illustrated in **Figure 6.A7a**.

From a qualitative point of view, the key feature of this graph (Figure 6.A7) is that the prey isocline now may actually be ascending at the point where the predator isocline intersects. Looking at a close-up of that intersection (Figure 6.A7b), we see dynamic results that are similar to Figure 6.6, except the isocline is not strictly linear. The change in the prey isocline from a horizontal line to an ascending function has caused the dynamics to destabilize, and the intersection point is actually an unstable oscillatory point. The predator and prey have expanding oscillations, coming ever closer to the origin, which is to say, ever closer to the extinction of the predator from the system. Because of the precise way in which the predator/prey equations are formulated, it is actually not possible for the predator to become extinct, but it is possible for it to become "practically" extinct in that its numbers go so low that extinction through some random event is almost inevitable. Generally, the closer the predator isocline is to the origin, the lower are the low points in the oscillations of the predator, as illustrated in **Figure 6.A8**.

Recalling the previous graphical analysis, one can easily see that if the predator isocline crosses the prey isocline to the right of the hump of the prey isocline, the oscillations will

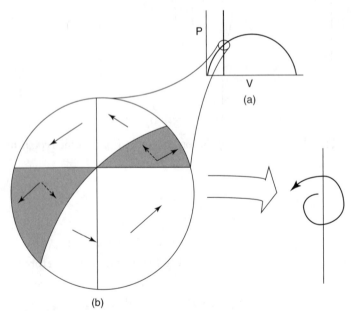

Figure 6-A7 Changing dynamics as a function of adding a functional response to the Lotka-Volterra predator/prey equations.

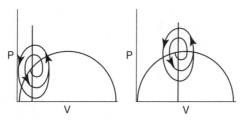

Figure 6-A8   Trajectories of the predator/prey system with predator isocline intersecting the prey isocline in its ascending limb.

dampen (i.e., the equilibrium point will be a focal point attractor). Also, note, however, that this powerful rule works only if the predator isocline is a simple straight vertical line. Finally, the overall dynamics of the system are such that the unstable oscillatory point is usually constrained to a limit cycle that is actually buffered away from the zero point on each axis.

Endnotes

[1]Unless otherwise noted here, as in Chapter 1, all notes about the Ireland tragedy are from Woodham-Smith, 1962.

[2]The treatise of Harris (1841) was undoubtedly an important influence on farmers and scientists throughout this period. Harris simply outlined a catalogue of specific techniques that had worked in the past for controlling specific pests.

[3]Beattie, 1985.

[4]Vanderplank, 1963.

[5]Cushing et al., 2002.

[6]Mundt, 1990.

[7]Mundt, 1990.

[8]Vanderplank, 1963.

[9]De Wolf and Isard, 2007.

[10]Harris, 1841.

[11]Peterson, 1967.

[12]Russell, 1993.

[13]Funes et al., 2002.

[14]Many recent examples have been cited. One of the more spectacular is the Tahitian land snail crisis inaugurated by the introduction of a predatory snail to control a molluskan agricultural pest (Lee et al., 2006).

[15]Arditi and Berryman (1991) were the first to note the general problem of biological control.

[16]Murdoch and Briggs, 1996.

[17]Vandermeer and Andow, 1986.

[18]Mundt, 2002.

[19]Hawkins and Cornell, 2001.

[20]This work is based on my own research, along with Ivette Perfecto and others in Chiapas. A summary of the work can be found in Perfecto and Vandermeer, 2008.

[21]Vandermeer et al., 2008.

[22]Liere and Perfecto, 2008.

[23]Vandermeer et al., 2009.

[24]Holling, 1959.

Chapter 7

Biodiversity, Imperialism, and Agricultural Landscapes

Overview

With a broad historical brush, agroecosystems have touched all spaces on earth, including all areas of the tropics. Whether speaking of centers or "noncenters" of plant domestication, the tropics of America, Africa, and Asia have all seen the mark of this human construction. In this context two transcendent features of the tropics merit comment. First, the industrial agricultural system is still in the process of penetrating tropical regions, having been first introduced through the export production dominated by the European powers, beginning in the 16th century, but only recently spreading throughout more traditional tropical agroecosystems from its birthplace in the United States and Western Europe. Second, the diversity of life reaches its peak in the tropics. The political and economic challenges posed by these two features are daunting—the poorest people in the world occur in the parts of the world that house the bulk of the world's terrestrial biodiversity. Consequently, we face a subject matter that merits analysis separate from other, perhaps more common, frames of reference normally considered as part of agriculture. This chapter reflects these two fundamental features.

First, the general topic of biodiversity as an ecological subject is discussed in its broad context, emphasizing well-known patterns to the extent they are relevant for agroecosystems. This is followed by an analysis of the way in which the industrial agroecosystem penetrated the tropical areas of the world, focusing on the relevant political structures that perpetrated that penetration. Then an extremely important background issue, ecosystem dynamics in space, is covered in some detail, followed by the contemporary reality of how agriculture, especially the most recent forms of industrial agriculture, impacts biodiversity through the creation of fragmented landscapes, including the political issue of how small farmers organize themselves within the context of that fragmented landscape. Understanding the consequences of such fragmentation on biodiversity requires knowledge of recent developments in the ecology of biodiversity, which is covered in the section entitled "the maintenance of biodiversity." Finally, with all of that background, a concluding section summarizes some of the current thinking on the interaction of agroecosystems and the current biodiversity crisis.

Biodiversity

We are currently facing a crisis—the world's biodiversity is being degraded at a phenomenal rate, perhaps far more rapidly than the popular press has reported.[1] Indeed, according to some estimates, we are currently experiencing a rate of extinction that is comparable to the Cretaceous extinction, casting us as a metaphorical asteroid. Let us recall that it took more than 3 million years to recuperate from that event. Yet the good news is that there seems to be universal concern over this surging problem. Who can be openly against the conservation of biodiversity? This *cause célèbre* enjoys universal approbation from all sectors of all societies. (Do you think biodiversity should be conserved? It is difficult to imagine a poll that would find anything other than almost 100% yes.) When constructing cost/benefit calculations, however, the answer changes dramatically. Local farmers in Nicaragua are not likely to conserve the poisonous snake that has slithered into their children's bedroom any more than the International Monetary Fund (IMF) is likely to revise its basic neoliberal approach to save a population of jaguars. The fundamental problem of thinking of biodiversity as nothing more than big charismatic creatures living in pristine environments remains a major obstacle to devising effective programs to engage this truly worldwide problem. Small things like fungi and insects do not get the same exposure as elephants and mountain gorillas.[2] It is becoming increasingly obvious that our great grandchildren will likely live in a world in which African elephants and mountain gorillas are known only in zoos and DNA banks, having been long extinguished from their original habitats in Africa. As sad as these losses will be, such large charismatic creatures represent a minute part of the world's biodiversity compared with the small things—mites, nematodes, insects, to say nothing of bacteria, the biodiversity of which we are only now beginning to appreciate. Furthermore, many of those small things live in nonromantic places like farms, old pastures, and cityscapes.

In light of common misunderstandings about the importance of biodiversity, it is not surprising that most efforts at biodiversity conservation have been failures. Although it is true that some national parks and other biological reserves function well, most are poorly managed or exist only on paper. To take a concrete example, I have been working in MesoAmerica in what is formally designated as an "extractive reserve." I have never met anyone in the area who knows that he or she lives in an extractive reserve—indeed, I have never met anyone in the area who even knows what an extractive reserve is. It looks good on paper, and beautiful colored maps can be posted on prize-winning websites but the problem of biodiversity destruction remains as large as it has ever been, perhaps larger than when the World Bank decided to spend huge sums of money on it.

Given the undeniable fact that we continue with the biodiversity crisis, even after many well-meaning intelligent and even rich people have become concerned, suggests that something is wrong. Although it is difficult to pinpoint the exact problem, one causal agent seems of overwhelming importance—an ignorance of the underlying nature of the problem itself. It is an ignorance that ranges widely, from farmers in underdeveloped tropical regions, to directors of famous botanical gardens, to CEOs of large international environmental NGOs, to officials in powerful international agencies. It is not the ignorance arrogantly assumed by some developed world conservationists that peasants simply cannot appreciate the importance of Quetzales. It is general ignorance of the subject matter itself. Biodiversity is actually an immensely complex subject. It is the fundamental issue that Charles Darwin was trying to explain in the book that defined the field of biology. It is the basis on which our dependence on domesticated organisms rests. It is the yet unanswered question of how many species exist on earth[3]—indeed, the even more fundamental question of what a species is in the first place.[4] It is a conundrum of theoretical biology since some hyperdiverse ecosystems seem to contradict fundamental principles of ecology.[5] It is the well-founded fear of indigenous people that first world entrepreneurs will once again reap gigantic profits off their biodiversity, leaving little behind.[6] It is the vexing question of biodiversity's functional role in ecosystems.[7] It is the question of bet-hedging with future needs for foods, medicines, and other useful items.[8]

When the popular media approaches the issue of biodiversity, the subject matter is almost always about charismatic megafauna— tigers, elephants, pandas, and the like. I too shed tears at the probable extinction faced by these evocative creatures. The world will surely be diminished as the last wild gorilla is shot by a local warlord beholden to one or another political ideology or even as a rare but beautiful bird species has its habitat removed to make way for yet another desperately needed strip mall or fast food restaurant. The irony is gut wrenching to be sure.

Such concerns, however, are a very small tip of a very large iceberg. If we simply take mammals as an estimate of the number of creatures that are likely thought of as charismatic, we are talking about approximately 4,500 known species. By comparison, there are currently about 500,000 known species of insects, and this is almost certainly a gross underestimate. We have no idea how many species there actually are, since estimates range from about 1 million to as high as 30 million. Even if the latter estimate is exaggerated, even if there are only a million species of insects, we see that restricting our attention to the 4,500 species that happen to look more or less like us is focusing on a rather small fraction of the Earth's biodiversity. To make the point even more dramatic, consider the biodiversity of bacteria. Microbiologists define two bacterial cells to be in the same species if their DNA overlaps by 70% or higher, which would likely put all primates in the same species if the same rule were applied. Simply from the point of view of numbers, the world of biodiversity is mainly in the small things, from bacteria to insects, leaving the charismatic megafauna as a rather trivial subplot to the main theme.

Apart from these dramatic taxonomic patterns, there are three basic patterns of biodiversity that are relevant to the subject of agroecology. First, species diversity tends to increase with decreasing latitude, a geographic pattern with profound political consequences. Second, species diversity tends to decrease on islands when the island is smaller and/or more distant from the mainland, an insular pattern with major implications for conservation in today's fragmented world. Third, species diversity tends to decrease as the intensity of management of the ecosystem increases, an intensification pattern. Each of these three cases is treated in turn.

Geographic Pattern: Biodiversity Changes with Latitude and Altitude

One of the most obvious patterns in the biological world is the dramatic difference between temperate and tropical worlds. In a 1-hectare plot of land in Michigan, we have identified 16 different species of trees. In a 1-hectare plot of land in Nicaragua, we have identified 210 different species of trees. Such differences exist for almost all groups of species. The bird guide of Colombia lists 1,695 species, whereas in all of North America (an area much larger), there are 700. Butterflies, ants, mammals, amphibians all show this same pattern. There are exceptions, but the general pattern is one of increasing numbers of species as you approach the equator.

The cause of this pattern has been the subject of an enormous amount of speculation and debate in ecology. More than 20 specific hypotheses that explain this pattern from an ecological point of view have been proposed,[9] yet there is no final agreed-on theory that explains the latitudinal gradient.

The practical political problems associated with this latitudinal pattern hardly need mentioning. It is in the tropics where the bulk of the world's biodiversity lies, and it is also in the tropics where the most destitute people reside, for reasons that stem from the history of the development of agroecosystems, as explained later here. Solving the crisis of biodiversity loss is closely related to solving the crisis of tropical agriculture, which in turn is closely related to solving the crisis of inequality—inequality of wealth, income, and power.

Insular Pattern: Biodiversity on Islands

One of the most noted biodiversity patterns is on islands. Generally, more species exist on larger islands and more species on islands nearer to the mainland. Nevertheless, exactly how many more species will be found on an island twice the size of another island remains highly variable. Despite this variability, it is a general rule that if you plot the logarithm of the number of species against the logarithm of the size of the island, you get what seems to be a linear plot. This suggests that the underlying relationship follows a "power law," where the actual number of species is a function of the area raised to a *power*. This relationship puts biodiversity in the same league with many other natural phenomena such as earthquakes and sand piles. Per Bak's delightful and modestly titled book *How Nature Works* fails to mention this natural phenomenon, but it clearly fits snugly with his catalogue of things that do follow power laws and perhaps suggests that some sort of "self-organization" is in play here. Exploration of this interesting idea is beyond the intended scope of the current book.

More mundane explanations of the pattern generally fall into three categories. First, it may be simply a sampling problem. As discussed later, this is certainly part of the explanation and is extremely important when it comes to the difficult issue of sampling biodiversity. Even today, the technical literature is filled with the elementary error of determining species richness in a series of samples and then taking the average of those determinations as the estimate of species richness in that environment. Because species accumulate nonlinearly with increasing sample size, this procedure is fundamentally wrong and can give highly misleading results.

The second category of explanation has to do with environmental heterogeneity. It seems obvious that a larger area, be it a sampling unit or an island, will likely contain more microhabitats than a smaller area and thus provide more "niche space" for species to fill. In the end, this explanation falls within the general category of differing extinction rates (if a species arrives at an island and fails to locate its habitat, it goes extinct)—part of island biogeography theory.

The third category is an equilibrium balance between migration and emigration rates, first fully elaborated by Robert MacArthur and E. O. Wilson in their classic work *The Equilibrium Theory of Island Biogeography*. The basic idea here is that populations of organisms are always dispersing in one way or another. Any area is thus likely to receive migrant individuals at regular intervals, whether in the form of a dispersed seed of a plant or a migrant individual bird or bat or the propagule of a microorganism. On an island, one could, theoretically, sample all of those incoming organisms and determine which ones were new to the island and which ones had been living on the island already. The rate at which *new* species arrive on the island is the immigration rate. Also, if you had enough helpers and technology, you could keep track of all the organisms on the island and thus know when any particular species went extinct on that particular island. Then, if the rate of extinction is greater than the rate of immigration, the total number of species on the island would decline. Similarly, if the rate of immigration were greater than the rate of extinction, the number of species would increase. The perfect balance, the equilibrium number of species, is when the extinction rates and immigration rates are equal. The details of this theory are outlined graphically in **Figure 7.1.**

This third category, the equilibrium theory, has been employed in many different guises in ecology and, especially today, is implicated in a variety of practical problems ranging from the ideal design of a nature reserve to the underlying reason that tropical rain forests have so many species of trees. It is also a major feature of the most recent refocusing of the conservation agenda, as described later in this chapter.

Intensification Pattern: Changes in Biodiversity with Management Intensity

More than 50% of the terrestrial surface of the earth is covered with managed ecosystems (i.e., the terrestrial surface not covered in ice). In the popular and romantic conceptualization of nature as a Garden of Eden, many conservationists think of agriculture as the defining feature of biodiversity loss. The world is divided into those areas untouched or minimally touched by *Homo sapiens* as contrasted to those areas "despoiled" by human activity. One of the main observations that caused a reevaluation of this bias was the correlation between the decline of populations of songbirds in the Eastern United States and the transformation of the coffee agroecosystem of Central America. As discussed in Chapter 1, the traditional method of coffee production includes a diverse assemblage of shade trees with the coffee bushes growing as if they were part of a forest understory. These coffee plantations appeared to be forested land if viewed from above, and as has now been convincingly demonstrated, they are important habitats for those very bird species from North America when they migrate south for the winter. This key observation

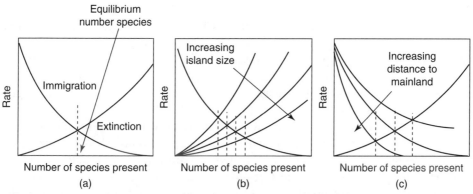

Figure 7-1 Elementary theory of island biogeography. (a) The basic curves. Immigration rate refers to the number of *new* species arriving per unit of time, which is why it is a decreasing function of the "number of species present" (if all possible species are already present, any species that arrives cannot be new, whereas if no species are already present, any species that arrives is new). Extinction rate refers to the number of previously extant species that completely disappear from the island per unit of time, which is why it is an increasing function of "number of species present" (if there are no species on the island, the number of species that are lost must be zero, whereas the rate is maximum when there is a maximum number of species already present). Where the two curves cross, we expect an equilibrium point (to the left of that point the immigration rate is greater than the extinction rate so more species are added, whereas to the right of that point, the extinction rate is greater than the immigration rate and species are lost—thus, the overall tendency is to approach the point where immigration is equal to extinction). (b) The effect of increasing island size on the equilibrium number of species on the island. The bigger the island, the lower is the extinction rate and the more species at equilibrium. (c) The effect of increasing distance of the island to the mainland. The further away the island from a mainland source for species, the lower is the immigration rate and the fewer species at equilibrium.

has been significant in demonstrating to the world that agricultural ecosystems can be critical repositories of biodiversity, but even more important, that the particular type of agricultural practice was a determinant of the biodiversity contained in the agroecosystem. Not all coffee plantations harbor high levels of biodiversity, and the characterization of what types of agroecosystems generally harbor greater or lesser amounts has only recently emerged as a serious scientific question.

When dealing with managed ecosystems, it is first necessary to distinguish between two concepts of biodiversity. First is the collection of plants and animals that the manager has decided are part of the managed system—rice in the paddies of Asia, corn and beans in the traditional fields of Native American Mayans, carp in the fish ponds of China, and so forth. This is referred to as the "planned" biodiversity. In contrast, in each of these ecosystems a great amount of biodiversity spontaneously arrives—the aquatic insects and frogs in the Asian rice paddies, the birds and bugs that eat the Mayans' corn and beans, the crayfish that burrow their way into the sides of the Chinese fishponds. This is referred to as the "associated" biodiversity. Frequently, the managers are determinedly concerned about the planned biodiversity, especially when dealing below the species level (i.e., genetic varieties of crops). However, it is almost certainly the case that the associated biodiversity is the most abundant in almost all managed ecosystems, and as such, it has received a great deal of attention in recent years.

The process of intensifying agriculture has come to provide a conceptual base for discussing the role of agriculture in biodiversity studies, especially when concerned with associated biodiversity. This conceptual tool derives mainly from anthropology and has to do with the presumed historical record of agricultural evolution. Recall the Haorani of the upper Amazon, previously discussed with respect to agricultural origins. They are mainly hunters, monkeys being their main source of protein, with a diversity of other animals and fruits gathered from the forest, providing a relatively rich diet. They also casually plant cassava, usually in naturally caused gaps in the forest, and they maintain at least a mental map of the position of their peach palm trees, almost all of which were planted by previous generations of Haorani. Thus, the forest contains scattered peach palm trees where the Haorani formerly made a small clearing, the forest having taken over that clearing. So part of their activities is "gathering" the peach palms from trees they had planted earlier. If this is agriculture, its intensity is certainly not very high, and we would expect the loss of biodiversity going from a natural forest to one with scattered peach palms would be minimal.

Other groups of hunters and gatherers are known to provide a low level of husbandry for the plants they gather, encouraging some species and discouraging others. From such austere beginnings, other technologies emerge—the provisioning of water in dryer areas, the burning of fields to eliminate weeds, and ultimately the mechanization and chemicalization that characterize our modern agricultural system.

With these observations, it makes sense to think of agriculture not as suddenly emerging from a pure hunter and gatherer existence, but rather as a slowly evolving intensification of cultural activities—from the casual incorporation of peach palm into a forested matrix, where it would be difficult to demonstrate any effect at all on the biodiversity of the region, to the establishment of a modern tulip plantation on land wrested from the sea in The Netherlands, where it would be difficult to find any biodiversity except for the tulips. It is a gradient of intensification, not a dichotomy of human versus nature.

Although the term *agricultural intensification* has a very specific and complex definition in anthropology, in the biodiversity literature the term *management intensification* is taken to be the transition from ecosystems with high planned biodiversity to low planned biodiversity. Thus, for example, in the case of coffee, intensification refers to the reduction of shade trees eventually ending up with an unshaded coffee monoculture. The ecology of agroecosystems is such that the final stages of intensification usually involve the application of agrochemicals to substitute the function of some of the biodiversity that is eliminated.

An obvious question that has only recently become important is: What is the pattern of associated biodiversity change as a function of the intensification of agriculture? This question remains largely unanswered for almost all agroecosystems and almost all taxa. The few studies that actually pose this question come up with results that depend on the taxon involved and even the definition of what constitutes greater or lesser intensity. Nevertheless, it is possible to make one central generalization. There are two basic patterns of associated biodiversity change that might be expected as a function of intensification (**Figure 7.2**). First, as tacitly assumed by many conservation practitioners, as soon as a natural habitat is altered by some management system, associated biodiversity tends to fall dramatically. Second, associated biodiversity declines by only small amounts with low levels of intensification and only after much higher levels are reached do we see dramatic declines. Which of these two patterns (or what combination of the two) exists in particular systems is largely unknown because this question has not been popular among those concerned with biodiversity—one of the largest intellectual lacunas in the history of the biological sciences.

Indeed it is frequently the case that conservation practitioners pay little attention to agroecosystems or forestry systems or aquatic managed systems. The assumption seems

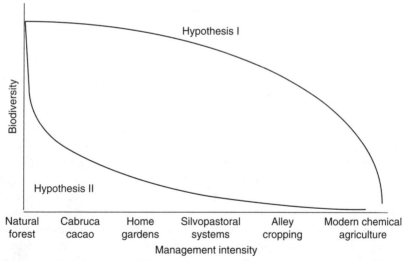

Figure 7-2 Two hypotheses about the relationship between management intensity and biodiversity. Some classic systems are represented on the abscissa, but they have been arbitrarily chosen. Other possibilities would include a range of coffee production systems as suggested in Chapter 1, from rustic coffee to full sun coffee or maize production from informal planting in natural forest light gaps to classic slash and burn agriculture to modern industrial production.

to be that once management activities are initiated in an area, the question of biodiversity becomes irrelevant, a point of view arising from a combination of romanticism and ignorance. Yet it is a point of view that could have devastating consequence when it comes to biodiversity conservation itself. If a high percentage of the world's surface is covered in managed ecosystems and if managed ecosystems contain even a small fraction of the biodiversity contained in unmanaged ecosystems, ignoring them will be counterproductive, to say the least.

The Problem of Sampling Biodiversity

As described previously, a plot of the log of the number of species as a function of the log of the area of an island usually results in a straight line—that is, the relationship is a power function of the form

$$\ln(S) = c + b \ln(A) \tag{7.1}$$

where a and b are constants, A is the area of the island, and S is the number of species. The parameter c refers to the log of the number of species in a unit area (because $\ln(1) = 0$), and the parameter b refers to the multiplicative effect of area on number of species. That is, rewriting Equation 1 as

$$S = S_1 A^b$$

S_1 refers to the number of species found in a unit area. Although this relationship seems to describe the pattern seen on islands fairly well, its true importance is in estimating procedures. For many reasons, it is of interest to know the biodiversity of a sampling area. It is not difficult to see that this problem, the biodiversity of a sampling area, is similar to biodiversity on islands, where different size islands have different numbers of species. Indeed, the pattern seen on islands could be nothing more than a reflection of a sampling program, where the species on an island represent a sample of a larger ecological community the biodiversity characteristics of which will be known only after a very large area is sampled. If the basic rule of nature is Equation 7.1, we face the problem that species diversity is related to area in a nonlinear fashion.

When I first started working in the tropics, I was told that one of the exceptions to the rule that tropical biodiversity is greater than temperate diversity is the biodiversity of aquatic insects in running water. Indeed, it is remarkably easy to see this pattern. If you pick up a rock out of a stream in northern Michigan, you find it covered with all sorts of aquatic insects: caddis flies, mayflies, stoneflies, and so forth. A single rock picked from a tropical stream frequently has nothing and usually at most a single species of some aquatic insect. It thus seems true and obvious that the tropics are depauperate with respect to tropical insects. However, picking a rock from a stream is similar to asking what is the value of S_1 (the number of species in a unit area, the unit area in this case being the rock).

A detailed examination of the way species accumulate as area (number of rocks in this case) increases shows that the parameter b is indeed greater for the tropics than for the temperate zone.[10]

Generally, if we wish to compare the biodiversity of two treatments (e.g., comparing associated biodiversity in an organic versus an industrial agroecosystem), if S_1 in the first (i.e., the industrial) is greater than S_1 in the second (the organic) but b is greater in the second than in the first, the size of the sampling unit becomes of critical importance. This point is made clear in **Figure 7.3**, where taking a large number of samples of size 1 and then taking the average of those samples would suggest that assemblage 1 had on average 1 species, whereas assemblage 2 had 1.3, even though assemblage 1 has a much larger number of species than assemblage 2.

The basic power law of species diversity seems to hold reasonably well in practice,[11] but it does imply an inconvenient truth—that species number increases indefinitely with increasing area. Most ecologists do not feel this to be reasonable. Consequently, alternatives to the specific form of Equation 7.1 have been used, sometimes with great success. Recently, computer-based extrapolations have proved to be extremely useful[12] and have effectively avoided the need for inventing a special equation (although it is still remarkable how closely nature approximates Equation 7.1 in most situations). These new techniques are effectively based on a simple proposition. If you have already sampled all of the species in an area, a further sample will include no new species. Indeed, further sampling will result in every species represented by more than a single individual. If one of the species in the sample is represented by a single individual, take another sample and another and another, and if indeed all of the species have already been sampled, eventually you will encounter another individual of that species represented by a singleton. This means that the singletons (species represented by a single individual only) will no longer

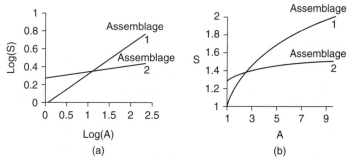

Figure 7-3 Basic relationship between area sampled and number of species in an assemblage (community). (a) Log–log plot illustrating the power function associated with two distinct assemblages of species. (b) Same data on an arithmetic plot. Note in the case of assemblage 2 that the curve seems to be "leveling off," whereas in the case of assemblage 1, it does not seem to be leveling off, suggesting that the sample size of nine is more or less adequate for assemblage 2 but not at all so for assemblage 1.

be singletons if enough sampling is done (and because all species are now represented in the sample, no new singletons can appear). Without going into details, relating the way singletons are converted to doubletons, doubletons to tripletons, and so forth, and the way these transformations are related to the accumulation of new species (usually singletons) with further sampling, is the basis of most of these new computer-based techniques.

Nevertheless, it warrants repeating that the fundamental nonlinear pattern of accumulation of species with area means that normal sampling, such as what one might do with forest tree biomass, or agricultural productive units, is fundamentally wrong when it comes to biodiversity. This error is frequently made when making biodiversity comparisons, at least in the past. A particular sampling unit size may give the wrong results, no matter how many times it is replicated (see Figure 7.3).

How Industrial Agriculture Penetrated the South

Background Structures

Biodiversity has long been a sort of *raison d'etre* for biology. Indeed, the Darwinian program is a response to the question of where all the biodiversity comes from. The tropics have always been a magnet for this interest. Tropical explorers from Alexander von Humbolt to Alfred Russell Wallace to Darwin himself gained inspiration, both spiritual and scientific, from their observations of the amazing biodiversity contained in tropical regions. Obviously, and a bit ironically, the peoples and cultures of the tropics have long related to that biodiversity in complicated and sophisticated ways, even if Western science has developed its own somewhat exclusionary framework that tends to exclude the peoples and cultures that have evolved in the context of that biodiversity. The most direct contact that tropical cultures have had with biodiversity is through agriculture. Indeed, one of the more important contributions to the move toward a modern sustainable system comes from lessons learned from Indian traditional farmers by the Englishman Albert Howard, as discussed earlier in the context of the process of decomposition.

Although the framework of sustainable agriculture includes an important suite of intellectual inputs from traditional farmers in tropical zones, the main features faced by today's agroecosystem transformations stem from the penetration of the industrial system into the tropical areas of the world, usually referred to in political economic circles as "The Global South" (in the recent past called the Third World). The complex sociopolitical tapestry into which this system penetrated needs to be understood, at least at a basic level to comprehend fully both the causes and consequences of the current state of affairs. Those not familiar with Wallerstein's basic framework of the "World System" and the fundamental ideas of articulation and disarticulation in socioeconomic relations should review the information in the appendix to this chapter before proceeding.

Like the rest of the modern capitalist economy, industrial agriculture is really an international affair, and a critical feature of that international structure is the relationship between the developed and underdeveloped world. The decolonizations of the past century created what in effect remains colonialism, but of a dramatically different form. The former colonies are now members of the Global South and retain important remnants of their colonial structure. As difficult as life may be for some citizens of the developed nations, one can hardly fail to notice a dramatic difference in conditions of life in the South. The South is where dangerous production processes are located, raw materials and labor are supplied to certain industries at ridiculously low cost, air and water pollution run rampant, people live in desperation, and talk of bettering the state of the environment is frequently met with astonishment—"how can you expect me to worry about tropical deforestation when I must spend all my worry time on where I will find the next meal for myself and my children." Such obvious qualitative differences are reflected in all major statistical measures. About 80% of the world's population lives in the Global South, yet more than 70% of the world's energy is used by the 20% that live in the North. Income distributions, standing wealth, and almost any other measure you care to contemplate show similar patterns.

The contemporary world is undoubtedly engulfed in a major transformation, with China and India growing at unprecedented levels, petroleum prices at record volatility, world food prices first being so low that small farmers could not survive and then 1 year later (2008) becoming so high that poor citizens the world over could hardly afford to eat. Under such obvious complications, the future cannot be predicted with certainty. Even here, in this volatile section of the world political system, suggestive patterns are already being revealed. Reports of peasant uprisings in China, an epidemic of suicides by Indian farmers, and other indicators suggest that the breakneck speed of development in these two countries entails internal contradictions that have not been worked out. Oil price swings suggest that alternatives to the industrial system of energy supply may soon become attractive. Peasant agriculturalists in the Global South are forming political organizations to confront their looming problems. The volatility in food prices has provoked a serious global debate about how food is produced.

While acknowledging that the current situation is likely to provoke substantial changes in the near future, certain consistencies are prerequisite to understand the current state of agriculture in the Global South. For the most part farmers in the Global South find themselves in one of two obvious social classes: those who produce cotton, coffee, tea, rubber, bananas, chocolate, beef, sugar, and many other crops for export and those who produce food mainly on their own small farms for their own consumption and local markets and, when necessary, provide the labor for those who produce export crops. Superficially, these groups appear similar to the factory owner and worker of the Developed World system (see appendix), but the similarity is skin deep. Yes, owners run "factories" (e.g., the tea or banana plantations) and the workers work (or have worked) in those "factories." But the owners are concerned with selling the products of their production process not to the workers and people like them, but rather to the workers (and owners) in the First World. The export producer of the South does not wear two hats with regard to the laborers on his or her farm, as his or her counterpart in the Developed World (see Appendix 7.A). The concern is not to sell the products to local laborers but rather to the laborers in the Developed World. This means that the dynamo of economic growth represented by the contradictory goals of the factory owner in the North simply do not exist in the South. The factory owner in the North must be concerned with the general economic health of the working class, and consequently there is social pressure to maintain consumption power in that class, even while trying to drive wages as low as possible in his or her own enterprise. There is general agreement that although such an arrangement tends to produce economic cycles of boom and bust, it also represents the mechanism of economic growth and has therefore been the base of development. That very base of development has not existed, to a significant extent, in the Global South.

Part of Wallerstein's overall framework includes the so-called peasant sector, that class of people who move in and out of the world system and effectively provide it with a convenient buffer. Move to the city to find a job in the newly industrializing sectors or closer to home in the large fruit plantations, but keep the grandparents on the farm. Then, when fired from the job, go back to the farm, thus providing the social safety net the global system requires. Although this system may be failing,[13] it has been an important historical force with regard to the large-scale development of agricultural landscapes, especially with regard to biodiversity, since most of the underdeveloped world is located in the tropics and most of the world's biodiversity is in the tropics.

Most regions are highly variable, and thus, it is difficult to generalize about either the ecology or the politics, but certain general patterns do seem to exist. First, the soils in tropical areas tend to be quite poor, with the main concentration of the world's oxisols and ultisols located there. The most productive of tropical soils are the andisols, mainly located at slightly higher elevations, and the entisols and inceptisols based on alluvial primary material. However, when initiating a farm, it is not always obvious what the character of the soils might be. Agricultural expansion tends to occur along rivers, which means that earlier farms tend to be located on alluvial plains. As they push further away from the river, the soils tend to get poorer, with high acid reaction, aluminum toxicity, problems with phosphorous, and low organic matter. When they are exposed, the frequently high concentrations of iron oxidize and form concretions that range from small pebble-sized structures to so-called ironstone formations that form a sort of cap on the surface of the soil.

In areas of moderate to high relief, agricultural expansion is also limited by access, with ravines and peaks difficult to farm and thus retaining natural vegetation in a patchwork located in an agricultural matrix. As discussed in an earlier chapter, more traditional modes of agricultural development in such areas employ fallows for "recuperating" or "resting" the land. In its most traditional form, the fallow land is allowed to return to a vegetation type that is far from what the climax would be, and the entire agricultural system includes extensive areas of natural vegetation, albeit not in a "pristine" form. This form rarely occupies the frequently extensive areas of primary habitat, mainly because through tradition they have discovered where all of the good soils are and simply do not undertake to farm on the poorer soils, thus leaving extensive areas of natural vegetation.

Modernism has generated a new pattern. Farming communities that have given their land over to large export-oriented agricultural concerns, such as large fruit companies, are forced to move on to lower quality soils, which means they need a longer fallow time for resting the soil, which means they need a larger area, which means they convert larger areas of natural vegetation. But, in general, this new kind of "migratory slash and burn" is highly unpredictable, and necessary fallow times are truly huge—far greater than had come to be the norm when they were generally located on andisols or alluvial entisols and inceptisols. Such farms tend to fail, and thus we see a new class of people facing a kind of double jeopardy. On the one hand, they are forced into ecologically marginal areas, and on the other hand, they are faced with unpredictable employment opportunities on the very plantations that usurped the good land to start with. The seemingly stable situation of leaving grandparents on the family farm as backup security while you go and make your fortune in the banana plantation is extremely unstable over the long haul. Even if political circumstances did not intervene to make things far worse (as has happened first with the political requirements of the Cold War and more recently with the expansion of the neoliberal model, as discussed later), the situation of displacing traditional peasant agriculture to marginal soils would have been unsustainable in the long term anyway.

The Green Revolution

In Chapter 2, the origin and spread of the industrial system was presented as effectively a response to a looming overproduction (or underconsumption) crisis that presented itself at the end of World War II. The appropriated inputs (e.g., chemical fertilizers, insecticides, improved seed varieties, machinery) became dominant elements of the overall agricultural system, as did substituted products (e.g., peanut butter for peanuts, tomato concentrate for tomatoes, corn syrup for sugar). Thus was created that famous agricultural system in which "manufacturing peanut butter from petroleum" replaced "growing peanuts."

As part of this same trend, industrial enthusiasts looked to the Global South for yet further market outlets. This need dovetailed quite well with the global political climate. The Western capitalist countries legitimately feared the promise of communism. The option for the poor that communist ideology provided, at least in theory, was seen to be seductive for the downtrodden. Food and agriculture were essential in the struggle for hearts and minds as the Cold War gained traction. Hungry people could be the "victims" of communist propaganda, as could farmers unable to support themselves. Projects to counter this possibility emerged as early as 1954 when the Eisenhower administration initiated Public Law 480 (PL-480). As U.S. agriculture was undergoing the massive transformation initiated by the underconsumption crisis that emerged from the wars, it became evident that the oversupply of grains resulting from the expansion of the industrial model would lead rapidly to another form of overproduction, that of the grain itself. This clear politicoeconomic necessity fit well with the new fear of communist ideology. PL-480 solved both problems. Grain would be shipped to "cooperative" countries. As insurrections emerged repeatedly in the Global South, the joint application of food aid through

PL-480 and military intervention (both overt and covert), along with strategic cartel formation, such as the famous coffee cartel,[14] provided an effective package. Exporting food thus became a central feature of U.S. foreign policy, along with military intervention, and by 1956, food aid accounted for more than half of all U.S. foreign assistance.

Thus, by the early 1960s, the full industrial model was applied in the Global North to produce an enormous surplus of food, much of which was exported under PL-480, solving both the underconsumption crisis for companies like Cargill and serving the interests of the West in the global political struggle of the Cold War. This would all change with a new technological package. Recall the history of the tomato mechanization program undertaken by Jack Hanna and Toby Lorenzen of the University of California. Hanna was breeding a tomato to be harvested for a machine that did not exist, and Lorenzen was designing a machine to harvest a tomato that did not exist. The philosophy of technological development in agriculture is well illustrated by this program, where complex and interconnected technologies are developed in consort with one another to create a "technological package," as discussed more fully in Chapter 3.

Operating within this developmentalist framework, agronomist Norman Borlaug working at the International Maize and Wheat Improvement Center outside of Mexico City, began a breeding program that took chemical fertilizers, pesticides, and plenty of irrigation water as underlying assumptions about the ecological background in which improved varieties of wheat needed to be developed. The degree to which any of the individual specific inputs contributed to the dramatic increases in production is debatable.[15] What is not debatable is that the technological package was potent and yields increased by as much as fivefold compared with production technology that used none of the new inputs of the technological package. This package, not all that different from the package now commonly used in much of the Global North, became a weapon in the Cold War.

The trend of exporting the basic industrial model to the Global South was a more-or-less continuous process as manufacturers of fertilizers, pesticides, machinery, and seeds independently moved their marketing operations there. However, as a political movement the Green Revolution was born in July of 1965 when political tensions between the United States and India flared.[16] A new Indian Prime Minister, Lal Bahadur Shastri, took a critical view of the U.S. bombing of Vietnam, to which the Johnson administration responded with a threat to cut off PL-480 food aid. This was a major threat since India had already lost much of its food-producing capacity due to the massive PL-480 distribution, which had effectively created an oversupply and dramatically lowered local prices, always a disaster for local farmers. PL-480 contracts were now renewed on a monthly rather than yearly basis, and it was made clear that India's attitude toward U.S. foreign policy in general would be a litmus test for continuation of the program. This was the stick. But the United States also offered a seemingly unassailable carrot. If India would become a more pliable and cooperative political partner, the United States would provide not food but the new technology to produce it, namely the fertilizer/pesticide/machinery/irrigation/seed technologies, in short, the industrial system, to select farmers in India. This massive transfer of technology, first to India and then to the rest of the Global South, became known as the "Green Revolution."

The actual performance of this new technological package remains a matter of considerable debate. Yields up to five times pre-Green Revolution technology have been reported, but it is often obscure as to what is being compared. To be sure, if we begin with seed not necessarily adapted to conditions on the farm, with low nitrogen content in the soil, with low cation exchange capacity, with insect and disease pests having built up in the local environment, and with a lack of water in a drought-prone area, adding the fertilizer/pesticide/machinery/irrigation/seed technologies to the operation is almost certainly going to result in massive yield increases. If those industrial inputs are greatly subsidized (which they were), economic yields will undoubtedly also increase dramatically. It is similar to comparing advanced technology to no technology and becomes a no-brainer. In contrast, if the comparison were to be made to an alternative technological package where agroecological techniques were used to promote soil fertility and pest management, where small-scale irrigation projects were initiated at a local level, where traditional varieties of crops adapted to local conditions were used, and where machinery was adapted to local small-scale farming operations, it is not at all clear that biological yields would be greater with the industrial package, to say nothing of potential economic benefits. Because it is now common knowledge that organic production in fact provides biological yields equivalent to the industrial model,[17] accepting the notion that the Green Revolution was a success remains, in my view, a controversial position.

On the other hand, from a strictly historical point of view, that revolution was one of the more successful in history. The industrial system did indeed penetrate the Global South. It is now the dominant ideological form of agriculture throughout the world, regardless of its actual performance. The average farmer in Latin America, Asia, and Africa does indeed want the fertilizer/pesticide/machinery/irrigation/seed technologies that constitute the industrial system. However, as we shall see, the new farmer's movements have slowly but surely developed not only a political position toward land ownership and economic security, but a critical view of the Green Revolution/Industrial System and are ever more frequently openly talking about agroecological technologies.

Neoliberalism and the Political Response

The global system continues evolving, even as the basic power structures remain relatively constant, ever consolidating their political power, yet perhaps facing ever more serious crises, as has so frequently happened in past global power shifts. Since the Cold War, the tendency has been continual penetration of global markets, using trade treaties to gain economic advantage when it is not there to start with and to protect that advantage when necessary. The vehicle that replaced the fear of communism during the Cold War has become the mantra of "free trade" and "democracy," with the added spice of "terrorism" since 2001. In all cases, the terms are slogans not intended to convey any serious philosophy, other than the continued operation of the World System as it has operated at its foundations since the beginnings of capitalism.

Unfettered from the necessities of maintaining a bulwark against the Soviet Union, the United States and its allies began a system of economic expansion based on so-called free trade and democracy that became known throughout the Global South as the "neoliberal model," an ideological framework based on a rejection of Keynesian governmental regulations and an unbridled faith in the theory of unfettered market forces.[18] Ideology aside, the practical plan has three components (see Appendix 7.A). First, the World Bank is used as a carrot to provide developmental capital for development projects, inevitably using U.S. (sometimes European) companies as outsourcing, sometimes covertly, frequently quite openly as part of the loan package (the infamous support of Haliburton by the second Bush administration is only the most blatant of what had been normal procedure for many years). Second, the International Monetary Fund organizes "structural adjustment programs" that usually call, among other things, for the privatization of state-controlled industries, frequently making them available at fire-sale prices to U.S. and European corporations. Acceptance of the IMF structural adjustment programs is taken to be the signal for private banks to begin or resume lending to a now stable, or stabilizing, economy. Third, the World Trade Organization holds out the promise of access to developed world markets, as long as its rules are followed. Litigation through the World Trade Organization can be lengthy and expensive, again providing developed countries with a natural advantage.

In the context of this neoliberal model, the Clinton administration took on the North American Free Trade Agreement (NAFTA) (creating a semifree trade zone among the three North American countries), initially crafted by the first Bush administration, and danced it through the U.S. Congress in the face of massive opposition from labor groups all over the country. The main concern of U.S. labor was the potential loss of jobs as Mexican labor became available to formerly U.S.-based manufacturing industries. From an agricultural point of view, however, the consequences of NAFTA could easily be seen to be even more devastating. To see this, we must return to two historical events that preceded the passage of NAFTA in 1994.

First, in the United States, the entire program of supply management that had been the theoretical basis of government subsidies to agriculture began to be eroded in the 1960s, and by the time of the Reagan administration, the system had been effectively reversed. Supply management subsidies were originally designed, at least in theory, to pay farmers to hold back production so that supply and demand could come to equilibrium at a price that would cover the cost of production and guarantee farmers at least a minimum profit. Whether the various attempts at supply management in the United States were ever successful remains debatable. A glance at the underlying theory behind supply management, however, leads to some obvious conclusions. Monopsonized grain and other product purchasers clearly would lose out on this arrangement, even though average farmers would be supported, with the market setting prices that at least covered their cost of production with a bit of a profit on top. However, this ideology changed, slowly at first, but finally completely under the Reagan administration, such that subsidies were offered to cover shortfalls in selling price compared with the cost of production. This effectively provided the farmer minimal insurance against low prices and encouraged fencerow-to-fencerow planting and increasing overproduction in commodity after commodity. As expected, prices collapsed in the face of excessive overproduction, and a state of massive overproduction in almost all agricultural commodities was the result.

One player benefits excessively from the arrangement, the giant grain companies—entities such as Cargill and Archer-Daniels-Midland. Now able to purchase grain at well below the cost of production, their grain silos filled and their profits soared. Eventually, however, they faced the underconsumption ceiling—excess grain that could not be sold (only so many cows to feed and corn chips to dip with). Following the now familiar script, they began looking for markets elsewhere and eventually turned their eyes southward, toward the countries of the Global South. However, before the triumph (still only partial) of the global neoliberal model, most countries of the South had tariff laws in place to protect their farmers from what would be effectively predatory pricing. To solve that problem, in stepped NAFTA.

The second antecedent was in Mexico, with the modification, or even effective elimination, of Article 27 of the Mexican Constitution. Article 27 had set in place the basic Mexican system of ejidos—collective land that guaranteed every peasant producer land security. Part of the arrangement was that land in the agrarian reform program could not be bought and sold. As part of the liberalization of the neoliberal model, this restriction needed to be eliminated to permit Mexico to enter into the NAFTA agreement. Then-president Salinas lobbied forcefully and eventually won the day, and Article 27 was effectively eliminated, paving the way for massive displacement of peasants from their land.

Many analysts, including many peasant leaders, could see in the pipeline highly subsidized grain (especially maize) from the United States and Canada, streaming into Mexico, displacing the peasant small farmers from their traditional markets. Because the giant grain companies of the United States were effectively receiving massive subsidies so that they could purchase grain at well below the cost of production, it could be expected that they would penetrate Mexican markets, if the traditional tariffs were eliminated, as was required in NAFTA. Selling maize below the cost of production of Mexican peasant farmers would destroy them.

In response to NAFTA, a remarkable peasant army, calling itself the "Zapatista Army of National Liberation" (EZLN, more commonly known as the Zapatistas), marched out of the mountains on January 1, 1994, the day that NAFTA came into effect, captured several towns in the Chiapas highlands, including the major tourist destination of San Cristobal de las Casas, and declared war on the Mexican Government. This particular revolution was remarkably different from the typical Central American revolutionary movement and captured the hearts and minds of a surprising number of Mexicans of all social classes. Its spokesperson, subcomandante Marcos ("sub" because "the people" represent the real commander), was a masked enigmatic figure who talked more in postmodern philosophy than spirited revolutionary jargon. At this writing, the EZLN and the movement it represents still exists and plays an unpredictable role in Mexican politics.

Regardless of the revolutionary protest of the Zapatistas, the predictions of the anti-NAFTA analysts played out almost perfectly. The highly subsidized maize from the United States began streaming into Mexico, undercutting the traditional Mexican maize producer. Yet the growth in employment opportunities that would have absorbed these newly displaced rural peasants was less than expected, such that the major cities swelled with surrounding shantytowns containing the rural migrants. Underground economies grew, and the usual social problems associated with growing urban poverty have become a major burden on the population. Pressure to migrate to the United States continues to increase as of this writing. As expected, the winners, the giant grain companies, have benefited enormously (e.g., The Associated Press reported that Cargill had earnings of 553 million dollars during the quarter ended February 28, 2007, up from 370 million during the same period the previous year). Mexican maize farmers are going out of business and are migrating to the shantytowns and now to the United States in unprecedented numbers. A coalition of Mexican farmers called "The Countryside Can Bear No More" formed in the late 1990s and staged strikes and even stormed the Mexican congress. And U.S. farmers face the threat that their subsidies will be cut at the same time that overproduction has driven the price they get for their grain to record lows.

The agrarian crisis created by the Industrial System also spawned an international movement of small farmers and farm workers, known as the Via Campesina (farmer's way). Via Campesina is an umbrella organization that represents 91 (at this writing) small farmer and farm worker organizations across the globe. Explicitly opposed to the neoliberal model, this organization has become a major voice in international agriculture, present at all major political demonstrations. The structure of Via Campesina is complex and under continual development, the details of which are beyond the scope of this text. However, a relevant, perhaps revolutionary, idea was spawned by this group of small farmers, a new philosophical framework that has the potential to completely transform global food and agriculture questions—"food sovereignty." Since the United Nations declaration of "food security" in 1994, the existence of hungry people has been an embarrassment for most governments. Accepting PL-480 and other food aid from the developed world was one way to meet the "security" needs of people's bellies. But the old adage that to give a man a fish means he eats for a day while teaching him to fish means he eats for the rest of his life rings true in most people's minds and forms the basis of this new agrarian philosophy promoted by the Via Campesina. Food sovereignty means not only the right of a region to sufficient food, but also to its own means of producing that food. The idea of food sovereignty, perhaps the most revolutionary idea in agriculture since the birth of the industrial model, is explored in more detail in Chapter 8.

The penetration of the industrial system into the Global South has led to a variety of forms of environmental degradation. On the one hand, displacing peasants to more marginal lands causes deforestation and massive biodiversity loss, the theoretical enemy of conservationists for many years (although too frequently conservationists see only the peasant farmers cutting forest rather than the overall system that causes them to do so). By global calculations, there is still enough food produced worldwide to feed every human being on the planet. At a global level, ignoring the politics, there really is no reason to convert another piece of forest for agriculture. The fact that we still see major deforestation of tropical forests in the Amazon, the Congo, and Southeast Asia is a consequence of the distribution of food, creation of poverty, and wealth and corrupt political policies, and has nothing to do with the need for more food production. On the other hand, farmers abandoning their lands, as is happening not only in Mexico but in many other countries of the Global South, while creating unprecedented social problems, has also created some land that is abandoned, regenerating, sometimes very slowly, the original natural vegetation. In a perverse way, one could say that the neoliberal model has been at least partly positive for biodiversity preservation, a controversial position taken by some conservationists.[19]

More generally, a set of complex problems, stemming from basic sociopolitical dynamics, needs far more attention than it currently gets. It is the problem of agricultural transformation at a landscape level and the spatially extended ecological dynamics that are inevitably involved therein. One of the major lacunae in the literature is the intersection of first, the sociopolitical problems that have their origin in the Wallersteinian international system but play themselves out at a regional landscape level, and second, the emerging principles of landscape ecology, including technical themes such as metapopulations, metacommunities, and spatial dynamics.

Spatial Ecology—Metapopulations and Metacommunities

Elementary Principles

Classic population ecology viewed the population as a homogeneous unit in space. Despite the fact that no environment in the world is homogeneous and thus we expect some sort of spatial structuring in almost all situations, classic approaches treated populations as if they were without such structure. The most important deviation from this orthodoxy was a very simple insight by Richard Levins. Noting that populations were frequently arranged in space in a patchy fashion, he suggested that the variable of analysis with which we begin should be, as it is in epidemiology, not the number of individuals, but rather the number of occupied patches (in epidemiology, the number of infected individuals). Thus, if we have a habitat that is divided into discrete patches (think of them as habitat "islands"), we ask what is the proportion of those discrete patches that are occupied by at least some individuals in the population of concern. Rather than beginning with N, the number of individuals in the population, we begin with p, the proportion of patches occupied. With that change of focus, the development of the idea is simple and elegant. The rate of change of occupied patches must be related to the rate at which previously empty ones become occupied (call it a migration rate) minus the rate at

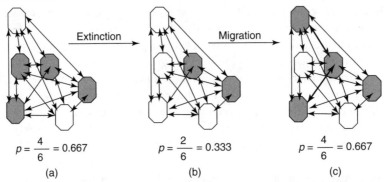

$$p = \frac{4}{6} = 0.667 \qquad p = \frac{2}{6} = 0.333 \qquad p = \frac{4}{6} = 0.667$$

(a)   (b)   (c)

Figure 7-4 Diagrammatic representation of a metacommunity. Each polygon represents a habitat or fragment of a habitat, and the double-headed arrows represent migration between habitats. Shaded polygons represent habitats that are occupied by the species in question, and unshaded ones represent habitats not occupied. (a) Sixty-seven percent of the habitats occupied. (b) Two of the previously occupied habitats become extinct, resulting in 33% of the habitats now occupied. (c) Migration, indicated by the bold arrows, reestablishes the species in one of the original occupied habitats and establishes the species in a formerly unoccupied habitat.

which subpopulations within a patch disappear (call it an extinction rate), and we expect a balance between the migration rate and the extinction rate. The basic idea is illustrated in **Figure 7.4**.

This is the most elementary level and is usually modeled with the simple equation:

$$\frac{dp}{dt} = mp(1-p) - ep \tag{7.2}$$

where p is fraction of habitats occupied, m is the migration rate, and e is the extinction rate. This is what has become known as the "Levins" form. It contains a logistic structure, $mp(1 - p)$ and an extinction term ep. At equilibrium (set the derivative equal to zero and solve for p), we have

$$p^* = 1 - \frac{e}{m} \tag{7.3}$$

Thus, the fraction of habitats occupied when the system is in balance is 1.0 minus the ratio between the extinction and migration rates. If migration and extinction are equal, the expected fraction of habitats occupied is zero and the population becomes extinct. If extinction is less than migration, the second term (e/m) is very small, and the equilibrium occupation is less than 1.0 but greater than zero. An important qualitative result here is that the equilibrium situation is such that not all habitats will be occupied. Thus, for example, if extinction is 0.1 and migration is 0.2, at equilibrium, we will find 50% of the habitats occupied. If we follow the population over time, we should see 50% occupation all the time. However, it is not that we expect each particular habitat to be either continuously occupied or not, but rather that there will be a turnover of which habitats are occupied at any one point in time, even though 50% of them will be occupied.

We are now in a position to ask what happens when a natural habitat becomes fragmented. It is clear that whatever spatial migration occurred, that is, however much organisms moved from place to place in the original habitat, that rate will be lower in the fragmented situation. It is equally clear that the rate of local extinction will also tend to increase, because by sampling a smaller fraction of potential habitats in the smaller fragments, the chance of including all the requisites for all the potential species will tend to decline. Thus, the expectation is that with fragmentation the parameter m will decrease and the parameter e will increase. The goal of conservation is to increase m and decrease e, obviously. It is easy to imagine planning production such that agricultural development is on a sustainable and biodiversity-friendly track, as much as preservation of remaining natural vegetation is a priority, especially in regions with significant disposable resources. Today, however, we must realize that we live in a world where the major concentrations of biodiversity are in those areas with the least "developed" agriculture and the most poverty, where such disposable resources are unavailable.

A fragmented world presents a series of other problems that are surprisingly complicated and presently only incompletely understood. For the most part, these problems are recognized only in theory so far. They could turn out to be major issues, as more fieldwork uncovers how populations of animals and plants actually respond to the new reality of fragmentation. In this chapter, only a few of the many problems that currently puzzle ecologists are noted.

Subpopulation Coherence

One of these new problems is coherence. All populations fluctuate over time for many different reasons—sometimes in response to seasonal fluctuations, sometimes because a predator and its prey do a sort of seesaw game where their populations alternate being common and rare, sometimes in response to forces that appear mainly random. Whatever the cause, populations tend to fluctuate over time.

Thus, the picture we must face in a fragmented environment is isolated subsections of a larger population, distributed in the fragments of the landscape, each of which fluctuates regularly or irregularly. The important part of this picture is the migration among the fragments.

The issue that arises theoretically is an issue that arose in physics in the 17th century. The great Dutch physicist Christian Huygens made detailed observations on two pendulum clocks affixed to the wall of his study. At first, he noted casually that their pendulums were swinging in opposite directions. When one pendulum was to the left the other was to the right and vice versa, what is normally called "antiphase symmetry." He then paid closer attention and experimented by placing the pendulums in near perfect symmetry so that they both started out in perfect synchrony. Even when set in motion from identical positions, in time, they came to oscillate with the same antiphase coordination. It turns out that the wall on which the clocks were mounted acted as a connection between them. The subtle vibrations set off by the oscillations mutually affected the motion of both pendulums through that wall,[20] and the result was pendulums that moved in perfectly opposite directions.

This phenomenon has been studied in all sorts of situations, from electric devices (in which currents are oscillating) to physiological rhythms to behavioral schedules. The issue is mainly one of in-phase synchrony, not the antiphase form originally observed by Huygens. The general phenomenon of things that fluctuate over time becoming synchronous when groups of them are connected together is the core idea around which much modern physical and biological research is organized, as so eloquently described in the book *Sync* by Steven Strogartz.

The significance of this idea for biodiversity conservation in fragmented landscapes is only now being fully appreciated, even though the structure of the facts on the ground is so clearly relevant. If subpopulations of a given species of animal or plant are living in a fragment and they are undergoing the normal fluctuations that characterize all populations, what happens if those subpopulations are connected by migration? The answer, at least when we examine the process in a theoretical form on the computer, is that the population fluctuations come into synchrony with one another. This fact could have enormous consequences for the potential of the entire population (i.e., all of the subpopulations). Most ecologists agree that extinction at a local level, which is to say the disappearance of a population from one of the fragments, is most likely to occur when that subpopulation is very small. The hopeful idea is that a metapopulation structure will exist, where locally extinguished populations are replaced by occasional migrations from other fragments. Thus, even though subpopulations disappear from fragments on a regular basis, they will be replaced by migrations from other subpopulations, also on a regular basis.

What if, however, those subpopulations are in synchrony with one another? Suppose that, for example, an environmental stress of some sort occurs with a frequency of once every 10 years. Now, if the subpopulations of a species are not in synchrony, every time the environmental stress occurs it will cause excessive stress on those subpopulations that happen to be in the lower part of their cycle, that is, those that happen to be rare at the time. Extinctions within fragments will occur, but in classic metapopulation fashion, reinvasions from other fragments will also occur. Now suppose that the subpopulations are all in synchrony with one another. It is probably unlikely that they will all be at their low point in their fluctuations at the precise time when the environmental stress occurs. Nevertheless, given enough time, with the stress occurring every 10 years, there will eventually be a time when the stress will occur at precisely the time when all subpopulations are at their low points, and each one of them will thus disappear. In this sense, coherence of the fluctuations of the subpopulations will cause the complete extinction of the population.

There is a seemingly inherent problem here for conservation. If the only way that a fragmented habitat can house a species in perpetuity is as a metapopulation, there must be migration among fragments. But migration among fragments seems to induce synchrony among the populations in the fragments, resulting in inevitable extinction of the whole population. It is a worrisome problem that deserves study.

If we make the situation a little more complicated, as it invariably is in the real world, the problem sort of disappears, although new and even more difficult problems may arise. No population occurs in isolation. Thus, when we connect fragments together, either with corridors or trampolines or higher quality matrices, we are not only generating interfragment migrations for a single species, but for much of the general community—that species plus the things it eats or nests in plus its natural enemies. Thus the question becomes: Which components of the biological community are migrating and how fast relative to one another? In a simple system of a predator and its prey in two fragments, on the computer, if the prey migrates faster than the predator, the two subcommunities (i.e., the two predator/prey pairs in each of the fragments) will oscillate in an antiphase fashion, whereas if the predator migrates faster than the prey, the opposite will happen, in-phase oscillations will occur.

Unfortunately, it is not possible to make a general statement about how migration affects coherence of the subpopulations, other than to note that which of the elements of the community migrates and at what rates is an important consideration. Much more work remains before we are able to say specifically which components will have a beneficial effect on the overall landscape when they migrate. We are at the point where ecologists acknowledge that the so-called metacommunity approach (where the migrations of all the species in the community are taken into account) is the approach that needs to be taken in future research.

Cascading Extinctions and Ecosystem Collapse in Metacommunities

Some recent computer simulations bring up another issue that is potentially disastrous for biodiversity conservation, that of extinction cascades, one extinction causing another causing another and so forth. A predator that specializes in only a single kind of prey will obviously go extinct if that kind of prey disappears. That is basically the idea, but because ecosystems can be so complex, the idea can become multiplied by many factors, and an underlying instability in the ecosystem can become exposed through fragmentation. A somewhat complicated but quite realistic example will illustrate the point.

Ecologists have come to believe that some communities of organisms exist in a state of neutral equilibrium, sort of a purgatory of negative interactions, as discussed later in this chapter. This happens when members of a group of species do more or less the same thing, when they have very similar ecological niches. It is similar to a group of gamblers in a card game, all beginning with a fixed stake. As we all know, most of those card games are largely a matter of luck, and thus, the basic setup is that no particular gambler has any advantage over any other one. We also know that one by one the gamblers will by chance be eliminated from the game until there is only one gambler left, having won all the stakes of all the others. A similar idea is thought to occur in some groups of species. For example, the community of floating photosynthetic plankton in the ocean is sometimes composed of many species, but they all do essentially the same thing. They float around and eat light. Which species gains an advantage over the others is as much a question of luck as in the case of the gamblers. Also similar to the gamblers, given this neutral setup, only one species is expected to survive. Similar situations are thought to exist in a variety of community types, such as tropical trees, or savannah grasses, or the organisms on the ocean floor.

We know, however, that many of these assemblages of species do in fact exist, sometimes for long periods, and consequently we are required to seek some sort of explanation (what if the gamblers, all with fixed stakes, continued playing the same game for years on end, with no one going broke—you would probably think something was intervening to help those with the bad luck or hurt those with the good luck). One explanation frequently proposed is that natural enemies—diseases, predators, parasites—may enter the fray, and when one of the competing species begins to dominate, the natural enemy comes into play and drives its population down to lower levels. It is as if a "cooler" arrives when one of the gamblers begins winning too much and cools down his or her luck (from the Hollywood movie *The Cooler*, referring to the legendary idea that some people naturally have bad luck and can be employed to stand near a casino customer who is experiencing exceptionally good luck to cool down the good luck).

If the natural enemies are there to stop the random process of one of the competitors dominating the situation, it is obvious that there must be at least as many natural enemies as there are competitor species. So, for example, if there are 200 species of tropical trees in a rain forest and they are all neutrally competing with one another, there must be a minimum of 200 species of natural enemies to make sure that the system stays in place. If just one of the natural enemies is missing (i.e., if one of the tree species does not have a "cooler," a natural enemy that can come around when it begins to get too dominant), that tree will simply take over from all the other trees, like the gambler who wins all of the stakes.

Thus, we have the neutral competitors with their natural enemies stopping the dominance of any of them. But there is a problem here in that such an arrangement is unlikely to last over the long run, especially if we bring in the idea of local extinction. If some outside process causes the random extinction of a species of natural enemy, the control over one of the competitors is lost, thus freeing that competitor to dominate the situation and take away all of the stakes. It would represent the complete collapse of the ecosystem—all but one species would be eliminated.

Casting these ideas in the context of a fragmented landscape, we have the process of competitors and their natural enemies (the gamblers and their coolers) along with the random extinctions of the natural enemies that inevitably operate in each of the fragments. However, the process of migration of the natural enemies operates among the different fragments. So, ultimately, if the rate at which extinctions occur is much greater than the rate at which the migrations replace those extinct species, we can expect that the point of overall instability in which one of the competitors has only a single natural enemy will eventually be reached, and the next extinction may leave that competitor with no control. This causes the extinction of all the competitors (which are, after all, the food for the natural enemies, so they will go extinct too). This general idea is schematically represented in **Figure 7.5**.

Given this setup (which is perhaps quite common in nature), we see that the migration rate among fragments is key to maintenance of the entire system. If migration rates decline, the process of local extinction can become sequential, leading to the collapse of the whole system. This prospect leads to the question: What causes migration rates to decline? At least one of the factors is a lowered quality of the matrix. So we see here the potential for a low quality matrix not only threatening the existence of single species at a time, but also threatening the integrity of the entire ecosystem at once.

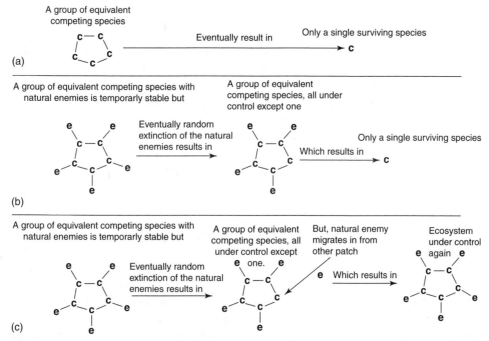

Figure 7-5   Diagrammatic representation of the possibility of ecosystem collapse rendered stable by migration of natural enemies. (a) The basic idea that a group of equivalent species will eventually result in a collapse of the system into a single species. (b) The maintenance of the system through the control by natural enemies on each of the competing species, exerting only temporary stabilization of the system. (c) As natural enemies migrate from patch to patch, they may provide a longer term stabilization of the ecosystem.

The sort of structure in this example is characteristic of a new focus of community ecology, the metacommunity. Taking a cue from metapopulations, a metacommunity is a multispecies collection of interacting entities that exist in a patchy formation in space, with migration among the patches. In this particular example, the parallel to a local extinction in the metapopulation context is ecosystem collapse, which similar to metapopulations is counteracted by migrations among the patches. Currently, there is great interest in the way metacommunities function, although there is little agreement on what are the major scientific questions involved.[21]

Fragmented Landscapes and Agricultural Development

Matrix Quality and the Landscape Mosaic

As elaborated previously, one of the main patterns of biodiversity in the world is a decrease with increased management intensity. The classic case is the coffee agroecosystem that ranges from the low intensity rustic system with high levels of biodiversity to high intensity sun coffee with dramatically reduced levels of biodiversity. This general pattern ought to be an important component of conservation thinking for two reasons:1) many agroecosystems contain very high levels of biodiversity, especially when the focus is on associated biodiversity rather than planned biodiversity, and 2) agroecosystems create matrices that are of varying degrees of "permeability" (i.e., different agroecosystem types present different probabilities of migration). These two issues are treated in turn.

The fact that many agroecosystems contain very high levels of biodiversity is now well established.[22] A certain historical momentum remains that still sees the agroecosystem as an enemy of biodiversity, but most modern analysts consider the kind of agroecosystem as important for biodiversity conservation, not its binary presence or absence. Thus, in conservation planning, it makes sense to ask not only how many hectares of preserved natural habitat exist, but also what is the "quality" of the intervening matrix. Generally, the biodiversity in the remaining fragments cannot really be improved on (although it can be degraded, as explained presently), but the biodiversity in the surrounding agroecosystem can indeed be manipulated by planning of the agroecosystem itself. For conservation purposes, it makes sense to advocate biodiversity preservation within the agroecosystem as part of the planning process of agricultural landscapes, beyond the argument that biodiversity provides ecosystem services for the agroecosystem itself.

Probably a far more important aspect of the matrix is its permeability. That is, in consideration of the fact that many, if not most, species living in fragmented landscapes likely exist in a metapopulation context, the migratory potential offered by the agroecosystem is far more important for regional survival than the nature of the fragments themselves. Generally, we know from island biogeography studies that the extinction rate on islands, and in fragments, is largely determined by the size of the island or fragment. Small fragments are expected to have a higher rate of local extinction than larger ones. Because for the most part fragmented landscapes are given to us, not planned by rational actors (at least not rational from the point of view of conservation), the size and distribution of fragments is a given, not something we are able to stipulate. Consequently, because the local extinction rate is mainly a function of the size of the fragment, there is not much we can do about local extinction rates. However, the migration rate is determined by the permeability of the matrix. That is something that we indeed can manipulate by careful planning of the agriculture that will be practiced in the matrix. A high-quality matrix in this context is a highly permeable one, permitting a high migration rate. Recall the basic metapopulation equation, $p^* = 1 - e/m$, which means that the larger the migration rate (m) the more likely it is that all fragments will be occupied by the species in question.

Thus, by combining the political reality of fragmented landscapes in almost all tropical areas of the world with fundamental theories of ecology, especially the theory of metapopulation biology, we arrive at a vision of biodiversity conservation that has a focus on the nature of the area between the fragments, the agricultural matrix. A matrix that is "friendly" to biodiversity (especially associated biodiversity) and is permeable to the necessary migrations that need to occur to maintain metapopulational structures over the entire landscape is what has been referred to as a "high-quality" matrix.

An example of these issues is the Atlantic Coast rainforest of southeastern Brazil. Popularly thought to be about 93% destroyed, this habitat is cited as one of the most threatened of the hyperdiverse habitats in the world. Although it is probably true that the biodiversity of the Atlantic forest is indeed highly threatened, it is definitely not because it is 93% destroyed. The figure itself is not correct, but more important, the landscape structure in the zone is the critical issue for conservation, not the percentage of original habitat remaining. The Brazilian government currently enforces, if imperfectly, a law that requires all landowners to maintain at least 20% of their land in natural habitat. Even if only half of the farmers actually respect the law, then at least 10% of the forest remains, albeit in a highly fragmented state. A recent study[23] using satellite information shows that in one river basin (the Rio Doce) over a million hectares remain in forest, almost 15% of the total area, but virtually all in small fragments (28,240 fragments larger than a hectare were counted). This same study, however, noted that almost all of the matrix between these fragments was either sun coffee or open pasture, neither of which contains trees as a part of the production system, and both of which likely preclude much interfragment migration of most organisms (**Figure 7.6**). The problem here is not in ensuring that the extant fragments remain. That has already been done (at least in theory). The problem is creating a matrix that will allow migration among the fragments to stem the inevitable tide of local extinctions from becoming regional. Research into this question of matrix quality is urgently needed.

As documented in previous chapters, the past several decades have seen a great deal of research that challenges the basic assumptions of the conventional agricultural system. Furthermore, a more ecologically sound form of agriculture has increasingly been promoted by many producers and planners, and much of the rhetoric of this alternative

Figure 7-6 Forest patch surrounded by low-quality matrix of pasture (to left) and sun coffee (to right) in Minas Gerais, Brazil.

agriculture movement has to do with creating structures within agroecosystems that are much more like the original habitat than is normal with the conventional system, as argued in Chapter 8. If this is true (and it is almost certainly true at least for tropical areas), the alternative agroecological matrix would be a "high-quality" matrix from the point of view of allowing migration among fragments. For example, in the Atlantic Coast rainforest of Brazil, the problem is the treeless landscape in the matrix, formed by sun coffee and open pastures. If the well-known technologies of organic and shade coffee[24] and silvopastoral systems[25] were to be pursued, which could easily occur with the proper political incentives, it is likely that the migratory potential of the matrix would increase dramatically.

This point then brings us back to the political issues raised earlier. What type of actually existing matrix provides the high quality that we require for landscape level sustainable management that preserves as much biodiversity as possible? An interesting case study is in the Pontal de Paranapanema,[26] in the westernmost edge of Sao Paulo state in Brazil, about as far southwest as the Atlantic Coast Tropical Rainforest extends. It is literally a point on the map of the state and is formed by the confluence of the Para and Paranapanema rivers and includes an area about the size of New Hampshire. As part of the most populated and developed state in Brazil, this region is a patchwork of pastures and sugarcane, of forest remnants and small farms. One very large conservation area (the Morro do Diablo State Park—about 30,000 hectares) occupies about an eighth of the area, but dotted across the landscape are about 20 small fragments of the Atlantic Coast rainforest formation.

What remains in the Pontal de Paranapanema is an irregular collection of large haciendas, small farms, and forest fragments. Brazilian conservationists have been in the forefront of understanding what this complicated landscape means for conservation. For example, recently they pressured the government to declare a protected "reserve" of four small forest fragments—the first declared reserve in Brazil with unconnected sections. These four sections, along with the large state park, comprise a sort of refuge for one of the most charismatic species I know, the black-faced golden lion tamarin monkey, a close relative of the famous golden lion tamarin, an eastern relative. In one of the small fragments, a healthy population of tamarins is currently in residence, and a population of perhaps a hundred individuals occupies the state park. The other two fragments do not contain any tamarins, but in historical memory, one of them is known to have formerly housed a small population—it presumably went locally extinct in the recent past. The eastern relative, the golden lion tamarin, is known to be a food item for the weasel-like tayra and undoubtedly many other predators feast on both of these species. A locally high population of any of these predators is likely a major threat to the survival of the species in any given fragment.

If we imagine the construction of corridors connecting these fragments, as is being currently promoted, we first think of those monkeys traveling through the corridors (or jumping the trampolines) from fragment to fragment, and much like Huygens' clocks, we might expect that the fluctuations of populations within the fragments might become highly correlated. This is, of course, worrisome, since whatever caused the one population to become extinct in the past could return, and if it returns at a time when all the populations in all the fragments are at a low population size, it could cause the entire population to go extinct. It could also be, however, that the tayras and other natural enemies will also use the corridors, and if they use them in a different way than the tamarins themselves, the population synchrony could be of the antiphase type. It is really not all that easy to say whether the migrations will cause in-phase or antiphase synchrony, but without doubt it depends on the details of how the food of the tamarins as well as the natural enemies of the tamarins use the corridors, just as much as how the tamarins use the corridors.

Exactly what will happen is a matter for future research. Even simple models of the system indicate that there is going to be a great deal of unpredictability involved, due to the ecological complexities we know exist. However, before such complexities become central, some more obvious issues require attention. The 20 or so forest fragments in the area are surrounded mainly by treeless pastures, land held in production according to the owners, but held for speculative purposes according to some critics. Yet there are also large areas that are occupied by small-scale farmers who live comfortable lives, not luxurious, but comfortable. The process whereby they obtained this land was through the political organizing efforts of the "Movimento de los Sem Terra" (the landless people's movement, or in its Portuguese acronym, MST). The MST organizes landless people into communities that take over a piece of land that is "not being used" and demand that the government intervene to take action under the agrarian reform law and provide compensation for the original land owner and transfer title to the now landed families that have occupied that piece of land. Depending on your political point of view, these are "land invasions" or "squatter settlements" or, as the MST frequently refer to them, "land rescues."

Although the politics and law are incredibly complicated, what is important for agroecology is the attitude of most of the new farmers that come out of the land rescues. These farmers are small scale in their planning and production operations, they normally employ organic or semiorganic methods, they usually have trees as part of their production system, and they seem to have a sincere conservation ethic. If we are to construct a high-quality matrix for conservation purposes and provide food security for people, the process as it is unfolding in the Pontal de Paranapanema seems an impressive example, at least for now. Indeed, one detailed study of birds confirms the conservation expectations.[27]

The Via Campesina

The MST is a member of the umbrella organization known as the Via Campesina (the farmer's way), as introduced earlier. The formation and continued evolution of the Via Campesina is a complicated issue,[28] beyond the scope of this text. However, the position of the organization on two fronts is particularly important to the subject of this text. First, the group has pioneered the idea of food sovereignty, an all-encompassing view of rural development that is emerging as a logical alternative to many of the crises of modern society, from food security to urban violence. I return to a more detailed discussion of the idea of food sovereignty in Chapter 8.

The second front in which Via Campesina is having a major global impact is in the expansion of the alternative, ecologically based model of agriculture, commonly referred to in the Via Campesina literature as agroecology. Almost all the member organizations of Via Campesina originated with popular struggles for land reform or other social justice concern. Within those political struggles, environmental struggles emerged, partly because of the participation of environmentally concerned individuals and groups alongside the social justice groups, but partly a consequence of the concrete conditions created by the industrial model itself.

As has happened in the developed world, a growing realization of the environmental and health problems created by the industrial system has been penetrating the Global South and undoubtedly has influence on the members of the member organizations of Via Campesina. It is not difficult to see a connection between pesticide poisoning and the conditions of social justice. It is not the corporate board rooms of Monsanto that are sprayed with pesticides, but rather those same small farmers and farm workers who are organizing the organizations that form the Via Campesina. Safe working conditions, long a fundamental demand of labor organizing the world over, articulate naturally with ecological forces when the labor is agricultural. Rural communities need safe drinking water, not water polluted by either nitrate or pesticide runoff. Individual farmers do not want themselves or their children exposed to chemicals that the world now knows are tied to cancer and teratological effects. Consequently, the emphasis on sound environmental management is perhaps a natural outgrowth of class consciousness when the class is rural.

Thus, Via Campesina's environmental projection has evolved as a two-pronged program, first to oppose the most environmentally damaging aspects of the industrial system and second to promote an approach based on ecological principles. Their opposition has taken the form of street protests, militant actions against the spread of transgenic crops, and a call to get the World Trade Organization (WTO) out of agriculture altogether,[29] arguing that food should be for the purpose of human nourishment, not corporate profits. The promotion has taken on many of the same ideas that emerged with the beginnings of the alternative agriculture movement. Much as Albert Howard "discovered" the Indore composting methods through interactions with Indian farmers, the Via Campesina approach has been a "farmer to farmer" approach, recuperating techniques that have been locally traditional and promoting their diffusion through interactions among farmers themselves. Articulating a technical agenda with this new approach to "bottom up" research is a challenge[30], an issue we revisit in Chapter 8.

The Maintenance of Biodiversity

Biodiversity Revisited

We return here to the topic that opened this chapter. Biodiversity is a transcendent issue, to be sure. And the vast majority of the terrestrial biodiversity of the world occurs in tropical areas, which is to say in the Global South, where the industrial agricultural system has not yet completely penetrated but is on full speed in its intention of doing so. The specific rules imposed by the Bretton Woods institutions with their new offspring, the WTO, threaten the small farmer with extinction, in much the same way that has been happening in the Global North. Given the underlying ecological patterns associated with islands and agricultural intensification, it is difficult to ignore the fact that this penetration will likely promote an even more rapid decrease in biodiversity with the gradual disappearance of small farmers, the majority of whom, in the Global South, engage in relatively low intensity (and thus biodiversity friendly) agriculture and increasingly have a biodiversity-friendly ethic. If the matrix within which natural habitats are embedded is mainly agricultural and the migration among those habitats is critical for biodiversity preservation, the conclusions are obvious.

Recent ecological research has emphasized the ideas of metapopulations and more recently metacommunities along with the importance of diverse landscapes. Few professional ecologists would argue with the idea that landscape processes are frequently, if not always, essential to biodiversity maintenance. This point has already been discussed in the context of metapopulations and metacommunities.

In addition, something of a pernicious political problem has emerged with the question of biodiversity preservation. It is a two-pronged problem. First, there is a sense that there are habitats with many species (usually called natural habitats) in contrast with habitats that have few if any species (usually called "destroyed" habitats). Such a point of view ignores certain ecological realities, such as the landscape determinants of migration rates, all of which were discussed fully previously. This question of good versus bad habitats remains a critical issue for both biodiversity preservation and agroecosystem development.[31] As I hope is clear from previous discussion, categorizing habitats as either good or bad ignores ecological realities and can have politically pernicious consequences.

The second prong of this problem is the assumption that the structure of biodiversity within the "good" habitats is somehow fixed and constant. Thus, the focus of conservation activities is on preserving the "natural" habitats remaining, tacitly assuming that the greater extent of natural habitats preserved the more species will be saved. Elementary ideas of metapopulation theory belie this point of view (the migration among habitats is critical to avoid extinction, as explained previously), but there is also a far more difficult question of how that biodiversity is structured within those natural habitats and how the details of that structure may affect the ultimate consequences of land use planning. Questions of competition, natural enemies, and dispersal, in the context of a spatially explicit framework, are all involved in what is yet an incomplete understanding of the organization of biodiversity in natural habitats. In this section, we explore some of the more recent developments in community ecology that may shed light on such issues.

The Importance of Gause's Principle

The basic idea "no two species can occupy the same niche" has come to be a central organizing principle in ecology, for better or worse. The intensity of competition is related to the degree to which the ecological niches are similar. This degree of similarity is frequently referred to as "niche overlap," and the basic principle is thus related to niche overlap. Two forms of Gause's principle exist. First, if competition between two species is too intense, one of the species will disappear from the environment. Second, if niche overlap is too large, one of the species will disappear from the environment. It is not necessarily the case that these two forms are identical, but for now, we assume they are.

In a very influential work, Richard Levins and Robert MacArthur expanded Gause's principle to a multiple species context and asked some basic questions about how evolution would be expected to change species niches in response to competitive pressure.[32] They asked two simple questions: 1) What would happen if two species existed with too much niche overlap? 2) What would happen if two species existed with very little niche overlap? Remember that the idea of the ecological niche had become closely associated with the process of competition, such that Gause's principle could be stated either as "no two species can occupy the same or similar niches" or "if two species compete with one another too strongly, both cannot coexist in the same environment." Thus, both of their basic questions were couched in this elementary tradition. Given Gause's principle, the answer to the first question, what will happen with too much niche overlap, was well known—one or the other species will disappear. The answer to the second question, what if there was very little niche overlap, was also well known—the two species could coexist in perpetuity. But Levins and MacArthur further explored the significance of this coexistence. What if there were so little niche overlap that a third species could fit in? Given that there are always species ready to invade an area, it made sense to suppose that if some "niche space" were available, a species would invade it (see **Figure 7.7**).

The conclusion was obvious as soon as the problem had been properly formulated. If niche overlap was too small, another species would fit in, and if niche overlap was too large, an extinction would occur. We could thus imagine a sort of dynamic equilibrium in which the arrival of new species would tend to balance the exclusion from competition such that the environment would be continually "packed" with as many species as could be accommodated according to Gause's principle.

Using this simple idea, it is not difficult to generalize. If you fix the maximum allowable niche overlap at some particular value, it should be possible to stipulate how many species would be allowable in the community. Thus, it should be possible to relate the maximum number of species in a community to the average competition coefficient (or niche overlap) in the community. Using a variety of mathematical tricks, Levins and later May devised some simple relationships between the average competition coefficient and number of species in the community. One such relationship is that the number of species in a community should be approximately equal to the inverse of the square of the average competition coefficient, which suggests that as the average competition coefficient approaches 1.0 there should be only a single species in a "maximally packed" community while at the other extreme, competition coefficient approaches 0.0, there should be an almost infinitely large number of species. Obviously, this approach can be only approximate, but does give a kind of underlying expectation, at least when there are few complicating features.

Competition and Biodiversity—The Paradox of the Plankton

Even before Levins and MacArthur presented their elegant theory, a problem had been recognized by MacArthur's mentor, G. E. Hutchinson. Hutchinson, assuming Gause's principle of competitive exclusion, noted that many communities contain species that would be difficult to characterize as having distinct niches. Consider, for example, phytoplankton. A given lake, or a given section of ocean, will contain many species, all of which more or less passively float and eat light and use, for all practical purposes, the same nutrients. It is difficult to imagine how so many species in such a uniform environment could be partitioning some aspect of the environment into separate niches. Here, then, was the paradox. All plankton seem to do the same thing, which is to say they all occupy the same niche, which according to the classic theory, means that they cannot coexist. Yet so many of them do.

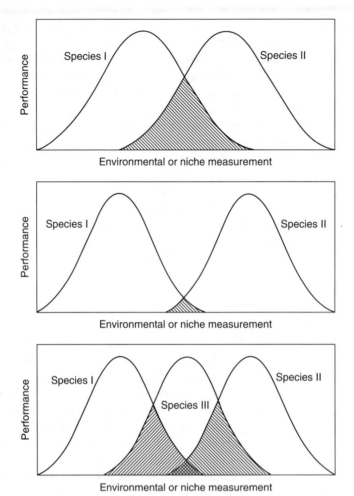

Figure 7-7 Diagrammatic representation of the species packing idea. (a) Two species with much niche overlap (hatched area), such that one or another must be eliminated through competition. (b) Evolutionary selection pressure reduces the niche overlap to the point that both species are able to coexist in the environment. (c) By reducing niche overlap between species I and II, however, a space is created for the insertion of species III into the system, resulting again in too much niche overlap. This process will continue until some limit on the narrowing of the niche is reached.

A variety of explanations for the paradox have been offered, but probably one particular hypothesis has received the most currency, that of "intermediate disturbance," proposed more or less simultaneously (although in slightly different forms) by ecologists Joe Connell and Mike Huston. In this hypothesis, it is assumed that indeed the competitive interaction among species is such that all but one species will be eliminated from an area, as the classic theory would have it, but that some outside event continually delays this inevitable outcome. Thus, for example, in the mowed fields of Darwin (it might be argued that Darwin was the actual originator of the hypothesis), the plants that were competing with one another would have eventually resulted in a single plant species outcompeting all others, but the periodic disturbance of mowing reduced all the species to the same level of competition. In other words, competition would proceed to elimination of all but the best competitor, but some sort of external disturbance periodically reduces the community to the point where the competitive process starts all over again. Put in terms of Gause's principle, all the species except one are expected to go extinct (because they all occupy very similar niches), but before the Gaussian equilibrium (of only a single species) is reached, some intervening force acts to stop the process and start it all over again.

The general line of reasoning that emerges from theories that are similar to the intermediate disturbance hypothesis has been referred to as the "nonniche" approach, whereas that based on the underlying assumptions of the competitive exclusion principle have been referred to as the "niche" approach. This debate was most active among researchers working in tropical rain forests during the 1990s. It is treated in a later section.

An historical note is appropriate at this point. The debate about niche versus nonniche approaches to understanding ecological communities is not new. In 1978, Peter Yodzis simply expanded the two-species Lotka-Volterra approach to multiple species and noted that there were two dramatically different structures that could be regarded as the ends of a continuum. At one end were those communities containing species that competed with one another at relatively low levels, in which case the species that dominate the community would be those that had the relatively larger competition coefficient (but all competition coefficients were relatively small). At the other end were those communities containing species that competed with one another at relatively high levels, in which case the species that just happened to gain the initial advantage at the time of founding would be the winner in competition (assuming all competition coefficients were relatively large, specifically, all were equal to or larger than 1.0). Thus, at one end were those communities that corresponded to the Lotka-Volterra case of permanent coexistence (dominance-controlled) and at the other those communities that corresponded to the Lotka-Volterra case of competitive exclusion with indeterminate outcome (founder controlled). Even though most communities would be found somewhere in between these two extremes, merely recognizing the continuum enabled researchers to ask where along that continuum particular communities lie. Subsequent debate has labeled the two ends somewhat differently. The niche versus nonniche debate or the equilibrium versus nonequilibrium debates are essentially about where to place communities along the Yodzis continuum.

Problems with the Classic Theory

After the critical experiments with fruit flies (see Chapter 3), Levins presented a complicated argument that pointed out some very basic problems associated with the classic theory.[33] The problems were varied, he argued, but could be classified into two general categories: 1) an unrealistic assumption of linear (proportional) responses and 2) an unrealistic assumption of deterministic (ignoring random effects) dynamics. In the end, he demonstrated convincingly with a generalized mathematical approach that there was effectively no limit on the number of species that could coexist when either of these two assumptions was relaxed. He suggested that a way of incorporating Gaussian thinking into these sorts of results was to think of one species as consuming the resource in proportion to its density, the second species according to the square of its density, the third according to the cube of its density, and so forth. Similarly, one species could respond according to the mean, another to the variance, another to the skewness, and so forth. It was a clever way of getting home the point of nonlinearity and stochasticity, but perhaps was too kind to the basic Gaussian framework, which effectively was being challenged.

At about the same time, Armstrong and McGehee published a detailed mathematical analysis of the two-consumer/one-resource system after relaxing the assumption of a proportional response of the consumer growth rate to the density of the resource,[34] effectively a detailed specific example of the more general points made by Levins. The classic theory had assumed this rate to be linear. Thus, for example, a lion would eat a certain proportion of the zebra population, no matter how many zebras were there. If that proportion were 1%, for example, then in a population of 100 zebras, the lion would eat 1 zebra per week. In a population of 300, she would eat 3. In a population of 1,000, she would eat 10, and in a population of 100,000, she would eat 1,000 per week, which would be difficult for even the most voracious lion. The assumption of a linear response of the predator to the prey is biologically unreasonable (as explained in the appendix to Chapter 6). The question arises as to whether it makes a difference for species coexistence.

Although this complication had been appreciated by ecologists at least since the late 1950s, it was largely ignored when studying species coexistence under competition. Thus, when Armstrong and McGehee decided to relax the linearity assumption, its significance became especially important. Basically what they found was that the fundamental result of Lotka and Volterra, that two consumers could not coexist on a single resource, was only true if the restrictive linear assumption was made. If it was allowed that the consumers responded nonlinearly to the resource densities such that permanent oscillations would result, it was possible to maintain two consumer species on a single resource. Thus, with permanent oscillations (which resulted from relaxing the direct proportion assumption), it was clearly possible to maintain more than a single consumer on a single resource.

Nonlinearity and the Mechanisms of Biodiversity Maintenance

A variety of recent papers has expanded on the theme originated by Levins and Armstrong and McGehee, and it may appear that the entire enterprise may be undergoing something of a paradigm shift. In a recent article, for example, it was shown that it was possible to maintain five species in competition on three resources.[35] Especially interesting in this work was that the coexistence of the five species occurred in a chaotic state. It has even been shown theoretically that under certain circumstances, it is possible to maintain more species in a community when competition is larger than when it is smaller, a dramatic contradiction of the basic Gaussian framework.[36] In another study, with only three types of resources and a continual input of new species, it was shown that more than 100 species could coexist at least for a while when small species migration rates were included.[37] All of these recent theoretical results on competition occur generally under conditions of oscillations, effectively multispecies versions of the results reported by Armstrong and McGehee, and not really all that surprising given Levins original analysis of nonlinearities.

A different approach to the question, but ultimately quite closely related, is through the examination of trophic webs. The ideas here began with a classic work of Hairston, Smith, and Slobodkin in which they noted that the effects of predators on herbivores would be such that plants would have to be controlled from "below." That is, if predators control the population densities of herbivores, the latter could not possibly control the population densities of the plants, meaning that the plants must be controlled through their own overutilization of their resources (commonly called control from below). Subsequently, after much debate, a spectacular experiment[38] showed that if you eliminate fish-eating fish from a lake, the other fish, the ones that eat the zooplankton (the principal herbivores in a lake system), are no longer controlled, and thus, they overeat the zooplankton. The plants in the system, mainly phytoplankton (the major photosynthetic organisms), thus had no control on them, and they dramatically increased in their abundance, turning the experimental lake green. There had been a cascade of effects from the top predator down to the plants, a phenomenon now known as a "trophic cascade."

Trophic cascades are just one form of a more general idea, that of "indirect interactions"—when the effect of a species (call it A) on another species (call it C) is mainly felt through the operation of a third species (call it B). So the predator (species A) has an effect on the plant (species C) through its action on the herbivore (species B). This sort of indirect effect has been shown in a variety of experimental situations and is thought to be a major mechanism involved in the structure of multiple species ecosystems. Indirect effects are not at all limited to trophic cascades, however. For example, in a classic series of experiments, it was demonstrated that a species of predatory starfish preferred the competitive dominant of a series of competing species, such that its removal actually resulted in the elimination of most of the relevant species in the intertidal.[39] By releasing the competitive dominant from predatory control, it was able to competitively exclude all other species in the community. Because the predator was effectively acting to keep the integrity of the intertidal community together, this sort of situation is referred to as the "keystone" species concept—the predator is the keystone because its removal causes a dramatic loss of species from the system.

The Janzen/Connell Effect

Two prominent ecologists, Dan Janzen and Joseph Connell,[40] independently and virtually simultaneously, suggested a generalization that has come to be profoundly important in questions of biodiversity maintenance, even though in its original formulation its significance was not generally appreciated. Consider, for example, a population of some species of tropical tree. A variety of natural enemies, from insect herbivores to pathogens, is invariably involved in the dynamics of that tree population. If we explicitly consider the dynamics of this species in space, it has some way of dispersing its propagules (seed dispersal), which will result in some degree of nonrandom distribution of propagules. If, for example, the seeds from the tree simply fall to the ground, a concentrated grouping of seedlings will be found beneath each seed-producing tree. The basic principle states that the natural enemies will tend to congregate where their food source is itself concentrated, not an especially unusual expectation. However, when looking at the joint dynamics of dispersal and natural enemy concentration, a pattern can be expected. The basic idea is presented in **Figure 7.8**.

Although the basic principle seems to be important in spatial processes, its importance seems to be more general, as discussed later. On the other hand, it is one of the few principles in ecology that has seen a great deal of empirical support. A variety of plant species seem to conform quite well with the expectations of the basic idea.

The fundamental idea that a consumer will tend to concentrate on concentrations of its food source is a relatively old idea. Indeed, one form of the basic parasitoid/host theory generates a rather uncomfortable conclusion that all parasitoid/host systems are unstable. Adding the idea that the parasitoid tends to concentrate on higher local densities of the host stabilizes that particular theory.[41] The special place of the Janzen/Connell hypothesis is in its spatially explicit nature.

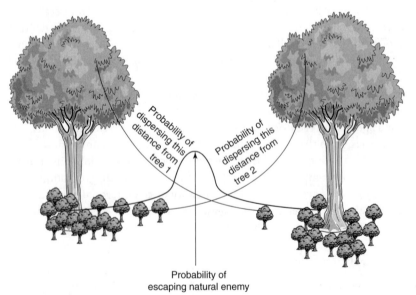

Figure 7-8 The Janzen/Connell hypothesis. As seedlings are dispersed from the mother tree, they tend to be concentrated near the source, but natural enemies can more easily locate them (or build up local population densities) when they are so concentrated. Thus, the final fitness of a seedling is located somewhere intermediate between two individual trees of the same species because of the concentrating effect of the natural enemies.

Nonlinearities in Space

A very powerful theoretical formulation recently cast light on the issue in a new and elegantly obvious fashion, recapturing some older ideas and casting them in a modern framework.[42] The paradox of the plankton has been an important metaphor for ecological communities for many years. As described previously, a debate emerged during the 1990s surrounding the idea of the ecological niche and the maintenance of biodiversity. One of the most important advances was due to Hubbell, who proposed, effectively, that the biodiversity is determined by the way individual populations interact with one another in a spatial context. Although the argument is complex and has a rather polemical history, the basic idea that seems to be emerging stems from a variety of formulations, all of which are cast in a spatial form, either explicitly or implicitly,[43] and involves the basic ideas of metapopulation theory coupled with dispersal phenomena with a Janzen/Connell topping.

If we consider a patch of land as being divided into unit areas in which each unit can be occupied by one and only one individual tree, we can basically apply the ideas of metapopulation theory, in which each patch is either occupied or not by an individual of a particular species. A very powerful theory simultaneously developed for epidemiological and community ecology applications concludes that if there is a balance between competitive ability at a point and the dispersability to that point, a certain level of stability in the entire community will be sustained. That is, if we presume that there are S species available in the species pool, we can order those species according to their competitive dominance. If species A is dominant over species B, when two individuals, one A and one B, come to occupy the same spot, species A will eliminate species B from that spot. If we now presume that the species have different dispersal capabilities and that there is a negative correlation between dispersal ability and competitive dominance, a very large number of species can be maintained in a particular space. Although the competitive dominant species keeps eliminating other species from the places it occupies, over the long run, the less efficient competitors, because of their better dispersal ability, are able to persist by continually occupying places before the more efficient competitors. One of the obvious consequences of this arrangement (although it seems surprising at first) is that if a habitat is modified such that dispersal becomes overall less efficient, the dominant competitor may be the first to disappear.

This viewpoint of the dynamics of biodiversity maintenance carries important implications for landscape planning, specifically for the type of agriculture that is planned in the matrix in a fragmenting landscape. If dispersal is the countervailing process to competitive exclusion, modifying it carries important implications. Unfortunately, it is not at all clear what to expect. It could be that reductions in dispersability across the board will shift the competitive balance toward the competitive dominant, such that there will be a general decrease in biodiversity. It could also be that a particular pattern of dispersal limitation, a consequence of land use planning generally, will have the reverse effect, such that the distribution of individuals among the species present will become more equitable than before. It all depends on the details, not only the details of the agricultural activities, but also on the details of competition and dispersal of the species in question.

Whatever the particular result, however, the important point is that the underlying rules of biodiversity maintenance in the undisturbed habitat could determine which agricultural practices will preserve most biodiversity.

An alternative, but closely related formulation, sees the relationship between dispersal and the Janzen/Connell effect as key. All species are limited to some extent in their ability to disperse propagules in space, but all species are also subject to the Janzen/Connell effect of the natural enemies that attack them. Thus, we can imagine clusters of individuals forming in space through the simple process of dispersal (more clustering from more dispersal-limited species), but then the clusters themselves being subject to the Janzen/Connell effect. The result is that all the available space cannot be completely occupied by an individual species. Again, the consequences of the internal organization of the biodiversity maintenance processes could have important consequences for the way in which biodiversity will be affected by the expansion of agriculture in the matrix. Indeed, the quality of the matrix depends to some extent on the rules of biodiversity maintenance that exist in the natural, undisturbed state.

Agroecosystems and the Contemporary Biodiversity Crisis

In light of all the material presented in this chapter, a vision of both biodiversity conservation and agroecological development seems to emerge, identical with the ideas that seem to be emerging from other quarters, namely the world's poor farmers who have consistently argued some issues (land reform) and are now increasingly adopting a modernist focus toward the ecological foundations of their art (biodiversity conservation as part of environmental sustainability). The sociopolitical aspects of these visions are further discussed in Chapter 8. For now, it is worth summarizing how the ecological theory, the political realities, and the nature of agroecosystems intersect to form what has been called a new paradigm of biodiversity conservation.[44] Summarizing this new paradigm is as difficult as summarizing any issue that is composed of complicated and intersecting elements. Yet, presenting all the complexities (as has been my intent in this chapter) frequently leaves a confusing jumble of trees where one seeks to view the forest. Thus, in this last section, I summarize briefly what I see as a skeleton of the new paradigm, using a three-part classification scheme: 1) what ecological science has to say about biodiversity, 2) what the current political and economic realities are on the ground in the tropics, and 3) what contemporary small farmers and rural communities are currently doing.

Regarding the current state of ecological thinking on biodiversity, it first must be understood what the nature of the problem has always been. The paradox of the plankton has been a centerpiece of thinking about biodiversity in areas where there is much of it. How can so many species coexist when fundamental theory says they should not? Resolving this apparent contradiction (i.e., asking the question of how biodiversity is maintained) has been a holy grail for many years. Most recently, the problem has been approached through the lens of metapopulation theory (or metacommunity theory when applied to the community as a whole), which brings the issue of extension in space and spatial dispersal of species to the forefront. A key element is the way in which species disperse themselves in space (through pollen dispersal, seed and spore dispersal, or simply physical movement when possible). Most practitioners now agree that patterns of dispersal in space are key to understanding biodiversity maintenance.

The current political and economic realities are based on simple observations—the tropics are generally fragmented, with natural vegetation occurring in small fragments that are embedded in a matrix that is largely agricultural. The elementary ideas of island biogeography (that small islands tend to have higher extinction rates) support a large literature claiming that local extinctions will be common in these fragments. Given this reality, we should expect that the organisms in the fragments will be subject to the basic rules of metapopulations: Since the fragments are generally of a size that imposes high extinction rates, it is only migration from fragment to fragment that allows the population to persist in perpetuity. Most important, that migration must occur through the agroecological matrix, leading to the question: What sort of agroecological matrix will permit this necessary migration?

Finally, the current needs, desires, and demands of the world's small farmers and rural communities link to the question of what sort of agroecological matrix is needed for biodiversity conservation. In growing opposition to the excesses of the industrial model combined with intense pressure to quell some of the problems exacerbated by the neoliberal model, small farmers and rural communities are increasingly voicing their demand for more ecologically sound methods of agriculture and against the social and economic dislocations that have resulted from decades of neoliberalism. Most recently, this tendency has taken on the political slogan of "food sovereignty," which is discussed more completely in Chapter 8. From the point of view of the general new paradigm, this social movement is precisely the sort of movement that is necessary for the construction of a high-quality agricultural matrix.

Appendix 7.A: The Political Economy of International Trade and World Agriculture

The patterns of both rural marginalization everywhere and economic stagnation in the Global South have been persistent problems even before World War II. Certain structural details go a long way toward understanding the origin and maintenance of that pattern and set the stage for understanding the basic issues discussed in this chapter. The outline presented here is particularly located in the work of Immanuel Wallerstein, with a strong focus on the dynamics of the articulated versus disarticulated economic arrangements that evolved slowly with the Industrial Revolution.

The World Systems Approach

Analysts of the late 19th and early 20th century generally agreed on the basic mechanism causing the pattern of rural marginalization and persistent poverty in the Global South. It was thought that the South was simply behind and that the nations of the South would develop in time, just as the developed countries had done before. This "time" hypothesis remained the received wisdom for some time, but gradually, it became obvious to even casual observers that something was dramatically wrong with this interpretation. Had it been correct, one should have been able to observe a gradual improvement in conditions of life in countries of the South as time passed. Yet by the end of World War II, it became evident that not only was the South not developing, the trend was actually the reverse—things were getting worse not better.

This very simple and obvious observation led to a great deal of analysis, beginning in the 1950s, but with especially vigorous discussion and debate during the 1960s and 1970s. The general field of analysis became known as "dependency theory" and in its most general form holds that the underdevelopment of the South is neither an accident of history nor a product of bad real estate, but an organic outgrowth of the development of the North. Development in the South was seen as dependent on events in the North and somehow maintained a lower level to benefit that North. A variety of flavors of dependency theorists has seen the light of day, and hundreds of books have articulated various points of view within the general idea of dependency. The one and only thing on which they agree is that the underdevelopment of the South occurs because of the development of the North. The contentions lie at the point of trying to explain how and why that happened, a discussion that remains alive today in its details. Nevertheless, the general outline of how the developed world developed and how that is related to the underdevelopment of the rest of the world is more or less agreed on. Here I follow the insights of Wallerstein, focusing on the idea that the capitalist system is not something that exists in some single locality, but rather is a world system, and has been since its inception.[45] Wallerstein's approach is conditioned by the initial historical framework provided in the 1940s by economic historian Polanyi in his influential book *The Great Transformation*.

Polanyi sought to explain the late congealing of the capitalist system from its early state of isolated parishes, focusing on the importance of incorporating free labor into the system as a whole. Much as nation-states have separate rules of governance, commerce, and movement restrictions today, parishes in West Europe had separate systems, a historical relic stemming from late medieval times and lasting to the dawn of the Industrial Revolution. Within this framework, the efficiencies expected of the emerging capitalist system could not be realized, largely due to the new bourgeois class's failure to appreciate the principle of collective demand (i.e., as all capitalists sought to drive workers' wages to starvation levels, a goal that left workers unable to buy the products of the new factory system). This is why the elimination of the parish system and the freeing of labor, including freedom to migrate from parish to parish, was key to unleashing the capitalist system to realize its promised efficiency.

For purposes of the present text, Polanyi's categorization of market types is of special importance. He noted that at the dawn of the Industrial Revolution there were three very distinct types of markets—agricultural markets, artisan markets, and international trade markets—that had little if any relationship to one another. Agricultural markets, certainly the most common and historically most persistent form, existed throughout Europe and were largely rural to rural or, especially important, rural to urban affairs. Their importance at the beginnings of urbanization is evident. Less evident are the artisan markets that emerged only with the move toward a social division of labor, itself only possible after the development of agricultural surplus. Although these two market forms undoubtedly involved the bulk of the population of Europe and certainly dominated the lives of average people, the most important market form for the present text is that of the long-distance trade markets.

Perhaps the most influential long-distance trade system was that dominated for a couple of centuries by the East India Company of Great Britain, with its rule over three of the largest entities in what would later become the "Third World" and eventually the "Global South"—China, India, and Egypt. However, the basic form was invented by the Dutch with the formation of their Dutch East India Company (known by its Dutch initials, VOC). The VOC emerged from the collective negotiations of several different shipping companies, effectively as a lobbying organization, seeking governmental protection from the then-common practice of piracy on the high seas. Seeking protection from a government that hardly existed (today's Dutch were formally under the administration of the Spanish until 1648 while the VOC was established in 1602), the formation of the VOC was, in a structural sense, the first giant corporation in the world. Its general framework was to establish on-the-ground trading posts in far off corners of the globe, retain the protection of national armies and navies, and work in close consort with local political powers. In this sense, the VOC set the stage for the emergence of the modern international corporation. The rules of the game were thus set in stone as a government/military/trade complex that has been extremely successful ever since and, in many ways, is the foundation of all of today's modern industrial societies in the Global North.

To understand the dynamics of this system as it operates today, it is essential, as Wallerstein noted, to conceive of capitalism as a worldwide system. Focusing, to start, on the local level, the system can be simplistically but heuristically represented as consisting of two social classes: the owners and the workers (capitalists and proletarians). During the late stages of development, the so-called Fordist accumulation phase, which occurs mainly in the industrialized North, the owners have to wear two hats in making decisions, first as owners of the factories and second as members of the social class that wants to sell products. On one hand, they try to pay their workers as little as possible, to minimize production costs. On the other hand, they want workers in general to make as much as possible so that their products can be purchased. This contradictory position represents the "engine" of economic growth in an advanced economy. As worker salaries increase, there is pressure to mechanize production. As production is mechanized, average worker salaries go down (relative to constant purchasing instruments) and consumer power declines. As consumer power declines, enterprises must close (this was referred to by Marx as an "underconsumption crisis"). As they close, there is more downward pressure on worker salaries, leading to further deterioration in consumer power. This downward spiral can continue into a recession or even a depression. The countervailing pressure is that the availability of a cheap workforce stimulates further growth in the economy (i.e., opening up of new economic sectors), which then causes lower unemployment and upward pressure on wages. This "oscillatory" behavior is referred to as the business cycle and is commonly assumed to exist, although its nature is quite irregular. Nevertheless, as a fundamental idea, sort of a Newtonian frictionless basis, it seems to explain a a great deal about how the developed world developed under what is effectively anarchistic modes of organization.

In the Global South or in those areas especially under the influence of modern European Colonialism, the dominant economic situation is quite different. Partly through the extraction of precious metals and harvesting of other raw materials (e.g., lumber), and most important for this text, partly through the expansion of a particular form of agriculture, European powers cemented their control over their spheres of influence. The form of agriculture was largely a monocultural production form, aimed at export to the European "center." Beginning with sugar cane production in the West Indies, and adding coffee and tea planters in Asia, banana production in Latin America, cotton and grain production in India and Egypt, and a variety of other large-scale endeavors, the large-scale monocultural production of grains, raw materials, and luxury crops formed the main economic activities of many countries in the Global South. Superficially, such endeavors may have resembled the factory system that had been developing in the Global North, with owners of plantations hiring workers, but the underlying structure was dramatically different. The workers on the plantations were not the consumers of the products of the plantations, meaning that the social requirement for consumer demand was absent (a worker on a banana plantation does not consume the bananas produced—they are produced for export to the factory workers in the Global North). There is thus no incentive of any sort for the wages of the workers to rise (other than from pressure from the workers themselves). If the dynamic relationship between the conflicting demands on the capitalist in the Global North is an important part of the engine of economic growth, we see how that engine is unavailable to countries of the Global South.

The other feature of the Global South that completes the overall picture is its dualism. On the one hand, there is the evident large-scale agricultural operations described in the previous paragraph. On the other, there is a major peasant-like economy that exists mainly in rural areas and the informal sector. That is, two sorts of economies exist side by side—a dualism. This dualism is useful for maintenance of the system as a whole, since there has been a dynamic interchange at the level of worker/farmer. Export agricultural production may go through bust and boom cycles, just as factory production. However, when factory workers in the North are put out of work, there must be social concern because those workers also participate in the functioning of the system (as consumers).

When plantation workers in the South are out of work, there is no parallel social concern, as those workers do not participate in the functioning of the system (they just go back to the farm and sustain themselves there).

Within this dynamic construct of contradiction in the Global North (desiring workers to make as little as possible while, at the same time, desiring workers to make as much as possible), the normal word used to describe the situation is "articulated" (not articulated in the sense of a speech, rather in the sense of an elbow or knee). The similar economic structures in the Global South are not, to any great degree, so articulated. Thus, the typical arrangement in the North is referred to as an articulated economy, whereas that in the South is disarticulated (in the sense that the two main sectors of the economy are not connected, articulated, with one another). Furthermore, because the production systems of the traditional agriculturalists and the export agriculturalists are not connected (at least not in the way the similar processes are connected in the developed world), the economy is dual—the traditional part operates relatively independently from the export part.

This dualism, or disarticulation, goes a long way toward explaining the differences between the similar classes in the North and South. Factory owners in Mexico, whether the private farms that produce export crops or the government company that exports petroleum to the United States, need not overly concern themselves over the fact that the general populace lives a physically deprived present and a hopeless future. They do not buy oil. They do not buy beef. They do not buy any of the multitude of other export products that make the factory owners' profits. On the other hand, they do work in those factories (sometimes), and the incentive to pay them as little as possible in return for their labor remains. The principles of Fordism (workers should make enough money to purchase the products made in the factories in which they work) simply do not apply within these societies as much as they do in the Industrialized North.

On the other hand, as noted previously, the owners in the United States, whether private enterprises or government owned and/or subsidized industries (e.g., defense contractors), care quite a lot that workers in general have purchasing power. Indeed, this is the basis of so-called Fordism, the phase of capitalist expansion roughly corresponding to the period after the initiation of Henry Ford's assembly line. Factory owners always try to pay their workers as little as possible, but that desire is balanced by their wish for the workers in general to be good consumers.

Thus, the different structures of economies (articulated in the North, disarticulated or dual in the South) go a long way toward explaining the relative conditions of the lower classes in those areas of the world. But in addition, there is an important connection between the two systems. Indeed, following Wallerstein, it is not possible to understand either system without understanding how the system functions as a whole. It is a system whose dynamics are derived from a worldwide web of connections, and any attempt to understand it from a smaller perspective is doomed to the metaphor of the blind men and the elephant.

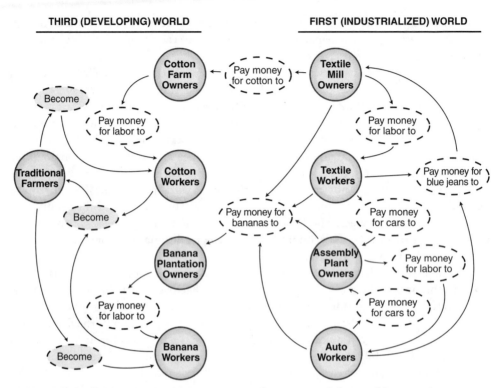

Figure 7-A1 Diagrammatic representation of a section of the world system.
Adapted from: Vandermeer, J., & Perfecto, I. *Breakfast of Biodiversity.* Oakland, CA: Food First Books, 2005.

In **Figure 7.A1**, the basic idea is represented as a simplified flowchart, assuming that the developed country produces only automobiles and blue jeans and that the underdeveloped country produces cotton and bananas. Just follow the arrows (e.g., the auto workers and textile workers "pay money for cars" to the owner of the auto factory who "pays money for labor" to the auto workers). From the diagram we see that in the developed world the money that goes to labor eventually goes to purchase the products, thus providing the machinery of economic growth. The South does not include an arrow that connects the export producers (the factory owners) to the traditional farmers (the workers) in the process of consumption. It is the workers and factory owners of the developed world that are the consumers of the products of the South. Thus, the "owners" in both countries are concerned with the consumer power of the "workers" in the developed country, but neither needs to be concerned with the consumer power of the "workers" in the underdeveloped country.

What remains to be explained, and this is a somewhat more difficult concept, is how this dualism in the South is maintained. On the one hand, supplying raw materials, food, and luxury products at extremely low cost (e.g., cotton, timber, grains, sugar, coffee, bananas, bauxite, nickel) is clearly beneficial to the developed part of the world. On the other hand, an additional, more systemic dynamic is involved. The developed world, because of its basic structure especially during the period of Fordist expansion, tends to go through cycles of bust and boom, sometimes severe and other times just annoying. One of the underlying bases of those cycles is the classic business cycle resulting from periodic underconsumption crises, as described previously.

Whether the real basis is an underconsumption crisis or not, the reality of cycles cannot be ignored, at least during the roughly two centuries we might refer to as modern (i.e., 1800 to today). Thus, capitalists must face periodic downturns. During those low economic times, investment opportunities are scarce. Clearly, without the presence of a South, the crisis would be of a qualitatively different sort, in that the South provides a sink for those investments during rough times in the North. In this sense, the dualism of the South is functional for the preservation of the system as a whole. It functions to provide an escape valve for investors from the developed world. The Global South, in addition to its function as a source of raw materials and luxury products for the North, also provides a production and marketing outlet, especially during times of crisis in the home country. Union Carbide located its plant in Bhopal, India, not Great Neck, New York. U.S. pesticide companies export insecticides banned in the United States to countries of the South. U.S. pharmaceutical companies pollute the ground water in Puerto Rico because they were not able to treat the ground water as a commons in the United States (or the court costs would have been too great). In all cases, the people of the Global South must accept such arrangements—it is part of how the world system works.

With this framework, it is evident that the developed world has been so successful at economic development, first because it has an articulated economy and second because it is able to weather the storm of economic crisis by looking for investment opportunities in the South, in addition to counting on cheap raw materials from there. The South in contrast has been so unsuccessful because its economy is disarticulated and thus dual, lacking the connection that would make it grow in the same way as the developed world. At a more macro scale, the dualism of the South is quite functional, in that it maintains the situation in which investors from the developed world can use the South as a source of raw materials and as an escape valve, especially in times of crisis. Indeed, the developed world is developed at least in part specifically because the underdeveloped world is underdeveloped.

This picture is very general and certainly does not strictly apply across the board. Many countries simply do not fit into the picture. For example, Taiwan and South Korea have recently experienced remarkable levels of economic growth. Indeed, many economists now categorize South Korea as a developed country. Hong Kong and Singapore have few peasant farmers. Finnish capitalists have few investments in the South, and none of the former command economies (the eastern European countries) had significant investments in the South. Indeed, with the new globalization that emerged at the close of the

20th century, many aspects of the dynamics of the world system are still being rewritten. Communism fell in the old Soviet Union, but the new Russia has not been the expected compliant tenant as it looks for its new place in the international order. Yugoslavia broke up, and past wounds were exploited by local warlords as much as international players. Brazil signs commercial contracts with China at a rate unheard of just 10 years ago, whereas India and Russia coddle favor with both in the new BRIC block (Brazil, Russia, India, China). Although Wallerstein's vision of a total world system is certainly still relevant, his notion that the rules under which that system operates change continually is perhaps what is important at the time of this writing. Indeed, Wallerstein more recently argued that the world system is coming apart and a major shift in how it operates is on the horizon, perhaps involving unprecedented levels of violence.[46] This prescient writing was before September 11, 2001, or the subsequent invasion of Iraq and Afghanistan.

With hundreds of independent countries each clamoring for a place at the world economic table and struggling to use their own natural resources for economic gain, environmental and social problems continue emerging. It is is not exactly a well-tuned orchestra that constitutes this World System, even though its underlying structure may be at least qualitatively understandable through Wallerstein's basic framework. At a more practical level, things have become extremely chaotic, especially during the post-World War II period of massive and uncoordinated growth. Managing this global economic and ecological cacophony has been difficult. Understanding the myriad of international relations relevant to this management requires an understanding of the tools available to the manager in the modern era.

Concerning the Definition of Capitalism

There is an unfortunate tendency to conflate the ideas of the "free market" with the definition of the Capitalist system. Wallerstein's framing insists that the defining feature of Capitalism is its functioning as a system worldwide, that the complex structures of markets are certainly part of that system, but nevertheless, it functions as a world system, not as a "theory of the firm," which is sometimes taken as the basis for understanding not only price formation but also as a definition of capitalism itself. Nevertheless, the world systems approach maintains the fundamental idea of social classes defined by internal (within class) self-interest and focusing on where those self-interests are contradictory. It is common to assume that those classes are more or less the classic classes of capitalist (owner) and proletarian (worker) and proceeding to analyze the system based on the contradictions between them. Such a vision has certainly been useful in the past. However, in the case of applying social theory to the agroecosystem, it may be the case that an alternative formation will be useful.

Of particular importance is the work of French historian Fernand Braudel. Braudel's writings were undoubtedly influential in many ways in the development of the world systems approach to social history, but his contributions that most specifically impact the understanding of agroecosystems are two: first, his notion of three scales of dynamic change, and second, his analysis of capitalism as the monopoly takeover of "free" markets.

Regarding time scales, Braudel recognized the distinct dynamics inevitably involved in three different scales. First is the scale of geography and environment, where forces largely, but not completely, independent of human activity are the main determinants. Today we recognize global climate change as an important force that drives major changes in the environmental background to agroecosystem development. Changes in soil, water, biodiversity, and climate all operate at levels that, although influenced by human activity, also have their own internal dynamical rules and change normally quite slowly. Nevertheless, their changes may be determinate in transformations of sociopolitical structures.

The second scale is the so-called long duration (*longue durée*), in which fundamental socioeconomic structures are seemingly impervious to change. Although the particulars of what he sees as these "stages" of world development are interesting, what is important for understanding agroecosystems is the general concept. The long duration systems represent something like an underlying structural arrangement on which individual events take place. Taking this point of view avoids the pitfalls of assuming that underlying social structures are akin to physical laws, inviolable and tinkered with only at the peril of attempting to change some "natural" course of events. Yet, at the same time, it permits an analysis that is more than just historical narrative, an underlying structure that indeed may inform us in the same way that physical laws do. In European Feudalism, decisions of the monarch or landed gentry were inviolable, and it makes sense to seek understanding of particular socioeconomic dynamics by studying the forces that impact those centers of power, whereas in modern capitalism, the interests of centers of financial power effectively constrain socioeconomic developments. We no more seek to understand 13th century France by appealing to the machinations of the international banking industry than we contemplate the world Capitalist system today as driven by the needs of Queen Beatrice of the Netherlands. Yet, in their day, those seats of real political power would have been part of an underlying sociopolitical structure that provide meaning and understanding to particular events.

Third is the scale of particular events (*histoire événmentielle*). The Great Depression of the 1930s was of historical significance and is best understood as part of the underlying dynamics of the capitalist system, but is certainly different from the Central American wars of the 1980s, even though they too were part of the capitalist system. Particular events may carry the weight of the underlying socioeconomic structures but frequently must be analyzed in light of their own unique circumstances.

In addition to this tripartite classification of time scales, Braudel's approach to capitalism was influential and most important for the analysis of agroecosystems. In opposition to classic Marxian scholars, Braudel rejected the grand deterministic trajectory of capitalism inevitably overtaking feudalism and the central place of the bourgeois revolution. In particular, the Marxian interpretation suggests that extraction of surplus value from labor is the central defining feature of capitalism, distinguishing "petty capitalists" from true capitalists, the former being small scale, the latter being large and usually monopolized, but effectively rendering them both in the same analytical category of "capitalist." Braudel's categories recognize (as did Marx) the productive potential of petty capitalists and the associated free markets they participate in, as generally a good thing, a step forward from the feudal peonage such actors used to be involved with. However, Braudel recognizes (as did Marx), the tendency to monopolize. Braudel parts with Marx on the basic definition of capitalism in that he insists that capitalism is that system that is dominated by monopolized markets. This distinction is especially important for understanding agroecosystems in that many of the actors (i.e., small farmers) remain in the category that Marx would have called "petty capitalists," or perhaps "peasants." Standard Marxian theory, classic economics, and neoliberal ideology all share the same assumption that the elimination of this class is inevitable (and largely positive). Braudel's definition of capitalism is effectively the point where monopolization comes to dominate, thus violating standard economic assumptions of a large number of small independent producers participating in free markets. His framework permits at least a thought experiment involving political power focused at the level of a true entrepreneur (i.e., one who seeks technological and social advancement as a fundamental goal, rather than the pure accumulation of capital).

From Bretton Woods to the Asian Tigers

World War II left Europe and Japan totally devastated. The challenge of the Soviet Union and its allies was announced and palpable. In face of this challenge, the last thing that the West needed was an unstable world economy, which a devastated Europe and Japan could have generated. There was a general realization that the origins of the war were mainly economic, including the Great Depression, the German malaise after World War I, the complete collapse of the German economy, and the expansionist desires of Japan. To analyze all of these problems with an eye to averting them in the future as part of the new Cold War strategy, the allies convened a conference at Bretton Woods, New Hampshire, in 1944. The importance of this conference cannot be overemphasized.

Realizing that much of the chaos that led to World War II was economic in origin, the Bretton Woods conference emphasized economic arrangements that would stabilize the world economic system. Central to the main goals was the idea of exchange rate stabilization. Fixed exchange rates were established to stabilize currency fluctuations so that investors and traders would face less of a risk. All currencies were pegged to the U.S. dollar and that was fixed at 1/35 of an ounce of gold. The basic ideas were seemingly sound if you looked at the system from a short-term point of view but had serious drawbacks if viewed from a longer term perspective. For example, in order to maintain the credibility of the fixed exchange rates, the actual exchange rate (the one people actually trade for on the street) could not be too different from the theoretical fixed rate. To manage this need, central banks were forced to spend U.S. dollars to purchase their own currency (to create a bigger demand for their currency when its value began to fall on the street). Financial speculators could thus anticipate particular countries running out of U.S. dollars, which effectively placed great downward pressure on domestic money outside of the United States. Because the U.S. dollar was pegged to gold, however, there was no mechanism for the United States to lower the value of the U.S. dollar (which, of course, would have been another mechanism to maintain the credibility of foreign exchange rates).

To implement this foreign currency arrangement, the Bretton Woods conference created the IMF whose original mandate was 1) to promote international monetary cooperation, 2) to facilitate expansion of international trade, and 3) to promote exchange rate stability. The mechanisms at the IMF's disposal were basically two: structural adjustment programs (SAP) and bailouts. Since its inception, the IMF has come under severe criticism from many quarters. The critics have noted that the SAPs in general have been devastating to many countries, mainly those countries in the South that began the post war process in an economically depressed condition. The second, more structural criticism of the IMF has been that its bailout programs have mainly benefited the international banking industry, forcing country after country to forego development of social infrastructure (e.g., schools, healthcare facilities) so as to repay international banks the loans that were, sometimes, given out to corrupt officials who used the money for personal gain. The critics of the IMF say, "Why bail out banks that have made bad loans?" The defenders of the IMF say, "If the loans are not repaid, future loans will never be made." According to Chalmers Johnson, the IMF has become "the premier instrument of deflation, as well as the most powerful unaccountable institution in the world. The IMF is essentially a covert arm of the U.S. Treasury, yet beyond congressional oversight because it is formally an international organization."[47]

Although IMF critics certainly have a point, much of the structure of SAPs makes sense if viewed from the constructs of IMF thinking. For example, consider interest rates on loans to farmers. Given a particular rate of inflation, if a farmer takes out a loan at an interest rate that is lower than the rate of inflation, he or she would be better off converting the money to dollars, waiting until the loan comes due, and then reconverting to local currency. The local inflation would have made a profit without using the loan for any particular productive purpose. This obvious problem causes the IMF to require governments that receive aid to charge interest rates at something above the local inflation rate. This clearly makes sense from the IMF's point of view. However, in many of the countries that must confront these SAPs, inflation rates are so high that they are required to charge interest rates on agricultural loans that are well above anything that would be acceptable to any farmer, thus facing increased political instability. Thus, although the IMF is making decisions that are perfectly consistent with "rational" economic policy, that policy becomes devastating to the stability of the country receiving the aid.

In addition to the IMF, the Bretton Woods conference also set up the World Bank (formally titled the International Bank for Development). The World Bank was originally created for the reconstruction of Europe and Japan, so devastated from World War II.

But the Bank rapidly evolved into the primary financier of development projects in the South, with a formal goal to reduce poverty worldwide. Projects are funded through grants and loans throughout the Global South with the intent of creating infrastructure that will lead to economic growth. Currently, the countries of the South owe the World Bank more than 300 billion U.S. dollars (Indonesia alone owes $132 billion and India over a billion).[48] Furthermore, poverty in the Global South is comparatively worse (excluding China) than when the World Bank started. Many current officials at the World Bank acknowledge failure at their most basic task of reducing poverty. However, most recently the major criticisms of the Bank have been associated with the environmental problems created by its projects. For example, it has been necessary to resettle at least 5 million people because of the World Bank's program of building large dams.

The Vietnam War injected a fundamental change in this overlying structure. Reflections on the political and moral aspects of this particular war would take up too much space. However, for understanding the world system, it is necessary to understand that the Vietnam War ushered in several important global changes. First, there have been enormous political restrictions on the use of U.S. military forces ever since. It is even suggested that the senior President Bush was unable to "complete" the work of the Gulf War because he was politically unable to commit ground forces to invade Bagdad, and as I write these words, support for the occupation of Iraq by his son has dwindled to less than one in three U.S. citizens. The massive propaganda campaign unleashed in preparation for that occupation would not have been necessary before Vietnam.

At a global level, Vietnam caused an international delegitimization of the United States. Because of its central role in the defeat of Hitler in World War II, the United States had gained enormous moral authority. That moral authority was dramatically squandered in Vietnam (and subsequently all but eliminated with the occupation of Iraq). Also, there was an attitude change in the South as a consequence of the war. The CIA's victories in Iran and Guatemala were more than canceled in Vietnam, and countries of the Global South regained a notion that they too could struggle for independence. Finally, Nixon was forced to eliminate the gold standard on which the entire world system set up in Bretton Woods was based. The cost of the Vietnam War was so exorbitant that the only way to pay for it was by minting more money than the gold reserves would sustain, given the fixed standard of 1/32 ounces of gold for each dollar.

By eliminating the gold standard, the West was effectively returned to the monetary system of the 19th century, a notoriously unstable system. Indeed, the entire rationale for the Bretton Woods conference was to create economic stability in the world, and the principal vehicle was supposed to be the rock solid dollar, which was rock solid because of its pegging to gold. That advance (if indeed it was an advance) was eliminated in 1974 when Nixon was forced into his action. The worldwide economic instability seen in recent years (e.g., during the 1990s and early 2000s and then even more spectacular in 2008) is at least in part a consequence of that action (although some would argue that the original Bretton

Woods institutions were flawed to begin with and doomed to failure anyway—but those arguments are beyond the scope of this text).

Partly as a response to the failure of U.S. policy in Vietnam and partly because of the obvious success of the Japanese model subsequent to World War II, other Asian countries were encouraged to take up the Japanese development model. In particular, first South Korea, Taiwan, Hong Kong, and Singapore, and somewhat later Malaysia, Indonesia, Thailand, and the Philippines were the candidates for this new strategy. Collectively, they became known as the Asian Tigers. Mainly a product of the still-hot Cold War, these countries were encouraged to develop industries or services that aimed at exports, especially to the United States, coupled with the protection of local industries (high import tariffs on anything produced in the country). The consequences of this strategy were overinvestment and excess capacity in almost every new industry, a massive deformation of any sort of supply/demand equilibrium, and the world's largest trade deficits in the United States. This latter point has had an enormous effect on employment in the United States, changing the nature of employment from a dominance on secure high-paying generally union jobs to a dominance on low-security, low-paying nonunion service sector jobs. It has also resulted in the debilitation of the manufacturing sector in the United States. Asian historian Chalmers Johnson has noted that "to base a capitalist economy mainly on export sales rather than domestic demand . . . ultimately subverts the function of the unfettered world market to reconcile and bring into balance supply and demand."[49]

As Johnson explains, export-led growth depends on strong foreign demand—in the case of the Asian Tigers, an artificial demand that was created and sustained by the American imperialist power. The strategy worked because of the exceptionally strong American economy during the Cold War—but only while Japan and other dependent economies remained relatively small. Eventually the Japanese economy grew so large that it began to disrupt not only the American economy, but economies throughout the world. When its neighbors began to pursue the same strategy, they generated an immense level of overproduction that simply could not be absorbed. As a result, Johnson writes, "There were too many factories turning out athletic shoes, automobiles, television sets, semiconductors, petrochemicals, steel, and ships for too few buyers."[48]

The Asian Tigers were remarkably successful, however, even if that success was temporary. After the fall of the Soviet Union and the end of the Cold War, there was an attempt to maintain Cold War structures, since they had been so successful. From 1992–1997, there was an ideological campaign promoting "free trade" as a new religion. But starting in 1997, the unrestrained weight of global capital came to bear. Loans to the tigers were recognized as nonperforming (and probably nonperformable). And, worse for ideological reasons, countries that had followed IMF advice were particularly vulnerable. Loans were called home, and immense downward pressure on local currencies became the rule. Then, in 1998, the crisis hit. All of the Asian tigers went into economic free fall, and they have not completely recovered yet.

The Neoliberal Model

As the Asian Tigers were beginning their journey, a new politicoeconomic framework was in the works.[50] In 1947, an influential group of economists and political philosophers gathered at a Swiss spa to form the "Mont Pelerin Society," the founding statement of which includes

> The group holds that [the loss of freedom and human dignity] have been fostered by the growth of a view of history which denies all absolute moral standards and by the growth of theories which question the desirability of the rule of law. It holds further that they have been fostered by a decline of belief in private property and the competitive market; for without the diffused power and initiative associated with these institutions it is difficult to imagine a society in which freedom may be effectively preserved.

The members of the Mont Pelerin Society regarded themselves as liberal in the classic sense, and thus their new theoretical formulation came to be known as neoliberalism.

Attending this foundational meeting in 1947 was economist Milton Friedman along with Austrian political philosopher Friedrich von Hayek. Although von Hayek provided the philosophical framework, Friedman took that framework to the University of Chicago from where it extended tentacles into university economics departments throughout the United States and United Kingdom, gaining influence in powerful circles wherever those two countries were dominant. The basic idea is that unfettered markets are more able to make decisions about the distribution of goods and services than any planning board or citizen group or government. Consequently, any restriction on so-called free markets was anathema to efficiency. The role of government, then, needed to be completely transformed. It was to be nothing but the guarantor of the unfettered freedom of markets, mainly through the control of labor and restrictions on government or, for that matter, any other organized pressure groups. The undeniable tendency of capitalism to produce inequalities was countered with the now famous idea that a rising tide lifts all boats, that enrichment of the rich was a good thing because those riches would "trickle down" to the poor, and that everyone would prosper.

The context of the formation of the Mont Pelerin Society was, ironically, the success of a totally distinct economic model, the one constructed by John Maynard Keynes in which the state intervened aggressively to prevent market distortions and excessive drift in obviously negative directions, inventing regulatory structures in the 1930s that effectively steered the United States out of the Great Depression. Neoliberalism explicitly rejected Keynesianism.

However, neoliberal policies remained in the shadows for some 3 decades finally to be embraced with gusto by the administrations of Ronald Regan in the United States and Margaret Thatcher in the United Kingdom. In both of those countries, the assault on Keynesianism and any other remnants of the progressive decades post-World War II was vigorous. With distinct methods in the two countries (see Harvey[51] for details), an unrelenting policy was pursued with, of course, the major funding supplied by those whose boat would be raised highest. By the time of the Clinton and Blair Administrations (the 1990s), the neoliberal rules were so embedded in the U.S./U.K. system that even two leaders with progressive credentials were boxed in to work within its framework.

NAFTA (establishing a so-called free trade zone among the United States, Canada, and Mexico) was pushed through by Clinton over major objections from his own political party. Although it was not the only such agreement, it has become a major symbol for the penetration of the neoliberal model into the underdeveloped Global South. As the Asian Tigers were collapsing in 1996–1997, the full weight of the neoliberal model was brought to bear on the rest of the Global South.

The most recent manifestation of the trend toward neoliberal globalization has augmented the Bretton Woods institutions with the newly created World Trade Organization (WTO). The overall goal of the WTO is to promote prosperity through international "free trade." It is attempting to do what had been done earlier in Europe as part of the Industrial Revolution, but without the key factor of creating a free labor market. That is, in the same way that individual parishes in England had different economic rules and regulations and it was necessary to unify them all under the same set of economic regulations, so the current globalization movement is trying to do the same thing but at an international level, and with unrelenting neoliberal ideology. Today's nation/state occupies the position held by the parish in preindustrial Britain.

The WTO is the organ that is intended to supply the overall supervision and legitimization of the new global order. Unfortunately, its current manifestation is such that it has no accountability to anything resembling democratic institutions. It operates in secret, is not obligated to explain or justify any of its actions, and is accountable to no party at all. It is largely composed of economists allied with large corporate interests and has thus far acted so as to smooth the way for corporate activities to be carried out without disruption wherever and whenever they are so desired. Most vociferous complaints about the WTO have been from environmentalists and labor groups. The WTO, through its strict ideology of the market as religion, has acted and is expected to act repeatedly to thwart local environmental and labor laws in favor of corporate interests.

The fly in the ointment emerged in the 2000 meetings of the WTO in Seattle, Washington. A coalition of environmentalists and unionists came out in force and shut down the meeting. A great deal of world attention was focused on the WTO as a result of the previous fiasco in Seattle, and ever since that time, the WTO has been unable to meet without a retinue of security that makes the whole operation look at best a bit suspicious, at worst, according to some critics, like a new brand of fascism.

Many of those critics fear that the current situation of inequality in the world is going to be worsened by the new globalization movement. Today, the richest 10% of adults account for 85% of world wealth, and fully 40% of the wealth is owned by just 1% of the population.[52] Some people suggest that leaving 90% of the world with only 15% of the world's goodies is immoral, some suggest it is unfair, some suggest it is in the end self-defeating.

Endnotes

[1]Lawton and May, 1995; Wilson, 1986.

[2]Wilson, 1987.

[3]May, 1988.

[4]Ereshefsky, 1992.

[5]Hubbell, 2001.

[6]Shiva, 1997.

[7]Loreau et al., 2002.

[8]Yachi and Loreau, 1999.

[9]Palmer, 1994.

[10]Stout and Vandermeer, 1975.

[11]This relationship was first noted by Arrhenius in 1921 and has been discussed and criticized extensively since. See, for example, Ugland et al., 2004; Tjørve, 2003; Tjørve, and Tjørve, 2008.

[12]A number of computer packages are available. My favorite is *estimateS*, written by Robert Colwell of the University of Connecticut.

[13]Recent literature has emerged regarding the social dislocation of small farmers and biodiversity (Aid & Grau, 2004; Green et al., 2005). The migratory activities of peasant and former peasant agriculturalists are complicated and beyond the scope of this text. The hopeful premises of conservationists are not likely to be borne out in practice. Former agricultural land left "idle" by poor farmers does not generally revert to biodiversity-rich natural vegetation, but frequently falls prey to more "efficient" agricultural production, which to add insult to injury, is generally biodiversity poor (Garcia-Barrios et al., 2009). This issue is discussed in the last section of this chapter.

[14]During the years 1960 to 1989, coffee producers around the world were organized into a cartel that secured effective supply management by holding back on production at national levels. Not surprisingly, the formation and maintenance of this cartel were strongly supported by the West. As the Soviet Union fell, that support was removed, and the cartel was immediately disbanded (Pendergast, 2000).

[15]Although the success of the Green Revolution (in productivity terms alone) is frequently attributed to the genetic changes in the crops themselves, the fundamental evolutionary principle that adaptation occurs within the framework of particular environmental circumstances holds here as much as in other cases of evolution. The environmental "constraints" included irrigation, chemical fertilizer (especially nitrogen based), and pesticides. The fact that varieties evolved to be more efficient in this environmental background than in a background of low water, low nitrogen, and abundant pests is not in any way surprising. As a general evolutionary principle, the performance of a variety adapted to environment A is better than a variety not adapted to environment A. Suggesting that there is something remarkable about the fact that the adapted variety performs better in environment A than the nonadapted variety betrays a naivety about the basic structure of biological evolution. To suggest that the Green Revolution was a great success in terms of production makes this elementary error.

[16]This section follows closely the interpretation of Patel (2008).

[17]This statement will attract scorn in some quarters. For the interested reader, the argument of Badgley et al. (2007) is unassailable.

[18]Harvey, 2007.

[19]Aid and Grau, 2004; Perfecto and Vandermeer, 2008, Perfecto et al., 2009.

[20]Pikovsky et al., 2003; Bennet et al., 2002.

[21]Liebold et al., 2004; Holyoak et al., 2005.

[22]See literature reviews in Perfecto and Vandermeer, 2008; Schroth et al., 2004; Perfecto and Armbrecht, 2003; Vandermeer et al., 1998.

[23]Hirsch, 2003.

[24]Perfecto and Armbrecht, 2003.

[25]Ashton and Montagnini, 1999.

[26]Cullen et al., 2005.

[27]Goulart et al., 2007.

[28]Desmarais, 2007.

[29]Rosset, 2006.

[30]Nelson (1994) has been the only one to my knowledge that has approached this problem experimentally, documenting a "farmer-first" approach to the development of pest management in tomato production in Nicaragua. Morales (2002; Morales and Perfecto, 2000; Morales et al., 2001) has been one of the most persistent advocates of an integrative approach to understanding the mutual interaction between farmer and researcher.

[31]Vandermeer and Perfecto, 2005a, 2007a.

[32]MacArthur and Levins, 1967.

[33]Levins, 1979.

[34]Armstrong and McGehee, 1980.

[35]Huisman and Weissing, 2001.

[36]Vandermeer et al., 2002.

[37]Huisman and Weissing, 1999.

[38]Carpenter and Kitchell, 1996.

[39]Paine, 1966.

[40]Janzen, 1970; Connell, 1971.

[41]Hassell, 1978.

[42]May and Nowak, 1994.

[43]Hubbell, 2001.

[44]Vandermeer and Perfecto, 2005b, 2007b; also see our book *Nature's Matrix* (Perfecto et al., 2009).

[45]Wallerstein, 1974, 1980, 1989.

[46]Wallerstein, 2004.

[47]Johnson, 2004.

[48]www.democracynow.org, 2005.

[49]Johnson, 2006.

[50]Harvey, 2007.

[51]Harvey, 2007.

[52]Cornia and Court, 2001.

Chapter 8

Toward a Sustainable Future

Overview

This chapter begins with some observations of several agroecosystems around the world to remind the reader that this book intends to cover agroecosystems in their rich interdisciplinary diversity. The section that follows presents a critical analysis of the current state of the one agroecosystem that dominates the world today, the industrial agroecosystem. As a counter, the next section outlines some of the issues associated with the development of an alternative. This is followed by two sections devoted to some details that must be considered in pursuing that alternative—first focusing on the more technical issues and then on the more socioeconomic issues. Finally, in the concluding remarks is a vision of the attitude required for a productive search for that alternative.

Agroecosystem Viewpoints

This book began with a series of vignettes designed to explore the broad range of subjects that should be included in a proper framing of the science of agroecology, writ large. The human tragedy of the Irish Potato famine, sun coffee's contribution to the global biodiversity crisis, and the national experiment with alternate forms of production in Cuba, together illustrate the broad range of concerns that define agriculture as an ecological subject. Although the book has had a technological bias, I have tried to continuously reflect the interdisciplinary nature of the subject matter with a tapestry of subjects as diverse as anthropology, chemistry, sociology, and history. That tapestry defines, for me, the ecology of agriculture. It cannot be about nutrient cycling alone, about predation and parasitism alone, about plant competition alone. The agroecosystem is the most fascinating ecosystem of all because of its intimate connection with the keystone species, *Homo sapiens*. Studying it without incorporating that species would be intellectually naïve—similar to studying the prairie ecosystem without considering fire, the intertidal ignoring tides, or the African savannah disregarding the large mammalian herbivores.

On the other hand, there is no doubt that agroecosystems are under management by incredibly distinct human cultures with incredibly distinct operating procedures. Even so, certain commonalities are evident. Plant competition is the same wherever you go, and herbivores can sometimes become common enough to be classified as pests, to say nothing of soil falling off devegetated hillsides. Although human relations are always involved, they are not able to cancel natural ecological forces, despite the dreams of some technocrats.

The three vignettes were chosen arbitrarily. For me, they bring to mind a plethora of personal experiences in agroecosystems the world over, illustrating both the convergences and divergences of the human influences. Standing in the middle of an organic strawberry field in California, the farmer, under contract to a giant berry corporation, talks of the problems with organic production in strawberries (**Figure 8.1a**). For him it is not a question of maintaining sustainability or harmony with nature or any such lofty ideals. It is a question of satisfying the requirements of organic production, implying that the particular standards of California organics are too strict. It is the problem of workers' rights, which for him are too lenient in California: "In this state, the field workers dictate agricultural policy." It is the problem of timing, with strawberries needing to get to the processing plant as quickly as possible, with a workforce that cares little for either the quality or quantity of strawberries.

(a)

(b)

(c)

(d)

Figure 8-1 (a) California organic strawberry farmer in his strawberry field. (b) Nicaraguan farmer in his silvopastoral system. (c) Indian farmer in his mixed agroforestry system. (d) Farm Labor Organizing Committee and supporters march in 1978 in Ohio.

My conversation could not be more different as I speak with a farmer friend in Nicaragua (Figure 8.1b). His cattle have become lazy because they have learned to hang out under the orange trees waiting for oranges to fall—they seem to prefer the sweet orange pulp to the grass that they actually are adapted for eating. He has a problem with a butterfly that bores into his orange trees, but has devised a method of controlling them with cow manure plastered to the base of the tree. He heard about this from a neighbor. He still has trouble getting his oranges to market because of the impassability of the road in the wet season, but is disconcerted about government plans to pave it, because that will bring all sorts of city problems to his idyllic farm. He even suggested he might have to use an assault rifle to protect his family. Worse of all are the migrants who come from the Pacific side of Nicaragua (he calls them Spaniards—he himself is a black English-speaking "Creole," one of the three main ethnic groups of Nicaragua's Caribbean coast). They do not understand local ecological conditions and are treating the land badly.

Even more different is my conversation with an Indian farmer in the state of Kerala, India (Figure 8.1c). Farming coffee and about 20 other crops on an area of 1.5 hectares and marketing only locally, he has produced an upper middle-class lifestyle. He has a two-story house that could be located in any middle class neighborhood in Ann Arbor, Michigan—satellite dish, automobile, motorcycle, kids with shoes, and access to free health care and public education. And he does it all on 1.5 hectares. He claims to be not exceptional, that most of the farming families in the area have a similar operation. His farm is organic, but he does not regard that as anything special and certainly not certified. He simply does not need chemicals. I imagine asking him what he thinks of my friends in Mexico who would claim that farmers simply cannot make it on 1.5 hectares, but I refrain.

Yet a different point of view comes from talking with a migrant farm worker on a picket line in Ohio (Figure 8.1d). She migrates up from Texas to pick pickles and tomatoes (or did so a few years back) in Ohio, Indiana, and Michigan. She understands the problems of soil fertility and the devastation that can come with tomato blight, but that is far from her main concern. Picking tomatoes and pickles is backbreaking work, and she wants better compensation, which she thinks comes only with union recognition, which a local union organizing effort (Farm Labor Organizing Committee) has been working on for many years. She recalls her childhood when her main concern was whether her family would get back to Texas in time for her to start her regular school or whether she would have to begin school again in Ohio and withstand the racist taunts of local Ohio whites who always referred to her as "Mexican" (she was born in Texas). It was not so much the experience itself, but rather the constant insecurity. Her mom had to balance the educational quality for her children, which began with school in September in Texas, with the earning potential for picking tomatoes, which was in September in Ohio.

In these and many other casual experiences with farming people across five continents, I have developed an appreciation for the diverse functionality of agroecosystems: for the way they function well, or not; for the way in which what I had always referred to as the natural world was intimately connected with humans and their varied forms of sociality. Even so, I also saw the consistencies—soil fertility carries the same implications in

California, Nicaragua, Kerala, and Ohio. Insects can destroy a crop whether its purpose is to feed your family, sell on the local market, or harvest for a fee. There are certain ecological consistencies. And those consistencies are what I have tried to describe in detail in this book. My approach has been as nonnormative as possible, but I have not tried to hide the fact that I advocate a change in the dominant paradigm of industrial agriculture, that I do believe we need to change the way agroecosystems operate in the contemporary world. "Business as usual" will not do if we seek a better future.

In 2008, the United Nations' International Assessment of Agricultural Science and Technology, an international collection of several hundred scientists and other people associated with the agricultural enterprise, completed their report. The executive summary notes:

> This model [industrial agriculture] drove the phenomenal achievements . . . in industrial countries after World War II and the spread of the Green Revolution beginning in the 1960s. But, given the new challenges we confront today, there is increasing recognition . . . that the current AKST [Agricultural Knowledge, Science and Technology] model requires revision. Business as usual is no longer an option. This leads to rethinking the role of AKST in achieving development and sustainability goals; one that seeks more intensive engagement across diverse worldviews and possibly contradictory approaches in ways that can inform and suggest strategies for actions enabling to the multiple functions of agriculture.

Although the report runs the gamut from genetically modified organisms to traditional knowledge and provides a forum for all opinions, it nevertheless reveals a spectrum ranging from those who seem to see little problem with the way things go according to the industrial model and those who perceive the need for an alternative. This text tends toward the alternative side of that spectrum. If the overall situation is judged fairly, in my view, there is little doubt that the industrial system has reached the point where it creates more problems than it solves and some sort of alternative is desperately needed. In this final chapter, I ask two basic questions: What precisely is the state of the current system? How can an alternative evolve?

State of the Late Industrial System

Appropriation, Substitution, and Oligopolies: Reviewing the Past

With characteristic acumen, Lewontin crystallized the defining feature of industrial agriculture: "Farming is growing peanuts on the land; agriculture is making peanut butter from petroleum."[1] This simple statement summarizes much of the state of modern agriculture as it has evolved since the industrial revolution and especially since World War II. Because of this structure, farmers, those who actually have contact with the land, are caught in a great squeeze. Their metaphorical petroleum suppliers are large monopolized corporations whose selling prices are set artificially high, and their metaphorical peanut butter makers, who buy the peanuts they produce, are large monopsonized corporations whose purchasing prices are set artificially low. The farmer is frequently caught in the middle. Powerless, both economically and politically, he and she can only hope for government handouts to make up the difference between capital outlay and gross income. How this came to be is complicated to be sure, but some general ideas can guide thinking about the problem. "Appropriation" of some aspects of ecosystem function and "substitution" of agroecosystem outputs are two such general ideas.

Three general cases of appropriation have been discussed at various points in this text: mechanization, fertility of the soil, and management of pests. In all cases, an inevitable consequence of the appropriation was the ultimate control of farm inputs by oligopolies. Names such as John Deere and Caterpillar dominated the farm machinery industry; Monsanto, Syngenta, Dow, and Dupont the pesticide industry; BASF and Bayer Cropscience for fertilizers. Of course, there are others, but few would deny that such oligopolization is characteristic of the industries that supply inputs to agriculture in the industrial world.

Substitution has had a similar effect on the outputs. In Lewontin's aphorism that modern agriculture is manufacturing peanut butter from petroleum rather than growing peanuts on the land, there was a substitution of peanut butter for peanuts. So-called substitutionism[2] has had a somewhat different evolution than the appropriationism that characterized farm inputs. Consider, for example, the grain trade, in which selling grains of wheat to consumers was transformed into selling bread to those consumers—wheat was substituted with bread. How did that substitution come to be?

On the eve of the Industrial Revolution, long-distance trade in agricultural products had been largely restricted to luxury goods such as chocolate, tea, coffee, and sugar. Cotton and wheat, the main agricultural goods that fed the Industrial Revolution, were hardly traded internationally at all before the late 18th century. There was not enough demand because wool supplied the textile industry and local production of wheat fully satisfied carbohydrate needs: for porridge, the staple of the lower classes, and bread, a luxury for the rich. To make bread, wheat had to be ground into flour, mixed with water and yeast, kneaded into dough, allowed to rise, and finally baked. Such a large input of labor could be afforded only by the well-to-do.

Toward the end of the 18th century, a change in the social organization of work caused a dramatic change in this pattern—the working class throughout industrializing Europe began to eat bread. Improved milling technology had caused the price of bread to decline, making it more available to the masses. This, coupled with the movement of large sectors of the population out of the countryside into the urban factories, resulted in a dramatic increase in bread consumption and a concomitant increase in the demand for wheat. Local agricultural production could not keep pace. This led to an increase in the international trade in wheat, starting with expropriations from colonies in Asia and Africa (to say nothing of Ireland, one of the vignettes that began this book). This ended with the international wheat trading business complex, dominated worldwide by a few oligopolistic corporations—Bunge of Argentina, Continental of New York, Cargill of Minneapolis, André of Switzerland, and Louis-Dreyfus of France. These five giants dominated the world grain trade much like the Seven Sisters dominated the world petroleum trade for so many years. By the time the modest grain trader Louis-Dreyfus, founder of the company that still bears his name, arrived in Odessa, the serf farmers of Russia had little knowledge of or control over what happened to their grain after it left their farm. Those who purchased it, however, were involved in high finance throughout the industrial world, assessing wheat shipments as they passed through the Bosphoros (the narrow canal between the Black Sea and Mediterranean) and traveling overland to trade in future shipments of grain in the big trading houses in Western Europe. The selling of grain by farmers to consumers was forever transformed, and large grain companies came to dominate the grain business. From the point of view of consumption of agricultural products, bread substituted for wheat in the pantry of the consumer.

This pattern had been most highly developed in the grain trade, the remnants of which remain today with companies such as Cargill and Archer Daniels Midland. However, virtually all farm products delivered to the consumer saw similar substitutions—canned replace fresh vegetables, ketchup rather than tomatoes, corn chips rather than corn, in short, all processed foods. The substitution has been driven to miniscule levels of industrialization, with virtually everything a modern family in the developed world eats containing products of maize. Beginning with the Industrial Revolution and continuing with only minor variation today, the buyers of agricultural products grew large rapidly, and most financial decisions came to be made far from their source, the farmer.

The Political/Economic Structure of the Industrial System

Given the appropriation of inputs for capital in the form of machines and chemicals and the substitution of outputs for capital in the form of processed food, we see a general trend toward incorporating agriculture into the industrial capitalist system. Visionaries could see the inevitable transformation of agriculture as perfectly parallel to other forms of production, as represented in Samuel Copeland's famous 1865 *Agriculture, Ancient and Modern* in which the future was vividly portrayed (see Figure 2.16). Recall the problem. At the time of the great changes induced by the Industrial Revolution, agriculture remained something of an exception due to its ecologically enforced extensiveness—it was impossible to put all the farmers under a single factory roof. It is certainly true that expansion of the industrial system was inexorable because of the need to invest capital. Yet this tendency to incorporate all productive activities within the fundamental paradigm of factory production came face to face with the exigencies of ecology. If the critical inputs were the soil and light and the critical outputs were food and fiber, there was little available for the creation of profit for capitalists (family farms tended to remain in the product-money-product [PMP] system, whereas capitalists seek the money-product-money [MPM] route by definition—as discussed in Chapter 2).

This basic problem was solved, not in a revolutionary way as in the production of textiles, but in a slow and fragmentary way through the substitution of inputs and appropriation of outputs. The inputs involved machines, chemical fertilizers, pesticides, and seeds, the outputs milled grain, canned vegetables, and a host of other processed-food products. The extensive production of peanuts on land became replaced with the production of peanut butter from petrochemicals.

Because of this peculiar history, unique forms of political power developed simultaneously with the penetration of capital. Appropriation came from the efforts of already powerful economic entities. Tractors came from John Deere, fertilizers from BASF, pesticides from Shell, seeds from Pioneer. These entities were frequently powerful because of their monopoly or oligopoly status. At the other end of production, value added off the farm substituted products such as cloth for fiber, bread for wheat, ketchup for tomatoes, and so forth. Again, giants such as Louis-Dreyfus and Libby McNeil Libby dominated the scene. Yet the production process itself remained mainly in the hands of relatively small and politically weak family farms. Although there were many examples of farmers organizing into politically powerful entities (some of which remain today), for the most part they remained divided and politically weak.

This peculiar history gave rise to a structure that almost guarantees the development of a farm crisis. At the input end there are suppliers who wish to maximize the prices of their inputs, whereas the farmer seeks to minimize them. At the output end, there are buyers who wish to pay as little as possible for what they regard as raw material for their production processes, whereas the farmer seeks to maximize them. Thus, the farmer faces a double contradiction, one at the input end, the other at the output end, as illustrated in **Figure 8.2**.

Because of the dramatic difference in political power between monopolized and monopsonized industries on the one hand and the farmers on the other, it is almost inevitable that over time the farmers should face crisis. The farm crisis, in its modern form almost 70 years old in the Global North and somewhat younger in the Global South, is largely driven by this underlying structure.

The Industrial System in Crisis

The economic situation of the average family farm, ranging from serious to catastrophic, is only a small part of the consequences of the modern model. Rural society has been dramatically transformed to the extent that in some industrialized countries it can be said to have virtually disappeared. Especially evident to the average citizen is the deterioration of the environment associated with this model. Even more disturbing, the condition of rural society, the environment, and economic life in the Global South has been and continues to be dramatically altered through this model. Although governments in both North and South are trying to reverse some of the more obvious of these problems, the structure of the industrial system, especially the assumption that it is the only one possible, makes positive modifications extremely difficult.

Although the problems associated with the industrial system have been and continue to be severe, critics of that system have also been evident, going back at least to Albert Howard in India in the 19th century. But the publication of Carson's *Silent Spring* (1962)

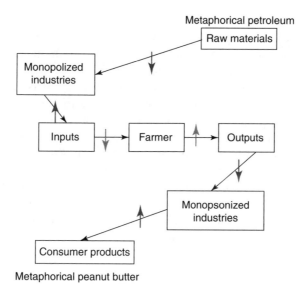

Figure 8-2 Diagrammatic representation of the modern conventional system, illustrating the plight of the farmer caught in the input/output squeeze. The vertical arrows indicate the "desires" of the purchasers and sellers of materials (e.g., the monopolized industries wish to make the price they have to pay for raw materials as low as possible and the price they get for inputs to the farmer as high as possible, whereas farmers want the price they pay for inputs to be as low as possible and the price they get for outputs as high as possible).

marked the beginning of what might be called the modern critique. This critique has become diverse and eclectic but can probably be best summarized as having two general tendencies. The industrial agroecosystem may be critically viewed as a system with problems or as a problem system, depending to some extent on whether one's focus is on "farming" or "agriculture." As noted previously, farming is the process of growing crops, whereas agriculture includes all aspects of the agroecosystem, from the manufacture of pesticides and tractors to the mixing of corn extracts with synthetic flavors and emulsifiers to produce corn chips. Repeating again Lewontin's one-liner: farming is growing peanuts on the land; agriculture is manufacturing peanut butter from petroleum.

That industrial farming is a system with problems is hardly contestable,[3] even reaching the consciousness of those whose self-interest would be obviously served through a defense of that system. For example, a survey in Washington State found that among both modern agricultural practitioners and their critics, there was general agreement that modern agriculture faces serious problems.[4] Some researchers are now referring to the

1980s as the "decade of awareness," and even deans of agricultural colleges openly admit that the modern system has spawned significant problems.[5] The expansion of hypoxic zones in the world's oceans, the continual problem with contaminated ground water, the loss of aquifers, pesticide residues causing health problems, pesticides directly and indirectly increasing mortality and morbidity, low prices paid to farmers, rising prices paid by consumers, the obesity crisis, the cancer epidemic, contribution to greenhouse gases—all of these have been tied in one way or another to the industrial agricultural system. It is increasingly evident that our modern cornucopia has been a Faustian bargain.

However all of these evident problems have caused some analysts to question whether framing the issue as a system with problems is correct in the first place. Rather, they contend, the problems we see with industrial agriculture are symptoms of a deeper crisis—that we are embedded in a problem system, rather than a system with problems. Because of the particular historical trajectory agriculture has taken, it has become inserted in an industrial system that drives it. As such, it will continue generating new problems as fast as old ones are solved. Consequently, searching for individual solutions to particular individual problems is seen as applying band-aids to a serious disease, and a total revamping of the system is necessary. In the North, such issues as commodity price supports and interest rates dominate most practicing farmers' concerns, whereas land tenure and market stability are the sorts of things a farmer in the South tends to think about, suggesting that there is something about the overall system that causes the evident problems. Critics who frame the problem in this way view current agricultural problems as derivative of a problem system.

Whether one views the modern system as a system with problems or a problem system, the sense that we are in a current crisis builds with each passing year. The economic and social problems faced by farmers the world over are now standard fare for international conferences and public policy debates. More relevant to the current text are those problems explicitly identifiable as ecological, most of which emerge from the industrialization processes. On one hand are the problems associated directly with biocides, fertilizers, and the like. On the other hand are the more indirect problems associated with the crude assumptions of modernization (e.g., assumptions of the necessity of factory-like production, excessive commodification, and monoculture), one of the consequences of which is a dramatic worldwide destruction of biodiversity, as described in detail in Chapter 7.

A key problem on which the post–World War II agricultural development had been focused was controlling pests. The methodology, much like the methodology in warfare, was total annihilation, achievable by using the new poisons that originally sprang from war research, as described in Chapter 6. It was not generally recognized that this could cause human health and environmental damage until the publication of *Silent Spring* in 1962, and it was not until 1978 that the general ecological and evolutionary mechanisms behind those problems were formulated so eloquently by van den Bosch as the now well-known pesticide treadmill (see discussion in Chapter 6).

By acknowledging the pesticide treadmill and taking some action to curb the most blatant negative effects of pesticides, the horrible environmental consequences envisioned by

Carson have been partly averted. Nevertheless, much of her vision did come to pass. Agricultural areas, such as those in California's San Joaquin valley, are largely devoid of life. A drive through the grape-producing areas, for example, is ominously similar to the opening sentences of *Silent Spring*, with the absence of bird song, the absence of flying insects, the feeling of desolation. The casual motorist in the central valley does not really get the whole picture. Pesticide residues are now known to occur in most food we eat. Although the levels of their occurrence are small and usually within government safety standards, our persistent exposure to them and the fact that two or more residues, even if at concentrations known to be safe alone, may have synergistic effects on our health that are a potential time bomb.[6] Particular incidents may stand out in public memory because of sensational press coverage, but far more ominous in the long term is the continual small dosage received each day by the majority of the population. Carcinogenic potential has long been acknowledged,[7] and more recently, the potential problems associated with hormone-mimicking pesticides have come to the public's attention.[8]

The problem cannot be solved by exporting the poisonous part of production to the South, as is generally done with onerous production tasks. Since the publication of Weir and Shapiro's *The Circle of Poison* in 1981, it has been difficult to ignore the human health effects of pesticides even though they may be applied far from home. Pesticide residues in tomatoes imported from Mexico are the result of exported pesticides applied by Mexican field workers. Frequently, those pesticides are banned in the United States, yet we must face the residues because of the international relations that allow producers to export dangerous technology. If political pressure is applied without an analysis of the entire political process, unexpected results may emerge. For example, because of consumer pressure in the United States, persistent pesticides were replaced by nonpersistent ones in vegetable production in the Tuliapan valley in Mexico. There is, however, a general relationship between acute toxicity and environmental persistence—short persistence time is correlated with high toxicity. So the less persistent but more acutely poisonous pesticides used by Mexican workers to avoid pesticide residues in the vegetables eaten by United States citizens were far more likely to poison those Mexican workers.[9]

Yet there was another problem probably started in the early 19th century when Justin von Liebig first discovered that you could put inorganic chemicals on the ground to increase crop yields, as described in Chapter 4. The problem of soil fertility was thus, in principle, solved, but soil management proved more difficult than had been expected. The Oklahoma dust bowl was for soil conservation what the Irish potato famine was for pest and disease control—something of a wake up call. The application of inorganic fertilizers on soils as part of the overall management system carried some strong negative consequences. Although not as well understood as the mechanisms of the pesticide treadmill, it is generally understood that provisioning the soil with large amounts of inorganic ions disrupts the normal nutrient cycling characteristics of the soil, thus making it even more necessary to apply inorganic fertilizer. Given the basic idea of having to apply ever-increasing amounts, it seems perfectly justifiable to refer to it as the fertilizer treadmill.[10]

Another ominous sign of a system in crisis is the loss of biodiversity so characteristic of the industrial agricultural system. Modern agriculture has emphasized production in large monocultures as opposed to the usually highly diverse traditional systems. The consequences of this trend are yet to be fully appreciated, but analysts everywhere are in agreement that the loss of biodiversity in general[11] and specifically in agroecosystems[12] is somewhere between severe and catastrophic. The way in which the industrial agroecosystem fits into this more general problem is complex and not completely understood, as described in Chapter 7.

Overproduction and the consequent fall in product prices for farmers in the developed world, coupled with the ever-increasing price of inputs and a continuing tendency for the capitalist system to appropriate inputs and substitute outputs, bode poorly for farmers. Already their numbers have dwindled to the point that in countries where they suffer from little political power (such as the United States), they are hardly recognized as a force at all—beginning with the 1996 census in the United States, "farmer" was not even recognized as a category under "occupation"; there are simply too few of them. The problems are different in the Global South, but no less serious—a lack of secure land tenure and disappearing markets. Couple these socioeconomic problems with the well-known problems of environmental deterioration and health risks, and the modern agricultural system indeed seems on the brink of crisis. On the other hand, development enthusiasts have been repeatedly predicting the complete elimination of the small farmer worldwide. As frequently maintained by those small farmers who refuse to disappear, there are still hundreds of millions of small family farms in the world today, and many of them are now making their voices heard as, for example, through organizations such as Via Campesina.

Toward an Alternative Program—Movements and Models

Problems with Identifying a Movement

The year 1977 was something of a watershed for organic agriculture worldwide with the formation of the International Federation of Organic Agriculture Movements (IFOAM). IFOAM has been an important organization for the dissemination of ideas about organic agriculture, bringing together what had previously been isolated and independently developing movements (obviously with a great deal of duplication of effort). Most important, the various national standards for organic produce have been codified into the "IFOAM Basic Standards,"[13] which incorporate the specific requirements of most organic certification boards around the world. Although IFOAM standards may not be specifically followed in all countries, they nevertheless represent a baseline to which individual national standards can be compared.

In only a slightly different context, the publication of *Silent Spring* in 1962 generated an entirely different, but ultimately related, movement in agriculture. Rather than challenge the entire modern industrial agricultural system, the vast majority of research workers, policy people, farmers, and agroindustry representatives scrambled to put as good a spin as possible on the watershed of information that Carson unleashed. The notion of integrated pest management is archetypical of this tendency and has as its foundation the idea that agriculture should be changed gradually, moving toward a more sustainable or ecologically benign system. The term LISA (low-input sustainable agriculture) was invented by the U.S. Department of Agriculture to encompass this basic philosophy, and LISA has now been incorporated into most conventional agricultural establishments (e.g., land grant universities in the United States, farmer organizations, agribusiness suppliers).

Today the nature of organic/ecological agriculture is not clear, especially if one wishes to incorporate the ideas of LISA. A great deal of ink has been devoted to the proper articulation of the concept,[14] and although all participants in this debate incorporate a critique of the industrial system, each has his or her own emphasis. Even what to call it is almost as diverse as its proponents, including at least alternative, biodynamic, sustainable, holistic, ecological, organic, regenerative, permaculture, agroecology, nonindustrial, natural systems, LISA, or the more recent low external input sustainable agriculture (LEISA), which fortunately can be pronounced the same as LISA. Each of these appellations may have different meanings in different places. Organic and biological agriculture are the same thing in Great Britain, whereas ecological or biological agriculture are the preferred terms in most of mainland Europe. Alternative agriculture is generally used in the United States but is eschewed in Europe because it can take on other connotations (including alternative lifestyles or systems that are simply not "usual," for example, hydroponics). Agroecology is common in Latin America (agroecología) and the United States, whereas in Spain, ecological agriculture (agricultura ecológica) is identical to the organic agriculture of Great Britain.

Today the debate about the definition of alternative/organic/ecological agriculture remains vigorous. It might be too early to be worried about formal definition. It is perhaps better to leave the definitional debate to active discourse until the problem and its solutions are better understood, as suggested by the National Research Council when they characterized alternative farming practices as

> not a well-defined set of practices or management techniques. Rather, they are a range of technological and management options used on farms striving to reduce costs, protect health and environmental quality, and enhance beneficial biological interactions and natural processes.[15]

Challenges to Ecological Agriculture

Contentious issues remain on the debate table. Within the community of proponents, there are significant, diverse, and complicated issues, many of which have been explored throughout this text. Nevertheless, the debate that takes center stage, at least in the popular media, is still the debate between proponents of the industrial system and the admittedly eclectic collection of people that form the ecological agriculture movement. Both sides, at times, can seem religious in their advocacy.

The modern form of the argument against ecological agriculture is certainly more nuanced than in the days when the chemical industry launched its dirty tricks campaign against Rachael Carson. Advocates of industrial agriculture have "greened" considerably and have accepted many of the arguments of their critics. Issues such as the loss of soil fertility, soil erosion, or the pesticide treadmill have all been accepted as real problems that industrial agriculture has spawned, and there is general admission that future agriculture must be organized so as to avoid them. Yet the Malthusian specter seems to be always waiting in the wings—the growth of the human population is such that in the near future we will face a challenge too big for ecological agriculture to solve. We will have more mouths to feed, and worse, those mouths will demand more food per person, since we expect the increased industrialization of the Global South. Even with no further population growth, if every Chinese and Indian person demands an egg per day, imagine how many chickens will have to be produced! Adding to the urgency of the argument is a moral position that urges a neoanti-imperialism on us—who are we to say that the Chinese and Indians should not have an egg every day, we in the industrialized world who have used and even squandered so much of the earth's natural capital!

The arguments against converting to more ecological forms of agriculture are eclectic and sometimes complex, much as the arguments in favor of the conversion. Although many people have come to accept the fact that the industrial system must be modified, there remains, even among them, a certain skepticism about the possibilities for any sort of transformation. It is worth examining where that skepticism originates. In the end, the argument against more ecological forms of agriculture, whether driven by disguised economic interests or sincere skepticism, seems to be based on a set of critical assumptions. Each of those assumptions is questionable.

Assumption 1: Ecological, especially organic, production always results in lower yield per land area.

This is the conclusion of the famous Landell Mills study, which was commissioned by the agrochemical industry.[16] Other studies, for example, the famous Lockeritz study of 1981, found no evidence that there was a difference.[17] Most recently, a study at the University of Michigan[18] searched the literature for production figures of actually functioning production units and compared production figures for organic versus conventional for a variety of commodities. After surveying almost 300 studies, they conclude that there is little support for the widely held assumption that organic produces less than conventional. Indeed, organic methods could easily provide for all calorie and protein requirements of today's world and even into the foreseeable future. They further note that in the Global South there seems to be an enormous potential for technological improvement along organic lines, suggesting that a serious research and development program within an organic framework could have great benefits for farmers of the Global South.

An additional fact is worth considering associated with agricultural productivity in the Global South. In an analysis of 15 underdeveloped countries, it was found that production (or productivity—amount produced per unit area) declined as farm size increased,[19] and another study found the same thing in the developed world.[20] That is, smaller farms tend to be more efficient than larger ones. A moment's reflection suggests why this is the case. Small farms, especially very small ones, tend to be managed "ecologically" at least in the sense that the farmer usually knows his or her farm very well, understanding the ecological peculiarities of each small corner of the farm. In contrast to the industrial mantra that large size yields more efficiency, when ecology is brought into the equation, the reverse seems to be the case.

Assumption 2: Future improvements in conventional production may be substantial, whereas ecological production will remain relatively static.

This assumption depends on a variety of factors. If governments continue subsidies, both direct and indirect (discussed later), for industrial production yet eschew such subsidies for organic production, it is natural that the research and development needed for improvements in ecological production are not likely to be forthcoming. The assumption itself becomes a self-fulfilling prophecy.

Nothing inherent in conventional agriculture suggests that it should outproduce organic production, when considering basic energetic and nutrient cycling facts. For example, there is no logical argument that could conclude that provisioning nitrogen, phosphorous, and potassium directly in ionic form will produce more than provisioning those same elements in the form of decomposable compost. The same amount of nitrogen will result in the same amount of nitrogen-induced production. Although organic proponents will argue that provisioning nitrogen with manure or compost brings other benefits to the soil, industrial proponents have no parallel argument for limiting fertilizer application to a restricted set of ionic forms. So many kilograms of nitrogen per hectare are, at the lowest common denominator, so many kilograms of nitrogen. Even if organic proponents are wrong about all they say, industrial methods can be, in principle, only as productive, never more productive, than organic methods, given optimal technology for both. Nitrogen is nitrogen, a dead pest is a dead pest, and a disease-free animal is a disease-free animal.

Behind this assumption lies an additional set of unexamined assumptions—that the sociopolitical and economic relations of production will not change in the future, and that production should be measured in terms of profit per input rather than kilograms per hectare. If a government subsidizes the production of an herbicide, for example, but refuses to subsidize worker benefits (as would be the case, for example, with socialized health care), replacing a worker who uses a hoe or machete to remove weeds with chemicals that remove them in the absence of workers obviously reduces the input costs. This assumption is closely related to Assumption 3.

Assumption 3: Conventional agriculture will be allowed to continue its practice of externalization, whereas more ecological forms of agriculture will receive no substantial support from governments.

This is a frequently hidden ideological assumption that may be true but, depending on political transformations, may very well change. As in many other industries, the industrial agriculture system is large and influential, promoting its own interests through official and extraofficial channels. The environmental consequences of industrial agriculture—from the creation of secondary pests to the elimination of wildlife, to long-term losses in soil fertility, to soil erosion, to secondary environmental contamination, and so forth—are currently allowed to be written off as externalities, a cost that will be absorbed by governments or the body politic. From the few studies that have tried to quantify all of the costs associated with industrial agriculture, it is clear that such externality costs are substantial.[21] If popular pressure builds for a more equitable accounting, industrial agriculture's components may come under more careful scrutiny and be forced to pay for some of the real costs of production that are currently absorbed by the body politic. In U.S. superfund sites, for example, many of which result from dumping the byproducts of pesticide production, cleanup costs are so extreme as to probably render the original production process a net drain on the overall economy, if anyone would do a serious complete accounting. Only through the indirect subsidy provided by a government subjected to intense industry lobbying are these products able to compete in an open market. If sufficient political pressure becomes manifest, such subsidies could become a thing of the past.

If awareness of the importance of conserving the environmental base of agriculture grows, political pressure could build for governmental support for organic conversion. If and when and how much are impossible to predict, but the possibility that such a change could occur cannot be discounted. In recent years, we have seen the beginnings of this transformation, somewhat reflected in a comparison between the European Union and the United States, the former taking a so-called government-facilitated growth model whereas the latter doggedly stuck to the "market-led growth" model.[22] By 2003, the European Union (EU) had almost 4% of its agricultural land in strictly organic production compared with 0.2% in the United States. The government-facilitated growth simply recognizes the need to switch the subsidy structure, whereas the obfuscating "market-led" model maintains subsidies for the industrial agriculture sector. The result is that there was almost six times as much land in organic production in the EU compared to the United States by 2003. Although growth of the organic sector is continuous and exponential in both EU and United States, the rate of that growth is much higher in the EU, suggesting that the maintenance of the industrial system is at least partly a consequence of subsidy structure and thus could be changed with the proper political pressure.

Finally, the 2008 report of the United Nations' International Assessment of Agricultural Science and Technology, referred to previously, strongly advocates for a change in approach of the science and technology involved in the industrial system, noting that the problems generated by that system leave us with the result that "business as usual is no longer an option." Such a consensus reached by a truly eclectic collection of interested parties could provide pressure for altering the focus of research and development activities, effectively altering some of the underlying subsidy structure, support for research and development being a form of hidden subsidy.

Assumption 4: Less affluent nations seek, and have every right to seek, the consumptive lifestyles already enjoyed by those in the developed world, which cannot be achieved except through high-intensity industrial agriculture.

Although the commercial messages beamed to the underdeveloped world by the industrial nations are highly successful and most in the underdeveloped world seek as much of the perceived good life as they can get, few, even those promoting the vision, believe the myth. The images of people "having more fun than you ever had" simply because they drink Coca Cola are only images in the end, and there is a limit to their ability to motivate. As populations become more educated, the tendency to adopt a knee-jerk response to advertising and other forms of propaganda may dwindle.

Furthermore, the assumption that all people deserve whatever they want is simply not a rational assumption. Some in Asia want powdered rhino horns. The physical reality of the world dictates that only a few could have all they want, and even then probably only for the short time it would take rhinos to go extinct. If consumptive desires are out of line with what the earth offers, those desires have to change. Although it is certainly true that people in the developed world have no business telling people in the underdeveloped world that they should not want a similar lifestyle, it is also true that people in the developed world cannot expect to maintain a lifestyle that precludes inclusion of the entire human family. It is not a question of whether every Chinese person can have two automobiles—it is a question of whether every U.S. citizen ought to have even one.

Assumption 5: The industrial system is necessary to provide more land for conservation purposes (frequently stated as, intensifying agriculture allows for more "land sparing").

This assumption became important in the late 2000s, especially as tropical deforestation seemed unstoppable and became intimately linked to global warming. Recalling the historical sequence of reforestation of temperate forests in North America, a similar model of tropical forest transition came to be seen as both inevitable and desirable, based on two simple propositions: 1) Rural populations will inevitably convert natural habitat to agriculture in proportion to their population density, and 2) the rural population along with its urban consumer base constitutes an inevitable demand on the land (for food). Both of these propositions are based on an underlying framework that has taken on an almost paradigmatic status[23] yet has little empirical support. The skeleton of that framework is a three-part logical sequence: 1) A given rural population density requires a certain land base to engage in productive activities, 2) The amount of food necessary to support that population plus its urban consumers, divided by current per-hectare productivity, gives the land area necessary for agricultural production, and 3) the total land minus that necessary for production is what is available for conservation. The two parts of the logical sequence are different only in their emphasis. The "rural/urban migration" argument focuses on (1) and notes that with the reduction in rural population, the conclusion (3) suggests that more land will be available for conservation (fewer rural people, less use of land for agriculture, and thus regeneration of forest or other natural habitat). The "productivity" argument focuses on (2) and argues that if per unit production could be increased, the required land base would be reduced, and consequently the conclusion (3) would again suggest that more land will be available for conservation (the same number of people, but higher productivity, thus more land for conservation).

Both arguments are wrong. Anglesen and Kaimowitz (2001a) have studied this issue in great detail, noting first that there is a fundamental contradiction involved in the basic proposition. First, "the belief that technological progress in agriculture reduces pressure on forests by allowing farmers to produce the same amount of food in a smaller area has become almost an article of faith in development and environmental circles." Second, ". . . basic economic theory suggest that technological progress makes agriculture more profitable and gives farmers an incentive to expand production on to additional land." Thus, according to the first assumption, technological progress would lead to less land in agriculture, but according to the second, that progress would lead to more land in agriculture. Anglesen and Kaimowitz (2001a) report on detailed studies that sometimes support one, sometimes the other point of view. As an "article of faith," it is simply not true that "progress in agriculture reduces pressure on forests."[24]

In a more extensive work, Anglesen and Kaimowitz (2001b) edited a series of chapters that include 17 case studies from Latin America, Africa, and Asia. Their conclusions from all of these studies is that the issue of intensification of agriculture and its relationship to deforestation is complex and that agricultural policy could be modified in such a way as to promote forest-preservative policies rather than policies that, however unintentionally, actually promote more deforestation with "improved" agricultural technologies. That "improvements" in agricultural technology *could* lead to forest protection does not imply that those improvements *will* lead to protection.

A close examination of the case studies presented in Anglesen and Kaimowitz (2001b) suggests that for the most part improvement in agricultural technology along industrial lines in fact causes more deforestation. The situation is complicated by many factors, and it is certainly sometimes the case that improved agricultural technology has decreased deforestation rates. Nevertheless, examining closely the 17 case studies presented, in 12, there was a clear indication that technological change had an effect on deforestation. Of those 12, 9 showed increasing deforestation as a result of industrial technification (3 of the 9 suggested it could go either way, depending on circumstances), and only 3 suggested a necessary decrease in deforestation with technification. All cases were treated with the complex analysis they deserve and seem to negate quite effectively and completely the assumption that increases in industrial agricultural technology automatically lead to land sparing. In Mexico, for example, the general trend for rural outmigration coupled with agricultural intensification has generally led to more deforestation, not less.[25]

Searching for Models

In searching for a general plan for the transformation of agriculture from the industrial system to an ecologically sound and sustainable one, we most naturally look to models, both for how the new agriculture will look and for how to transform the present system. Traditional systems present us with one form of model since by definition they were not subject to the same forces that produced the modern industrial system. Extant organic farms, isolated amid the ocean of industrial agriculture, provide another. Yet another is gleaned from the idea that the design of agroecosystems should be based on the ecological dynamics of local natural ecosystems, an idea that has come to be known as Natural Systems agriculture. All three of these are discussed later, but before proceeding to that discussion, it is imperative to note that in all three of these cases we are presented with systems in which some of the key social variables have been canceled out of the equation. Traditional systems, by their very nature, do not exist in the context of industrial agriculture but rather in isolated pockets where the industrial capitalist system has not yet fully penetrated. Extant organic or ecological farms are already run by people whose attitudes have been transformed and who largely are not concerned with promoting a global transformation, but rather with the survival and prosperity of their individual farms. Natural systems, sort of by definition, exclude the activities of humans.

Traditional Agroecosystems

Many authors have promoted the idea that traditional forms of agriculture, since they have not undergone the alterations of the industrial system, offer models wherein particular practices can be modified to fit into the transformation. Few would actually argue that we wish to "return" to the traditional systems and all live like Amish farmers or survive on only what we produce in our traditional backyard gardens. Rather, most claim that traditional systems are likely to contain rules of operation that have survived the test of time and may represent signposts along the road to an ecological agriculture. The general attitude is summed up by Altieri:

> Ecological principles extractable from the study of traditional agroecosystems can be used to design new, improved, sustainable agroecosystems . . . traditional farming systems have emerged over centuries of cultural and biological evolution and represent accumulated experiences of interaction with the environment by farmers without access to external inputs, capital, or scientific knowledge. Such experience has guided farmers in many areas to develop sustainable agroecosystems.[26]

Although this is certainly a good working assumption when studying traditional agroecosystems, it needs to be tempered with some skepticism. History is replete with examples of "traditional" agroecosystems that were devastating to the environment and were not sustainable. Examples exist from the salinization of croplands by the Hohokam in southwestern North America to the massive soil erosion that destroyed Greek ecology during the classic Greek civilizations.[27]

Traditional systems tend to share a surprising number of commonalities at a very general level:

1. Especially in the tropics, they tend to have a large number of species associated with them, both in the planned sense and the associated sense. Whether this diversity has anything to do with functionality is a hotly debated topic, with two special issues of major journals recently devoted to the issue.[28]

2. They tend to exploit all available habitats on the farm, including the habitats created by the crops themselves. Thus, for example, if trees are included in the system, usually some shade-tolerant crop is planted in the understory to take advantage of the space.

3. Nutrient recycling tends to be relatively closed, with kitchen wastes and animal manures recycled into the cropping system.

4. They use some kind of biological means for most pest control.

5. They have a very low level of off-farm inputs, relying mainly on human and animal energy and recycling within the farm.

6. They rely on traditional varieties of crops and animals.

Whether such principles can or should be applied to the development of modern systems remains to be seen. Some are relatively obvious and are already part of the movement toward sustainable agriculture (e.g., the entire philosophy of LISA is to maximize recycling to reduce inputs to as low a level as possible). Others evoke ridicule when suggested to developed world farmers, even the most ecologically conscious (e.g., try suggesting to a European farmer that he or she should revert to animal traction—indeed, some biodynamic farmers in Germany spray the required special preparations from helicopters!). On the other hand, specific technologies used by specific cultures represent, in a sense, the raw material of ecological development. It is not that we should try to convert all the wetlands of southern Mexico to the weed control system of a *Thalia sp.* fallow (see Chapter 2). Rather, we should view this method of cultural control of weeds in maize production as a principle that could be used elsewhere, perhaps with significant modification (e.g., could banana trees substitute for *Thalia sp.*, thus providing not only control of grasses but a valuable product as well?). Perhaps the general principles as previously outlined have already been indelibly etched in the minds of all those who seek to transform agriculture. Perhaps the real value of traditional systems as models is in the specific and particular techniques they use. Studying traditional agroecosystems then becomes similar to the study of traditional varieties of crops. Just as the classic crop improvement geneticist prospected for new genes, the new agroecologist will prospect for new technologies.

There is, as mentioned several times previously, another reason to look to traditional systems for guidance. Many, if not most, traditional systems rely on "autonomous ecosystem management," which is what seems to be emerging as a central goal of the agricultural transition. Recall the results of Morales with the highland Maya in Guatemala (Chapter 1)—they did not see pests on their farms, rather insects that never did become pests. Their management systems were examples of autonomous ecosystem management, a central goal of the ecological agriculture movement. While it is of great benefit to study these traditional systems from the point of view of how they solve the problems of production, it is also important to look at them from the point of view of how they avoid the problems in the first place. This brings up the all-important issue of local knowledge.

As early as Robert Boyle, perceptive scientists have recognized a historical truth: Academic knowledge is general but shallow, whereas local knowledge is specific and deep.[29] Local knowledge is frequently "flawed" in the sense of modern scientific understanding, yet is set in some other story or worldview that provides it with context. The fact that a local farmer in Nicaragua "knows" that grasses "burn" the corn is not ultimately different from the scientific ecologist's "knowledge" that the corn "dries out" because of "competition" from grasses. Whether we say burn or competition only identifies the discourse within which the actual observation of the corn's performance with grass is communicated from one person to another. The fact that the farmer knows the details of what happens to the corn when grasses grow around it is key to developing local alternatives that work. The fact that the ecologist uses the word competition to describe the phenomenon brings the rest of the experience of the world, to the extent it is catalogued, as possible knowledge for the planning process. The farmer knows what happens in his or her field. The ecologist can generalize that knowledge, compare it with what happens in other fields, effectively expanding the potential knowledge base.

The deep and local knowledge of the farmer, especially the traditional farmer, is essential to the development of the alternative model. The ecologist may help generalize and contextualize that knowledge to make it richer, but the local knowledge is imperative due to the unpredictability and locality-specific nature of many ecological processes. However, much traditional knowledge has been lost during the past 50 years, a result of the hegemony of the industrial model. We are thus faced with a contradiction faced by all utopians—it is difficult to construct an ideal system with less than ideal people. Today's world is filled with urban workers while what we need are rural ecologists. Although this problem is not as universally recognized as the lack of appropriate technology, it may turn out to be a more important one.

We thus are faced with a major contradiction in the move to ecological agriculture. On one hand, we do not have a complete catalogue of techniques that are tried and proven to work under all circumstances—the technical side of the contradiction. On the other hand, the destruction of rural society has taken with it the knowledge base and labor force that will be needed for the transformation—the social side of the contradiction. We are thus faced with the task of building a new system based on incomplete technical information and with a society designed to function only under the current system. As we strive to transform the current system, this is a fundamental contradiction that must be resolved.

Practicing Organic Farms

Extant organic or ecological farms offer probably the best model for the way individual farms will look after the transformation. There has already been an explosive growth of this form of agriculture in Europe as noted previously, and the famous University of Michigan Study[30] put an end to the previous propagandistic drumbeat that organic agriculture was a low-productivity option. The problems with further expansion of this sector have already been noted, although earlier warnings that the growth would soon halt[31] were clearly premature. Furthermore, there has been extensive recent criticism of the growth of the corporate sector of organic agriculture,[32] bringing to light an important inherent contradiction. The success of organic agriculture technically may lead to its incorporation and co-optation sociopolitically, thus diminishing its potential as a contributor to the more general transformation toward ecological agriculture. In this sense, it is difficult to maintain a posture of separating the technical from the social. Nevertheless, the practice of organic agriculture worldwide has developed into a technologically rich collection of practices that undoubtedly will inform the future system, even as the sociopolitical aspects evolve in a less than ideal direction.

To some extent, organic agriculture is largely based on what not to do and arises as a counterpoint to industrial agriculture. No chemical fertilizers (sometimes referred to as artificial manure) and no chemical pesticides are the two main principles. Reversing the tide of chemical usage requires very different management skills than contemporary farmers in the industrialized North generally have, including crop rotations and careful management of soil nutrients. In Europe, organic farms tend to be mixed farms, with animals providing an acceleration of decomposition so that nutrients can be recycled more quickly. Feedstock produced in one field can be cycled through a cow and provide nutrients to another field thorough the cow manure. In principle, it is precisely the philosophy that was the basis of the new husbandry system in Europe at the dawn of modernity.

In his influential book, *Organic Farming*, Lampkin[33] devotes considerable space to a useful discussion of what organic agriculture is *not*. Formulating his ideas in the form of common misconceptions, he notes that first, organic agriculture is not simply the nonuse of chemicals and rightly berates the idea. He notes, rather, that organic agriculture seeks to

> avoid the direct and/or routine use of readily soluble chemicals and all biocides whether naturally occurring, nature identical, or not. Where it is necessary to use such materials or substances, then the least environmentally disruptive at both micro and macro levels are used.

The second general misconception about organic agriculture, according to Lampkin, is the idea that organic agriculture merely substitutes organic inputs for agrochemical ones. He notes:

> Contrary to the dearly held ideas of organic "traditionalists," there is nothing magical about muck even if it is pushed in a heap and called "compost." The misuse of organic materials, either by excess, by inappropriate timing, or by a mixture of both, will effectively short circuit or curtail the development and working of natural or biological cycles.

He goes on to describe a third misconception, that organic farming is a return to the agriculture of pre 1939. Although much in this early agriculture is ecologically sound, it would be foolish to ignore the past 6 decades of agricultural research, advances in biology, and general experience of farmers in both the North and South who have continued experimenting with traditional techniques during those 6 decades. Organic agriculture should be just as modern (perhaps we should say more modern) as industrial agriculture, but with a different philosophical basis.

Much of the benefit of contemporary organic/ecological farms is in the favorable position of organic products in the marketplace. On one hand, we can probably expect an increase in demand for organic products as the populace becomes more educated about environmental and health issues, but on the other hand further expansion of organic production will inevitably lead to increased supply and a gradual erosion of the favorable market position. This strongly suggests that further refinements in organic methods, which will require serious ecological research, will be critical to the expansion of the organic sector in the future. Regardless, as a model, practicing organic farms remain important.

Finally, although beyond the intended scope of this book, the advantaged position in the market provided by the certification process creates, in some societies, a socioeconomic structure that has its own dynamic. In particular, as noted by Michael Pollan in his influential book, *The Omnivore's Dilemma*, there is a tendency to consolidate. This means that large corporate enterprises have become and are increasingly likely to become important as they continue penetrating the organic market. Indeed, the consolidation of the organic sector in processing, distribution, and retail has grown enormously.[34] Since the late 1990s, names such as Heinz, Coca-Cola, Cargill, and General Mills have become owners of organic processor companies. The Hain Celestial Group owns over 30 brands of organic products and is itself partially owned by Cargill, and recently, Heinz had a 19.5% equity in the group. This picture of continued consolidation is not likely to subside soon, at least in the United States.

Natural Systems Agriculture

It has been popular to suggest that an agroecosystem should reflect the natural vegetation in the region in which it is practiced.[35] Perhaps the most well-known proponent of this idea is Wes Jackson, who reasons that agriculture in grassland areas (which is where the most important cereal production in the world occurs) has gone wrong because the natural ecosystem in such an environment is a perennial polyculture, not the artificial annual monoculture that has replaced it.[36] The fact that we produce annual monocultures there is simply due to our inability to domesticate perennial grasses, which we now can do, given what modern science knows about genetics. Jackson's vision is to produce genetic variants of natural perennial grasses (or perennialized variants of annual grasses) that can be combined in a "permaculture" ecosystem that more nearly mimics the natural state of the grasslands that the modern annual monocultures replaced. The particulars of Jackson's vision are somewhat more specialized than the others mentioned in this section, but his logic warrants serious attention.

In one way or another, the idea of natural systems agriculture has been part of the organic movement since its popularization in the last century. For example, when Ehrenfried Pfeiffer examined the problems of Florida's citrus groves, he looked to the nearby native forests for insights into what was wrong with the agroecosystem, finding, not surprisingly, that monocultural production would likely lead to diseases. He suggested planting native trees among the oranges to protect against such problems, a clear example of learning from the local natural system. In fact, according to Conford (2001)

> This story fits well into the organic mythology in the morals that it offers: respect for the natural order as revealed particularly by the wilderness brings economic benefit to those who are not fixated on short-term gains; true science goes out from the laboratory and studies the ecological context, observing rather than trying to dominate; variety is more productive than monoculture; industrial products bring disease and waste.

The natural systems paradigm has been applied in the development of an agriculture based on the biology of the natural prairie of central North America at the Land Institute (a private research institute located near Salinas, Kansas, founded by Wes Jackson). Key elements in the prairie ecosystem (to be imitated in the planned agricultural system) include[37] (1) four major plant guilds—perennial C3 and C4 grasses, nitrogen-fixing species (primarily legumes), and composites (Asteraceae); (2) maintenance of soil through perennialness; (3) provision of nitrogen from within the system (the legumes); and (4) resistance to invasions of pests and diseases through internal ecosystem processes. The idea is to devise an agricultural system that mimics these four components of the natural system yet satisfies minimal conditions for agricultural production. The idea seems sound, and the research thus far is promising.

On the other hand, some other already extant agroecosystems correspond quite well with Jackson's vision of natural systems agriculture. As described by various authors,[38] traditional tropical agroecosystems tend to be complex multistoried polycultures that have certain features that resemble the natural forest that they replaced, including the so-called cabruca cacao of Brazil, the jungle rubber of Camaroon, and almost all traditional home gardens located in the humid tropics. Most impressive are the thousands of hectares devoted to coffee cultivation, one of the original three vignettes of Chapter 1. Traditional coffee production ranges from "rustic" to polycultures of overstory trees. In rustic production, the original forest remains except the understory has been replaced by coffee bushes. The entire ecosystem looks very much like a natural forest—indeed, aerial photos invariably categorize rustic plantations as natural forests. Decades of research into this system reveal that much of its operation is reminiscent of the ecological dynamics that occur in natural forests.

The New Ideology—Food Sovereignty

As stated on the Via Campesina web page,

> Food sovereignty is the RIGHT of peoples, countries, and state unions to define their agricultural and food policy without the "dumping" of agricultural commodities into foreign countries. Food sovereignty organizes food production and consumption according to the needs of local communities, giving priority to production for local consumption. Food sovereignty includes the right to protect and regulate the national agricultural and livestock production and to shield the domestic market from the dumping of agricultural surpluses and low-price imports from other countries. Landless people, peasants, and small farmers must get access to land, water, and seed as well as productive resources and adequate public services. Food sovereignty and sustainability are a higher priority than trade policies.

Food is produced by farming, and the purpose of food, ideology aside, is to provide nourishment for people. It is not, in any fundamental way, necessarily a commodity. Yet, prevalent ideology worldwide contends that whatever is not yet a commodity must be turned into one. Food sovereignty challenges this ideology at two different levels. First, human beings should have a right to food, not a right to choose to spend some of their money to buy food—a right to food as much as they have a right to *Habeas Corpus* or freedom of religion. This new model rejects the notion that food is nothing more than software or trinkets. Second, human beings should have the right to collectively decide, at a local level, how food is to be produced.

The first principle—that food is different[39]—is critical. Contemporary economic orthodoxy assumes as a basic principle that all goods and services may be viewed as tradable commodities. As part of the underlying historical transition from Feudalism to Capitalism, this idea is central. When the PMP model was replaced by the MPM model (see Chapter 2), viewing a product for its utility, its use value, was replaced with viewing a product for its potential to be bought and sold for money, its exchange value. Markets then are viewed not as places where products are bought and sold to increase the utility of those products in the society in general, but rather as places where profit is to be generated. How to "monitarize" products has become a buzz word in modern economic circles for everything from "economic service instruments" to water. Food sovereignty takes as a fundamental principle the assumption that access to food should be a human right, not a function of either the accident of birth or vicissitudes of economic fortunes. Like the air we breathe, societies should be organized so that no human being goes hungry. The UN declaration in 1994 that food security should be a basic human right is identical to this aspect of food sovereignty.

Food sovereignty goes further, however. Its second key element is the collective right to decide how to provide that security. This element, although simple in concept, generates important complications and contradictions. It means, for example, that a community must be self-defined and collectively see the food production system as part of community norms. Communities must reject dumping, for example. This implies a community consciousness that transcends the simple assumptions of *Homo economicus* that all individuals seek to minimize their economic output and maximize their economic input. It effectively views the past 500 years the way Dutch merchants might have viewed the seemingly intractable feudal system at the dawn of the Bourgeois Revolution, as a transitional period that eventually will give way to a more rational mode of organizing society. Although it was rational to replace the PMP system with the MPM system during the time of feudal domination, food sovereignty suggests that times have changed, that the modern world no longer needs to concern itself with overthrowing the assumption of royal authority, that the new system that needs to be overthrown is the assumption that corporate structure must decide how food is produced and distributed. The challenge of food sovereignty is to take control of that production and distribution process and organize it for the sake of producing nutrition for people rather than profits for corporations.

Technical Requirements for the Transformation

Strategies for the Technical Conversion

In recent years, despite ridicule from proponents of industrial agriculture and despite antagonism from even those who accept the need for a more sustainable agriculture,[40] the amount of organic production has increased exponentially. While certified organic per se is only a small component of the move toward a more sustainable agroecosystem, the formal certification process has enabled a relatively accurate assessment of the penetration of this sector into the food system. In the United States, for example, the penetration of certified organics into food sales increased from 0.8% in 1997 to 2.5% in 2005.[41] A recent report[42] claims that China, the largest country in the world, increased organic production to move from 45th to 2nd position in the world, contributing 62% of the annual increase in organic production during the first 6 years of the 21st century. Although the world total area under organic production remains a small fraction of the total agricultural area, the exponential trend is obvious. Clearly, much of this is driven by the extremely good market position of organic products, but certainly, an increasing awareness of environmental and health problems spawned by the industrial system is also a contributing factor.

On the other hand, it takes no more than a casual conversation with farmers in either the North or South to see that further expansion of the ecological sector (including, but not exclusively, organic) faces severe problems or will face those problems in the near future. In the developed world, the farming sector is relatively sophisticated about environmental problems but feels hemmed in by market constraints. Even though conversion to more ecological forms of agriculture is attractive both from an environmental protection standpoint and simply from market logic, it nevertheless is usually seen as a risky endeavor. The skills required for ecological agriculture are different from those needed for industrial agriculture, and the conversion time is unavoidably more than a single season, which means accepting considerable losses during the initiation phase. Furthermore, organic production is currently subject to the same market vicissitudes as industrial production, and there is no guarantee that commodity prices can be sustained at the high levels maintained for certified organics.

In the developing world, the situation is different. A significant sector of the farming sector remains illiterate or semiliterate, purposefully denied high-quality education, and thus easy prey to propaganda. The promises of advertisers are thus less skeptically viewed than has become customary in the developed world. Furthermore, the message of the environmental movement has made fewer inroads, and old habits of discarding garbage mix unfavorably with new technology of nonbiodegradable products. Noting that a pesticide could make you sick is frequently met with comments such as "I've been using it for the past 10 years and I never got sick." Noting that a fertilizer may contaminate the very soil on which the productive base of the farm is situated is less important when security of land tenure is fragile to begin with. Although there are certainly important exceptions to these patterns, they nevertheless on average combine to make the situation for ecological agriculture in the Global South substantially less favorable than in the developed world. On the other hand, probably the majority of small farmers in the Global South produce in a semiorganic way in the first place, mainly for economic reasons, and traditional ecological knowledge is typically sophisticated even when formal illiteracy is high.

Organic Versus LISA

One of the problems with the ecological sector at a more philosophical level is an inherent trap that seems inevitable, given world political realities. To prevent the co-optation of high organic prices, it is imperative that organic methods be well stated and rigidly adhered to with a well-policed certification system. Currently, such systems exist for most countries in the developed world. Indeed, in most of Europe, two certification systems exist, one for simple organic products (e.g., in the Netherlands the EKO label) and one for biodynamic products (the Demeter label). As previously noted, IFOAM has set standards that incorporate most all of the rules extant in most countries that have certification procedures and has become an international standard for organic products.

The trap that results from strong certification rules is that farmers must accept an initial period of reduced income (sometimes substantial reductions for substantial periods) and pay a significant fee for the certification process itself. Given the knife-edge of economic existence for so many of the world's farmers, it is difficult indeed to make a commitment to the pure ecological route. Thus, the need for formal rules in order to be certified creates a stifling entry barrier for many if not most of the world's farmers.

Because of this entry barrier, the recent rapid expansion of the certified organic sector may eventually wane, leading some advocates of ecological agriculture to promote an alternative route to the transformation. The popular notion of LISA has gained substantial credibility, although its susceptibility to co-optation by the very forces that have created the problem in the first place is great. Despite this problem, the LISA approach carries with it several very attractive characteristics. First, by assuming the underlying logic of current establishments, it is an easily arguable position to practical-minded farmers. One need not prepare the terrain with programs of education about environmental protection or philosophical approaches toward less domination of nature in order to promote LISA. The elementary notion that inputs represent a cost to the farmer and that he or she should try to minimize those costs is an extremely simple idea that is difficult for anyone to either misinterpret or misappropriate.

If coupled with proper research programs, input reduction ideology can ultimately result in input elimination reality. A farmer convinced that the cost of pesticide is too high and with a strong desire to reduce as much as possible his or her use of it will be continually moved in the direction of further reductions as technologies become available, such that eventually entire elimination of the pesticide is likely. In this sense, LISA takes on the characteristics of a transitional program, leading naturally from the industrial to the ecological.

Furthermore, in much of the Global South, agriculture is already low input, if not necessarily sustainable. Adopting a LISA strategy here would seem easy in that it asks farmers not to introduce something new into their farming practices unless that something can be clearly shown to be of benefit. Although I do not want to underestimate the strength of propaganda (i.e., advertising), the current reality suggests not that farmers need to adopt whatever changes are presented to them as new technology, but rather that they need to be sensible about what changes they in fact adopt. Considering the healthy skepticism many farmers have developed about "modern" technology, some confidence can be retained in their ability to be sensible about those changes.

These characteristics of LISA, combined with the obvious recent success of ecological agriculture in the developed world, lead to a sort of reformist conclusion. The ultimate transition to the next agricultural revolution, the complete conversion to ecological agriculture, will best be accomplished with the dual strategy of simultaneously promoting the expansion of ecological forms of agriculture per se plus the incorporation of LISA principles into the rest of the industrial agriculture system.

To some extent, both LISA and organic agriculture have as their basic *raison d'être* the elimination of some technologies (and the refusal to adopt some new ones) and thus are easy targets for *ad hominum* attacks (tree huggers, communists, idealists, etc.). This political battle needs to be faced squarely by those who wish to see agriculture develop in a more rational fashion. The details of that political battle are not subjects of this book, but acknowledging its existence is useful contextual information.

At this juncture, it is worth emphasizing a point made in Chapter 1. For the success of both ventures, expanding ecological agriculture and maintaining LISA as a transformational tactic, pushing back the frontiers of ecological knowledge about agroecosystems is key. The modern science of ecology will form the scientific basis of the new agriculture, but it is not yet prepared to take on that role. More and higher quality research is urgently needed, focusing the most recent results of pure ecology on the practical problems of the transition. For this to happen, a solid background in those ecological principles that impinge on agroecosystems is essential. Providing at least part of that background has been the intention of this book. Quoting Engels,

> We, with flesh and blood and brain, belong to nature, and exist in its midst, and . . . our mastery of it consists in the fact that we have the advantage of all other creatures of being able to learn its laws and apply them correctly.

which is, at its foundation, a call for more and more careful research into the ecological processes operative in agroecosystems.

Promoting a combination of LISA-type agriculture along with the expansion of the organic and agroecological sectors, as suggested previously, seems to be a rational plan. Some criticism of this plan exists, based on the failure of LISA to face up to some of the deeper and politically difficult issues involved. For example, it has been argued[43] that the LISA-style approach has actually allowed for the co-optation of the ecological agricultural movement and that the underlying structure of the industrial agricultural system has been allowed to remain, with simple substitutions of biological inputs for the previous chemical ones, a process referred to as "input substitution." If it is the underlying structure of the system that is the problem in the first place, focusing on input substitution is putting a band-aid on a cancerous tumor.

Others would argue that input substitution is an important, even essential, form of transition. Many farmers scoff at the idea of converting to more ecological forms of agriculture. Input substitution (or what is referred to generally in this book as LISA-type approaches) could be an important transitional argument for this sector. Although LISA can clearly be co-opted, it nevertheless can also be made to be transitional. "Don't apply unless you have to" remains an important principle of LISA and is an argument understood by any farmer in the world. The combination of promoting the expansion of more ecological forms of agriculture (e.g., organic, or agroecological) with the promotion of LISA (input substitution) seems to offer the best overall strategy for transforming the world's agroecosystems. Yet such a program seems cast in a vacuum. Who will promote it, and how will that promotion be facilitated both politically and technically?

Strategies of Knowledge Pursuit in Alternative Agriculture

Unfortunately, researchers in alternative agriculture have frequently adopted the same philosophy as more conventional agronomists, employing a simple-minded empiricism when a deeper understanding is called for. Thus, knowledge is sometimes restricted to knowing, for example, what crop response to expect from a given amount of applied matter, be it mulch, manure, or manganese. The alternative program requires deeper knowledge, acknowledging the complexity of the agroecosystem from the chemistry in its soils, to the interactions in its pathosystems, to the social structures that organize it.

An appreciation of complexity can lead to a different kind of error. The overwhelming complexity of ecological systems has generated an attitude that such a large and complex entity requires a large and complex theoretical framework to generate understanding. Unfortunately, the ready availability of high-speed computers and user-friendly simulation software has given this point of view a false sense of feasibility. It is relatively easy to construct a gigantic model and simulate it for a variety of parameter values. This general point of view attempts to include all factors and consequently loses tractability, either quantitative or qualitative. This super-model approach has not had much success, partly because of the difficulty inherent in measuring large numbers of parameters accurately and partly because of the tendency for large models to be intrinsically unpredictable. Ultimately, however, the large system models have failed because of their underlying philosophical message, a message that can be appreciated only by delving into its antithesis.

Historically, the general strategy of constructing models has been a key component, indeed a necessary step, in the process of scientific investigation. The point of having a model is to aid in understanding a natural phenomenon. The model also enables a prediction of what will happen in response to an intervention, a characteristic as important to the basic scientific method as it is to the development of technology. A model's ability to predict never has been the true measure of its utility in science. That utility is, as it always has been, its ability to provide understanding. In this context, both the conventional view of looking for the yield response to a soil amendment (for example) and the megamodel approach are misleading.

Ultimately, it's necessary to formulate a philosophy that is intermediate between the reduced empirical approach of the classic agronomist and the super-model approach of some systems-oriented scientists. It is a philosophy that utilizes models of intermediate complexity, sometimes quantitative, sometimes qualitative. They are models that purposefully and knowingly exclude many factors, even when those factors are known to be important. They exclude some factors in order to concentrate on the factors that are the "interesting" or transcendent ones. Deciding on which are the interesting ones is more art than science, as it always has been in the history of science. Throughout this book, I have tried to follow this general principle. The approach of using models with intermediate, or better said, appropriate complications can be applied to whatever categories seem suitable for analysis in pursuit of technical advances that promote the transition. Ultimately, the point is to create a metaphorical skeleton of understanding, fully knowing that the flesh on that skeleton is both complicated and necessary, but with the faith that, as in science throughout history, a model framework of understanding enables the details for engineering.

This then brings us to the question of what categories are suitable for analysis. Given the stated philosophy of this book, the question may seem odd, since my focus has been on confronting complexity, not ignoring it. Yet even when problems do not manifest themselves as fitting conveniently into a categorical scheme, in order to study them, even to talk about them, we are sometimes forced to create artificial boxes. Here I use what seem to me to be general schemes of ecological research that, although overlapping a great deal, nevertheless constitute what seem to be coherent groupings of research activities. The three are recycling, managing trophic cascades, and complexity. To be sure, problems may frequently manifest themselves as not fitting conveniently into one of these three categories, but as a vehicle for discussion, this tripart classification is useful.[43]

Recycling: Control from Below

Obviously a code word for the environmental movement, the fact of recycling is one of the undeniable tenets of modern ecology. Indeed, the two most fundamental principles of ecology are 1) energy flows and 2) nutrients cycle. Although pop ecology emphasizes recycling as a political principle, its importance in natural ecosystems cannot be emphasized too strongly, especially when focused on soil management. It is an unfortunate state of affairs that the industrial system has rather compromised the general vision of recycling nutrients in soils, with the dramatic introduction of simple chemical forms of nutrients effectively changing the soil from a semiautonomous ecosystem to a chemostat,[45] with nutrient inputs and outflows bracketing a black box. Technical research in agroecosystems should be focused on reestablishing recycling as the basis of the ecology of agroecosystems, probably most relevant when thinking about soils. In many ways, this focus corresponds to the ecological idea of ecosystem control from below; that is, plants and animals are affected mainly by their inputs, whether nutrients absorbed from the soil by plants or plants themselves consumed by animals. Soil management falls within a unique framework when viewed with this sort of ecological focus.

Nonetheless, as discussed earlier, management decisions must always be geared to the type of soil involved, since techniques that are productive and efficient in some soil types may be totally inappropriate for others. Conventional agronomists long ago realized this regarding conventional chemical fertilizers (e.g., applying K to a soil with low cation exchange capacity will not give the same results as applying it to one with a high cation exchange capacity). Furthermore, soil management involves separate management components that must be integrated. Application of organic matter adds nutrients to the soil but also contributes to the soil physical structure, which in turn affects the ability of plants to take up the nutrients available. Soil management must be approached as an ecological system, analyzing and synthesizing a myriad of interacting elements, but ultimately with the goal of recycling as much of the nutrient material as possible. The intermediately complex model of van Noordwijk and de Willigen, described in detail in Chapter 5, may be useful in this context.

A key problem, not to my knowledge controversial, is the problem of timing. For any soil input, the challenge is to supply the crops with precisely what they need at a given phenological state, thus minimizing loss from the system. Efficient management of the decomposition process is essential to this goal, and that efficient management requires sophisticated knowledge of the process itself.

Managing Trophic Cascades: Control from Above

The introduction of a biological control agent is the canonical example of this focus. That control agent is supposed to "cascade" down to the plant or animal we seek to protect. Thus, the vedalia beetle that controlled the cottony cushion scale in California had a cascading effect on the citrus trees—from the beetle, through the scale, to the plant. It is in its simplest form the notion that any enemy of my enemy is my friend. However, it is frequently complicated by many forces, as noted in previous chapters of this book.

Work in coffee plantations in Mexico is an example of some of this complication, as described in detail in Chapter 6.[46] A potential pest, the green coffee scale, is mutualistically associated with an arboreal ant, which itself is attacked by a fly parasite. The coffee scale is eaten by a lady beetle, but the adults of the lady beetle are unable to feed due to the aggressive action of the ant. The larvae of the beetle are covered with a waxy secretion that is impenetrable by the ants, causing the ants to patrol over the beetle larvae as if they were mutualists (while they patrol the scale insets). If the beetle larvae do not receive this protection from the ants, they are inevitably attacked by other hymenopteran parasitoids. The combination of the ants and the parasitic fly results in a self-organized spatial pattern of the ant nests, which means there are clusters of shade trees that do have ants, but many shade trees that lack the ants. The scale insects, which are the potential pests of the coffee, thus receive a great boost in those areas where ants are present, but survive at only low densities when the ants are absent. Flying adult beetles are able to find these smaller concentrations of scale insects and eat them. Thus, we have a system in which a species of beetle appears to keep a potential pest under control, but only because a mutualistic ant associated with the pest is distributed in space in a clustered fashion, something that arises as a self-organization through the attack of a parasitoid. Control of the potential pests is thus achieved in an "autonomous" fashion through a very complex web of ecological interactions.

The canonical example of the need to control trophic cascades is, as suggested in the previous example, pest control. Indeed, the basic principle of creating secondary pests is a direct consequence of a simple trophic cascade. Here the fundamental ideas of IPM[47] provide a general framework for management. Dividing various techniques into prophylactic versus responsive is a useful device for planners. The idea is to have the tools available to respond to a pest situation as it emerges, but also continually develop other tools that will provide a prophylaxis against that situation emerging so frequently. Generally, the responsive techniques may involve petrochemicals during the transition phase, although the environmental consequences of such use need to be fully assessed. The general idea remains, however, that as prophylactic techniques are perfected, there will be ever less need for the responsive techniques. Eventually, the tools of the responsive techniques will naturally become very rare, and we will hopefully approach the situation in which farmers in general respond to the question "what kinds of pests do you have?" with the answer given by traditional Mayan farmers, "none," fully understanding that a program of autonomous ecosystem management is underway.

Complexity

The previous example of autonomous control of coffee pests is a clear example of how ecological complexity is inherent in agroecosystems. Complexity is obviously a catchall category and includes a potpourri of issues, inevitably involving not only ecosystem processes, but also the intersection of ecosystem properties with social constraints. An example is provided by the inherent time lag involved in most agricultural production. Imagine a group of small farmers producing for a local market under the assumptions of classic economics (all are independent of one another and respond only to market signals). Because agriculture is constrained by ecology, planning must happen months, sometimes years, before going to market, even at the overly simplified level of standard economic theory. Thus, each farmer must decide how much of his or her land to devote to production of, for example, tomatoes. If only market cues are allowed, the fraction of land devoted to tomato production is dependent on the current price of tomatoes on the market. If supply and demand determine that price, assuming demand constant, the supply at last planting cycle will determine how much will be planted this cycle, which will eventually determine how much supply will enter the price-setting mechanism at harvest. The collective behavior of the farmers will result in high current prices generating vigorous planting of tomatoes, which results in high supply at harvest and consequently low prices, which in turn generates low planting of tomatoes, which results in low supply at harvest and consequently high prices. Thus, we see oscillations between high and low production and consequent low and high prices. But clearly the farmers do not respond to only local market signals (the current price of tomatoes) but rather to market expectations. Such necessary speculation involves predicting what other producers in the market will do, which generates, even at this extremely simplistic level of analysis, chaotic trajectories in prices and production,[48] effectively canceling anything that might seem like rational market choices.

If such unreasonably simplistic structures are capable of producing such extreme complexity, imagine what a more realistic assessment might do. Indeed, much like an ecosystem itself, the agroecosystem, with its complicated intersection of social, political, economic, and ecological components presents a daunting set of problems for serious planners. This complexity needs to be acknowledged, and the planning process needs to take it into account when developing programs, but with the humility that should accompany knowledge of the inherent complexity of the system we are trying to manage. All we can say for sure about the agroecosystem is that there is a certain inevitability of surprises, no matter how sophisticated our understanding. Indeed, the inevitability of surprise is a conclusion derived from sophisticated understanding of the nature of complexity. All planning of agroecosystems must be done within this constraint.

Planning within complexity does not occur on a single productive unit. Given that the agroecosystem is embedded in a larger ecosystem, its proper understanding is incorporated in that larger system. The entire landscape needs to be the ultimate focus of planning. If a one-hectare organic farm in India is sitting next to a large estate that sprays herbicides from an airplane, it would be foolish to ignore the effects of the pesticide drift. The consequences of landscape-level management are crucial for biodiversity preservation as much as for overall ecosystem function. Technically, this requires a great deal more research into landscape ecology, certainly an understudied subdiscipline, but also points to the absurdity of trying to maintain fixed barriers between scientific disciplines. Getting large numbers of individual farms to incorporate into a large-scale planning unit is a sociopolitical challenge as much as an ecological one.

Socioeconomic Strategies for the Transformation

Reruralization

As indicated earlier, there are major socioeconomic barriers to be faced in transforming agriculture. The industrial model has brought with it a debilitation of rural society in several important senses. First, the actual number of people who now live in rural areas is extremely small in most developed countries and decreasing in most underdeveloped countries. Second, the fabric of rural society, including both physical and social infrastructure, has been dramatically altered during the past 50 years, to the point that much of what might be needed to support the transition will have to be constructed from scratch. Reconstructing a rural society that will carry out the new agricultural plan is a major challenge involving a host of considerations.

Goldschmidt's hypothesis, intensely debated in the past decades, remains an important framework when dealing with this issue. Sociological research has shown that the change from small family farms to industrial agriculture leads, in rural communities, to "lower incomes and lower standards of living . . . lower numbers and quality of community services . . . lower community integration and greater psychological stress . . . and a less diverse economic base and higher unemployment."[49] In other words, the cornucopia of the industrial agricultural system as much as its profitability, comes at a great expense to the localities where food is produced. Workers in meat-packing plants or hand-harvest operations are generally not as well-off as members of small family farms. A vibrant rural society is a measure of human welfare, for the people who live there.

Nevertheless, there is a sort of conventional wisdom that regards the industrial agricultural system as an essential part of the overall industrial capitalist system, emphasizing the quality of life it has supposedly brought to the world. Indeed, under this model, one might conclude that massive urbanization is a good thing, relieving the people of the necessity of living in the degraded rural conditions created by the industrial agricultural system. Emptying the countryside seems, from this point of view, a positive goal.

If an alternative vision is to gain traction, it is not likely to do so without people to engage it. We thus face the important challenge that small family farms, in both the Global North and South, will likely be the agents to formulate the reality of the new agriculture, even as the industrial system pushes ever harder to deruralize the world. If the Goldschmit hypothesis has any traction at all, it would seem a sisyphean task to convince a new generation that rural life is attractive if industrial agriculture makes it not so. The call for "reruralization" clearly should be part of the call for a more rational agriculture. The call for a more rational agriculture clearly should be part and parcel of strategies for reruralization. With new information technologies (e.g., the internet, cellphones, DVDs) rapidly spreading across the world, some of the old prejudices about rural life become less convincing, and a more rational transportation network, called for in a totally different context, means that rural isolation is less isolated than it was when horses and buggies were the main form of transportation. The task of reinventing rural society is different today than it would have been in 1900.

Revising the Assumption of Cheap Food Policy

Much of what has produced the decline in rural population as well as rural society has been the economic assumption, made by planners throughout the world, that having cheap food is good politics and good economics:

> So-called "cheap" food comes with a huge hidden cost. Bills for diet-related ill-health, pesticide residues in food and water, motorways of which the supermarkets are disproportionately high users . . . when one takes account of these externalized costs, food isn't cheap.[50]

Cheap food policy initially helped solve the early problem of keeping the cost of labor down in the emerging factory system. Furthermore, by lowering the amount of money a family had to spend on food, a greater amount was left to spend on the products of industrialization, thus raising demand for other products. Clearly, this contributed to economic development in that workers had more disposable income that fed into economic growth, a prerequisite to the full functioning of the Fordist accumulation phase of industrial capitalism. In the post-World War II world, externalizing the production costs of inputs like fertilizers, pesticides, and machinery further enabled food to arrive at the consumer's table at an artificially low price.

Cheap food policy has been part of almost every active development strategy, especially in the Global South, subsequent to World War II. The basic logic is always the same—if people do not have to pay too much for their food, they will have money to demand industrial products. Development will then follow.

The policy has almost always failed. It had been a success in the early capitalist economies, largely because of dramatic growth rates in the industrial sectors, fueled not only by industrialization itself, but also by a lack of competitors and the ability to import food from exploited colonies. A handful of countries participated in the original expansion, and indeed, there were incredible demands for workers whose new-found wealth drove ever-increasing demand for industrial products. The 21st century is quite different, however. Countries of the Global South are not industrializing at the same rate as England in the 19th century, nor can we expect them to do so. Most of them retain a significant agrarian component to their economies with a large fraction of their population still involved in agricultural activities. If a large and relatively stable fraction of the population's

344 ■ Chapter 8 Toward a Sustainable Future

workforce is composed of farmers and those occupations that are directly related to the farming sector, a cheap food policy simply maintains a low effective demand in the industrial sector, and thus a low remuneration for the farming sector. So the farmers remain a significant fraction of those who must demand products because they remain a significant fraction of the population, yet their disposable income is low. Furthermore, the policy has added to the inequalities within these countries, placing an even bigger barrier than had existed before between rural and urban societies. The most obvious exception to this general rule in the postwar period is Japan. Because of special political considerations, postwar construction was based on an *expensive* food price policy. It is hardly necessary to dwell on the consequences.

Although the simple act of vertical integration was able to initiate a cheap food policy in 19th century Britain, more complicated forces are at play in food price policy in the postwar developed world. In the United States, for example, large food corporations have encouraged government policy that promoted extensive overproduction in almost all agricultural sectors (with some notable exceptions, such as cane sugar). Such overproduction drove prices very low, sometimes so low that the average farmer could not make ends meet. Even worse, the postwar emphasis on biocides, extensive monocultures, and inorganic fertilizers has in fact exerted an enormous environmental cost. That cost should rightly be part of the cost of production with the new industrial technology but in fact has been written off as an externality. Nitrate-contaminated ground water must be cleaned up, but those who are responsible for causing the problem are not expected to pay for it. Biocides eliminated natural enemies, requiring the spraying of more and more toxic biocides. Yet restoring the land with its normal concentration of natural enemies, an expensive task indeed, if ever actually pursued on a large scale, is an expense that those who caused the problem are not expected to assume.

At a very general level, the supply curve for any product can be maintained at a relatively high level if the producer is allowed to externalize much of the input cost. If there were a tax levied on every bag of fertilizer sold, to pay for the cleanup of ground water, or on every bottle of pesticide to fund the restoration of natural enemies and restore the health of workers and consumers affected, the cost of production would likely soar for most agricultural products. The politics of an "expensive" food policy, as advocated here, are complicated to say the least. The urban poor will be the ultimate victims of expensive food policy, something that is neither morally or politically acceptable. Yet having the rural poor as victims of cheap food policy is likewise not morally acceptable. Its political acceptance stems only from particular political power relationships that tend to pit urban poor against rural poor. Properly functioning governments are necessary to structure markets such that the urban poor see the system as a cheap food policy, allowing them to feed their families, while the rural poor see it as an expensive food policy, allowing the farmers to reap enough monetary rewards to support rural communities. Remember, it is never a question of whether agricultural subsidies are wise, it is always a question of exactly how subsidies are organized. Searching for an economic structure that is fair to both consumer and producer is nothing more than searching for a healthy agrarian economy.

The Health of the Agroecosystem

If revising the assumption of cheap food policy is effectively searching for a healthy agrarian economy, the simple idea of maintaining a healthy ecosystem is a central metaphor for the ecological aspects of the new agriculture. Recall that one of the major forces giving rise to the industrial system in the first place was the transformation of the medical metaphor (maintaining a healthy farm) to the war metaphor (vanquishing the enemies on the battlefield of the farm), as presented in Chapter 6. It is not without merit to consider challenging this transformation and seeking a new metaphor, and the idea of a healthy ecosystem is not a bad candidate.

However, such an idea within the science of ecology sounds strange. Indeed, a central organizing principle of modern ecological science is that ecosystems are not like organisms—that viewing the ecosystem in that way does more to cloud our vision than enlighten it. Because our notions of health are largely derived from a medical metaphor which implies the need for an object (organism) to query about its health, the idea of a "healthy" ecosystem is immediately suspect.

Because the environmental movement had gained such a prominent position on the world stage, it was inevitable that the idea of an ecosystem in either a good or a bad state would emerge. Indeed, the absence of songbirds in Carsen's *Silent Spring* was an indication of an ecosystem in trouble, an unhealthy ecosystem. Recently, there has been a flurry of discussion and analysis of this issue,[51] and a journal with precisely that title, *Ecosystem Health*, was launched in 1995 (more recently incorporated with other journals into *EcoHealth*). The idea is not really new, with no lesser a name than Aldo Leopold having clearly warned the world about sick ecosystems in 1941.[52] As is frequently the case with new and resurgent ideas, their birth or rebirth can be painful.

There has always been a problem with the health metaphor for ecosystems. Although alternative agriculturalists of various shades regularly use the notion of "a healthy ecosystem," professional ecologists would not likely speak with such emotionally laden terminology. It would be like talking about the well-being of a subatomic particle or the comfort of a molecule. It just does not make any sense. Or does it?

The arguments that it does not make sense are relatively easy to make. Ecosystems are collections of organisms and their interactions. They have energy flows and nutrient cycles. Our job is to understand them—what are the various organisms, how do they interact, in what channels do the energy flow and the nutrients cycle. It is a set of non-normative questions. To suggest that an ecosystem *should* behave in some way or another is thought to be what popularizers and pundits do, not something with which professional scientists would concern themselves. We are concerned with how they do behave, not with how they should behave. Furthermore, the notion that an ecosystem is a metaphorical organism has long ago ceased being popular, and the idea that a sick organism could equally mean a sick ecosystem is not thought to be defensible.[53]

The health sciences also regard themselves as sciences, equally committed to studying the world and avoiding normative judgments in their work, and they do not always treat "the organism" but more frequently are concerned with populations and communities. The difference is that the concept of a healthy body and even a healthy population is conceptually fixed in all of our minds because it affects us so closely. I am a scientist, but I have no trouble understanding that hypertension is not good for me, no matter how normative that understanding might formally be. I quickly add that this obviousness is not universally agreed on, as a quick sampling of various health states quickly reveals. Sumo wrestlers would certainly not get a clean bill of health from the American Heart Association. yet their bodies are wonderfully suited for their stated purpose, and the criteria for what made a healthy slave in the Caribbean sugar economies were clearly not the same as what made a healthy plantation owner.

Such caveats notwithstanding, speaking of a healthy body is not regarded as somehow unscientific in the health sciences, so why should speaking of a healthy ecosystem be so thought in the science of ecology? I propose that the appropriate way of looking at the idea is similar to looking at the science of physiology as it relates to medicine. Physiology is similar to ecology in that it seeks to be rigorously nonnormative. No physiologist says that the mobilization of melanin in the epidermis *should* be something or other, but rather, that melanin will be mobilized under certain conditions of genes and environment and that the mobilization may lead to other physiological consequences. It is society that defines, sometimes obviously, sometimes subtly, what is a "healthy" amount of melanin in the epidermis. In trying to mobilize large amounts of melanin, people of European descent sometimes develop cancer, a clear state of unhealthiness. But the "right" amount of melanin (the perfect tan) gives a "healthy" glow to one's complexion, and, in the United States, too much melanin fixed by one's genetic heritage is indeed unhealthy because of extant sociopolitical arrangements (one tends to be harassed by urban police).

Indeed, a close look at health, even in the case of the medical sciences, reveals some philosophical complexities that rapidly become difficult to sort out. Yet who would question the utility of the concept. "Are you healthy?" is not a question misunderstood by scientist or lay person alike. When a doctor says "the patient is healthy," few regard him or her as unscientific even though the judgment is in fact loaded with normative ambiance. The notion of health as a condition of an individual person is deeply embedded in our psyches.[54] The extension of the concept to a population of people is not difficult, even to plants and other animals. It seems only a natural progression then that it be applied to ecosystems as well.[55]

Ecosystems are different. Although most would agree to a clear meaning for the notion of having a healthy body, it is not completely clear what a healthy ecosystem would be. Unhealthy ones may be obvious (e.g., oil-soaked Prince William Sound, the canals of Venice, acid-rain soaked forests of the northern Czech republic, and so forth). Yet, are the pine forests of southern Spain "healthy"? They were never there before the Romans put them there, but they seem fine now. They clearly displaced other kinds of vegetation that had been there and perhaps had been "better" adapted. Are the Sahelian ecosystems generally healthy? The plant ecologist C. DeWitt once suggested massive aerial spraying of rock phosphate over the entire Sahel, since all of the soils there were extremely low in phosphate and such a spraying would have enriched, for a long period of time, those soils with phosphorous, thus clearly changing the nature of the ecosystem, making it far more productive for agriculture and grazing. Would that have been a healthier ecosystem than the ones currently extant?

Ecologists and others have recently tried to deal with this question. One response has been the recognition of the so-called Ecosystem Distress Syndrome (EDS), the symptoms of which are reduced productivity, reduced biodiversity, alterations in biotic structure to favor short-lived opportunistic species, and reduced population regulation resulting in larger oscillations and increased disease outbreaks.[56] An ecosystem exhibiting EDS is then thought of as an unhealthy ecosystem. Furthermore, according to Costanza (1992), "An ecological system is healthy and free from distress syndrome, if it is stable and sustainable— that is, if it is active and maintains its organization and autonomy over time and is resilient to stress." There are, to say the least, practical problems with all such views. Although superficially obvious, any attempt at operationalization runs up against words like "stable," a notion with a long history in ecology and not yet fully established as an agreed-on principle; "sustainable," which itself has a contentious literature far larger than that of ecological health; "active," which I have never heard of in the context of an ecosystem; and even "organization" and "autonomy" are subject to debate as to exactly what they mean. Nevertheless, the intent of such definitions is still clear. In a more ambitious attempt at characterizing the concept, Rapport (1995) tentatively offers seven "key properties" that indicate ecosystem health. An ecosystem can be thought of as healthy if it 1) is free from EDS, 2) is resilient to outside perturbations, 3) is self-sustaining (here there is exception made for agroecosystems and forestry), 4) does not impair adjacent systems, 5) is free of risk factors, 6) is economically viable, and 7) sustains healthy human communities. These features are presented as tentative ideas for delimiting a healthy ecosystem. Because the idea itself is relatively new, any attempt to fix its meaning is probably premature. Rather, questions about what is meant by ecosystem health need to be further debated.[57]

Such questions are not easy to deal with. I was shocked the first time I witnessed the central valley of California. It appeared as if the opening lines of Rachel Carson's *Silent Spring* had suddenly come true. No birds were singing, no insects flying, no ants crawling— a biological desert. For me, it was an unhealthy ecosystem. Other observers, for example, the vegetable producers and vineyard owners, would likely disagree.

Part of the philosophical problem is automatically resolved when we talk of the *agroe-cosystem*, for, unlike the tropical rain forest, it is an ecosystem with a purpose.[58] The agroecosystem exists for the purpose of producing food and fiber for humans. If it is in such a state that this function is jeopardized, it seems perfectly intuitive to refer to it as unhealthy. Thus, the problem of being normative by allowing a functional interpretation seems to be avoided in the particular case of an agroecosystem. Although this particular conundrum is avoided, there remains the more difficult question of what is healthy, in the sense of under what conditions is the ecosystem performing its function, a philosophical problem as much for ecosystems as it is for medicine. The extremes are easy to identify—an individual suffering from cardiac arrest, a desertified pasture in the Sahel. When less extreme situations are encountered, philosophical problems arise. According to some analysts, Albert Einstein was afflicted with attention deficit disorder, a recognized psychological malady, yet it was possibly this very malady that preconditioned him to his creative excesses—a diseased individual? Traditionally, European Americans regarded a high-fat diet as emblematic of prosperity and good health, yet today the famous meat-based diet is recognized as cause of poor health.

The same problem exists, probably more severe, in the case of agroecosystem health. The extremes are obvious, but the normal cases not so much so. What are the key features of agroecosystem health? First, the system must produce on a relatively sustainable basis. A farm whose production is clearly on a decline may be unhealthy (obviously in some cases this may not be the case, as when declines in production are sought as part of an overall sustainability criterion). The cause of that particular diseased state, declining production, may be a slow depletion of potassium from the soil, a slowly increasing pathogen load in the animals, a slowly increasing insect pest on the pasture, or many other factors. If the farm is producing less each year, however, despite the intention of maintaining production, it may be unhealthy.

Second, the system must be productive on a long-term basis. The heroin addict may judge his or her situation to be excellent when under the influence of the drug, but the long-term prospects for survival are less sanguine and most readers would agree that heroin addiction is a diseased state, regardless of the temporary feeling of well-being that shooting-up provides. Similarly, a cotton farm maintaining good profitability based on the requirement of ever-increasing quantities of pesticides could not be considered healthy over the long run.

These two criteria can be thought of as the underlying basis for determining the health of an agroecosystem: 1) sustained year-to-year production and 2) production over the long term. The first of these principles can be reasonably assessed from year-to-year experience. The second may require some more sophisticated analysis. That production was maintained from year to year may have seemed obvious to farmer and agronomist alike in the cotton fields of Nicaragua in the 1950s and 1960s. Yet the agroecosystem was obviously not healthy. It could have been shown (and indeed was in this particular example) that the development of pesticide resistance and the regular emergence of new pests caused this agroecosystem eventually to become extremely unhealthy, to the point of death. That is, despite early, sustained production, the agroecosystem was not healthy—as hooked on pleasure (profit)-producing pesticides as the heroin addict.

What then are the characteristics of a healthy agroecosystem? The following features seem to emerge repeatedly as issues promoting agroecosystem health,[59] both in the sense of year-to-year production and over the long run:

1. Closed nutrient cycles with minimal external input
2. Control of pests with biological or cultural practices, with a minimum use of synthetic pesticides

3. Maintenance of soil biodiversity, at least insuring that members of all functional groups of organisms remain in the system, including macro and micro faunal and floral elements

4. Maintenance of beneficials, such as parasitic wasps, spiders, ants, birds (the pest forms of these groups are covered in item 2)

5. Reasonable rewards to the practitioner (the farmer, or worker—the people who are in contact with the land in the production process)

6. Matching the needs of the consumer community (be it a local community or the international marketplace) to the capabilities of the land

7. Maintaining the health of farmers, farm workers, and consumers

8. Provisioning of landscape diversity with ecological interactions among patch types, both in space (e.g., mixed farming, intercropping) and time (e.g., rotations)

All of these items are offered here as general and tentative guidelines. Most seem obvious, and most are easily operationalized. I have tried to avoid use of popular yet vague terms such as stability and sustainability, although vagueness also may have a strong mind-of-the-beholder aspect to it (e.g., some readers may question what is "reasonable" in the reasonable rewards criterion). These seven criteria also fit snugly into Rapport's "dimensions" of ecosystem health.[60]

Concluding Remarks

Unless history has ended, we expect changes, some of them dramatic, in the way we do agriculture in this world. How practitioners and planners should approach these changes is not completely obvious, as both history and contemporary debates suggest. In the search for positive and progressive forms of change, a useful metaphor for understanding agroecosystems at a very general level is the idea of "syndromes of production."[61] Just as the conventional industrial system involves a host of factors that combine with one another to form the entire system, the alternative system combines a host of factors. That is, both ecological and industrial systems represent "syndromes of production," and it would be difficult to mimic either of them outside of their entire context. History provides us with some striking examples; the McCormick reaper, for example, was functioning and in production before the Civil War, but it remained largely unused for over 20 years because other conditions of production (labor costs) were not conducive to its adoption, as we saw in Chapter 3. A single new technology when simply added to a previously functioning system may not improve the performance of that system. The system as a whole is a "syndrome," and any small change in the syndrome is likely to make its performance worse, not better. A jump from one syndrome to another is required. This conceptualization can be thought of as a contour map with two mountains. One syndrome, as indicated by one "peak" in the contour map, may be our imagined sustainable system, whereas one of the other peaks is today's industrial system. The height of the position is proportional to the performance of the system. If we are fixed at either peak, a small change requires a decline in performance, even though the alternative peak may represent a generally higher performance.

As discussed previously, models for more ecological forms of agriculture exist in the many traditional and organic production systems. Nevertheless, a model for the actual transformation from the industrial system to the sustainable alternative is elusive to say the least. Theoretical models have been proposed, and many of them look good, but no country or region has yet undergone the transition. A useful tentative model may be Cuba, one of the examples that opened this book. The changes on the world political stage that forced Cuba to strive for a different model of production may provide us with a preliminary glimpse of the problems we will likely face in the upcoming transformation. As an example of an apparent jump in the potential map, a syndrome of production switch, Cuba is interesting.

High-Tech Expressways or Contour Pathways

From the Neolithic revolution to the post-World War II chemical revolution and the Green Revolution, it would be hard to deny that switches in syndromes of production occur, and unless history has ended, they will continue to occur. We now may be on the cusp of a new agricultural revolution. It is the revolutionary change to more ecological forms of agriculture. The consequence of the original agricultural (neolithic) revolution was that human cultural evolution invaded the natural world in an unprecedented way. Although any ecosystem is subject to the changes wrought by genetic change in component species, the agroecosystem is unique in that changes can be extremely rapid since they are driven by cultural rather than organic evolution. With cultural evolution as the driving force, changes in agroecosystems in the past have not been in the direction of "improving" the system, but rather emerged from whatever happened to be the forces shaping cultural evolutionary changes at particular moments in the relevant human population. Thus, for example, the expansion of sugar beet production in Napoleonic France was not really for the purpose of producing sugar for the French people. It was for the purpose of securing independence from English-controlled cane sugar markets as part of Napoleon's imperial strategy. Many other examples could be cited in which economic, cultural, and/or political forces, the forces that in fact shape cultural evolution, created the conditions under which agroecosystems underwent dramatic changes.

The contemporary world is no different, except perhaps in that change happens far more quickly than in the past, and its effects are felt worldwide. As agroecosystems change, part of the philosophy of the new agriculture is that changes in agroecosystem structure and function should be brought under control. Rather than allow skewed economic interests to dictate the direction of change, rather than allow nature to take its own course, rather than letting a philosophy born of other interests (e.g., the unrelenting search for increased profits) dictate the direction of change, there ought to be a concerted effort to design agroecosystems in a rational fashion. Here the ecological agriculture movement has been clear in its philosophical distinction from industrial agriculture.

A useful metaphor is the hunter in unfamiliar terrain with a topographic map for his or her guidance. To get from point A to point B, one might draw a straight line between the two points on the map and proceed to follow that line, climbing hill and valley, perhaps scrambling up cliff faces and rappelling down steep gullies, eventually getting to point B in the "most efficient manner possible"—most efficient because the line on the map was the shortest possible. Another way of getting from A to B would be to follow a pathway along the contours provided on the topographic map. Although this might be a longer absolute distance, one would likely arrive at point B faster than the straight-line approach. On one hand, we ignore the topographic contours and insist on our peculiar notion of efficiency that dictates a straight line is the shortest distance between two points, that we can ignore the contours. On the other hand, we view the contours as our signposts to guide us to the goal. The philosophy of industrial agriculture has been akin to drawing a straight line on a topographic map, blasting through mountains and draining waterways to construct a high tech expressway that follows that straight line. The metaphorical contours are the myriad ecological interactions that inevitably exist in an agroecosystem, the interactions that industrial agriculture has sought to ignore. The new agriculture philosophy acknowledges the contours not only as extant barriers to the straight line approach, but as useful signposts as to where to construct the contour pathway that will most efficiently get us to point B. In pursuit of the contour pathway, ecological knowledge is of utmost importance because it provides us with the necessary signposts.

Challenging Old Assumptions

Advanced industrial capitalism promotes certain assumptions, an ideology, about the natural world. First, there is a kind of triumphalist mentality about the industrial system. This triumphalism encourages the attitude that we can completely control nature. Spanning rivers with bridges and flying heavy metal birds in the air are certainly impressive engineering feats that seem to offer limitless possibilities for conquering any and every limitation nature seems to put in our way. Second, there is an extreme utilitarianism about resources, codified in an almost religious philosophy of commodification. That which can be commodified and traded, it is assumed, *should* be commodified and traded, and that which cannot be commodified and traded cannot be socially useful.

The triumphalism of the industrial system sometimes seems that it knows no bounds, despite the evident and well-publicized problems, from hypoxic oceans to a greenhouse atmosphere. It seems to an increasing number of people that the triumph has been a cruel façade. At least once a year, a major article in a major international newspaper recounts the continuing problems created by Union Carbide in Bhopal, India. It is not just an accidental gas leak that exposed half a million people to toxic gas, killing at least 10,000; it is the lingering effects of the toxic sludge that Union Carbide (purchased by Dow Chemical Co. in 2001) left in the ground around its plant. Those pesticides may have made production easier for the Green Revolution farmers in India and the industrialized farmers in the United States and Great Britain, but was it worth the death and destruction that went along with it? The pattern is repeated the world over. The secondary consequences of the industrial system make it look more like a Faustian bargain than a triumph of any sort.

The assumption that we will be able to triumph over the barriers placed by nature is one that needs to be challenged. The natural world is something we need to learn to cooperate with, not triumph over. It is a contour path that will get us there, not an autobahn.

The second major assumption to be challenged is the assumption of commodification—for a piece of nature to be useful, it must be commodified (in some quarters, the word is "monitarized"). Thus could, for example, a chemical industry representative at a conference on integrated pest management unabashedly list those techniques that were potentially "useful," based solely on the degree to which they could be turned into marketable commodities. Methods based on cultural control, according to him, had little potential in integrated pest management, simply because there was usually nothing that could be constructed and marketed by a firm. If it cannot be constructed and marketed by a firm, it simply could not be a part of future life.

This fetish about profitable commodities, along with the notion that nature is to be dominated by technology, has dictated the philosophy of the industrial agricultural system, at least since World War II, if not since the Enlightenment. It has led to the various problems that are now so evident, even if they were denied when early writers, such as Rachael Carson, pointed them out.

Alternative agricultural movements need to take the challenge to these two major assumptions as part of their overall agenda, as many of them do.

Boyle's Other Law and a Gentle Thought-Intensive Technology

Ecological processes are general, but their particulars are local. Weeds compete with crops, a general ecological phenomenon. In a Nicaraguan backyard garden the sedges are most competitive against maize, whereas morning glory vines and wild banana plants that may appear as weeds at first glance are actually beneficial because they maintain the field free of sedges over the long run. Predators eat their prey in all ecosystems, but which predator eats which prey is dependent on local conditions. It is itself almost a general rule that local particulars may override general rules in ecology, a fact that continues to frustrate the attempts by ecologists to devise meaningful general principles.

In much the same way that ecological forces are simultaneously general and local, our knowledge about those forces is also both local and general. The intimate experience of local farmers cannot be matched by the generalized knowledge of the ecologist, yet the sophisticated training of the ecologist cannot be matched by the experiential knowledge of the local farmer. Thus, for example, residents in a small valley in Cuba observe that trees grow toward the wind.[62] This particular valley is arranged such that the surrounding mountains block out the sun most of the day, except when it appears through the same mountain pass that allows the daily breezes access to the valley. Thus, the trees that strain for more light, according to ecological principles, in fact do grow toward the wind according to local knowledge. Such stories could be multiplied a thousand-fold across the globe. Local residents may have intimate knowledge about the ecological forces that surround them. However, their experience is limited to a relatively narrow geographical and intellectual setting, disabling them from seeing their knowledge in the same larger context that a professional ecologist may automatically assume. On the other hand, the ecologist may find the deviance of local circumstances baffling in relation to that same larger context and be unable to appreciate the rich texture that comes from detailed knowledge that the local farmer automatically assumes. It is not far from the truth to claim that general academic knowledge is broad but shallow, whereas local knowledge is deep but narrow. The trick is to combine the two in search of broad and deep knowledge. Robert Boyle, so famous for his law of gases, saw this as an important generalization in the development of science in general when he said, "As the naturalist may . . . derive much knowledge from an inspection into the trades, so by virtue of the knowledge thus acquired . . . he may be as able to contribute to the improvement of the trades," as important an observation as the famous law named after him.

Furthermore, if local knowledge is to be part of the process of agricultural transformation, a clear prerequisite to the development of a truly ecological agriculture, the people who own that knowledge must be part of the planning process. This implies a great deal about equality—equality in education, equality in economic and political power. It is, effectively, part and parcel of what the movement for food sovereignty seeks.

In response to the crisis, modern industrial agriculture researchers clamor for more funds to do more of the same, seeking a new technological fix to each problem that arises. At the other end of the political spectrum, and even more troubling, a surprisingly reactionary force seems to have emerged. If the modern industrial system is a consequence of "hard-nosed" scientific research applied to the problems of ever-higher production at whatever environmental cost, they argue, this hard-nosed scientific approach needs to be changed. This position frequently takes on a romantic air, and the techniques of our forebears become models to be emulated, even venerated, regardless of their potentially negative environmental effects. I have always been in opposition to those who locate the problems caused by industrial agriculture in the application of western science. Indeed, my position is not that we need less science and more tradition, but rather that we need more and better science.

Given that the industrial model still dominates, we might think of agriculture, as so many in the developed world yet think of it, as simply another form of industrial production, with inputs to and outputs from the factory, with expenditures, gross revenues and profits, and so forth. In this model, if agrochemicals are inputs to the factory and wheat or barley are the outputs from the factory, where is the factory? Is it the farm? Where are the machines in the factory? Are they simply the tractors, plows, cultivators, pickup trucks, and the like? Perhaps, but there is surely a more important collection of "machines" on the farm—the soil and its dynamic biological inhabitants, the insect pests and their natural enemies, the sun and shade and weeds and water, and all the other factors that are usually referred to as "ecological factors." These machines do the real producing on the farm, and the trucks and tractors only function to bring the inputs to and take the outputs away from the real machines.

But there is a profound difference between these machines and the machines of a regular factory.[63] The machines in a regular factory were designed and built by men and women, and their exact function and operation are known with great precision—they do pretty much what their designers intended them to do. However, no human being designed the machines in the factory that is the farm. They were molded by hundreds of millions of years of evolution, and their exact operation remains enigmatic to anyone who minimally understands our current state of scientific knowledge about how ecosystems work. We are, in effect, making products with machines we understand only superficially. It is as if we send a blind worker into a factory filled with machines, tell her nothing about what the machines are or what they are intended to do and then tell her to produce something useful.

The industrial model has been able to avoid understanding much about these metaphorical machines, at least temporarily, by a brute force approach. It is as if the blind worker began by knocking over all the machines that were not immediately familiar on first touch. It is increasingly obvious to most observers that the science of ecology must get more into the act. The machines in the factory are the ecological forces that dictate what can and cannot be done on the farm, and understanding those machines is the goal of agricultural ecology, whether applied to industrial agriculture or to a more ecological agriculture. The mobilization of the science of ecology in service of understanding agroecosystems is critical to future developments in agriculture if we are to avoid the mistakes of the past.

Most applications of ecological principles to agroecosystems remain either hopelessly naive or even mislabeled. The many "alternative," "organic," "holistic," or "ecological" agriculture texts and guidebooks are frequently guilty of substituting popular visions of ecology for what really ought to be concrete scientific knowledge. Not that popular visions of ecology are irrelevant, not that local experiential knowledge about particular ecosystem function is unimportant, but rather that there in fact is a body of scientific knowledge called ecology that should be mobilized in an effort to make a more ecologically sound agriculture. Just as physics represents the scientific foundation of mechanical engineering and chemistry that of chemical engineering, the science of ecology should, and I predict one day will, represent at least one pillar of the scientific foundation of ecological agriculture. The purpose of this book is to summarize at least part of that knowledge as it may apply to the new agriculture. In this way, I hope it will contribute to the continued development of a more ecologically based agriculture.

Richard Levins has provided a useful guiding philosophy for the future development of an agriculture based on ecological principles. We do not want the bull-in-the-china-shop technological developments we have seen in the past in industrial agriculture. We also do not want a return-to-the-good-ole-days philosophy that rejects modern scientific approaches. Instead, we want an approach to technological development that takes a holistic view, incorporates both the knowledge we have of academic ecological science and the local knowledge farmers have of functioning ecology, and seeks to develop an agriculture that produces for the good of people. This philosophy rejects standard economic accounting (assuming that what is profitable is what is good) and seeks to internalize all environmental costs. In short this philosophy looks to the contours on the map as sign-posts for development, gently maneuvering through the ecological realities rather than trying to bully them into submission—it is "gentle, thought-intensive technology."

To drive this point home, I end this text with some words of wisdom from the 19th century.

> Let us not flatter ourselves overmuch on account of our human victories over nature. For each such victory nature takes its revenge on us. Each victory, it is true, in the first place brings about the results we expected, but in the second and third places it has quite different, unforeseen effects which only too often cancel the first. The people who, in Mesopotamia, Greece, Asia Minor and elsewhere, destroyed the forests to obtain cultivable land, never dreamed that by removing along with the forests the collecting centres and reservoirs of moisture they were laying the basis for the present forlorn state of those countries. When the Italians of the Alps used up the pine forests on the southern slopes, so carefully cherished on the northern slopes, they had no inkling that by doing so they were cutting at the roots of the dairy industry of their region; they had still less inkling that they were thereby depriving their mountain springs of water for the greater part of the year, and making possible for them to pour still more furious torrents on the plains during the rainy season. . . . Thus at every step we are reminded that we by no means rule over nature like a conqueror over a foreign people, like someone standing outside nature—but that we, with flesh and blood and brain, belong to nature, and exist in its midst, and that all our mastery of it consists in the fact that we have the advantage of all other creatures of being able to learn its laws and apply them correctly. (Engels, *The Dialectics of Nature*, 1860)

Endnotes

[1]Lewontin, 1982.
[2]Goodman et al., 1987.
[3]Buttel, 1990; Soule et al., 1990; Patel, 2008.
[4]Beus and Dunlap, 1992.
[5]Lacy, 1996.
[6]One of the best treatments of this subject is due to Thornton's "Pandora's Poison," 2001.
[7]Steingraber, 1997.
[8]Colborn et al., 1997; Hayes et al., 2003.
[9]Wright, 2005.
[10]Although not as commonly cited as the pesticide treadmill, Hepperly and Wilson (2006) make casual reference to Robert Rodale having referred to it in this fashion long ago.
[11]Wilson, 1999.
[12]Vandermeer and Perfecto, 2007a, 2007b.
[13]IFOAM, 2009.
[14]For example, Brown et al., 1987; Carter, 1989; Farshad and Zinck, 1993; Francis and Youngberg, 1990; Hassebrook, 1990; Keeney, 1989; Lockeretz, 1988; Madden, 1990; Stinner and House, 1989.

[15]National Research Council, 1989.

[16]Landell Mills, 1992.

[17]Lockeretz et al., 1981; Andrews et al., 1990; Faeth et al., 1991; National Research Council, 1989; van der Werf, 1993.

[18]Badgley et al., 2007; also see Gomiero et al., 2008.

[19]Cornia, 1985.

[20]Rosset, 1999.

[21]Tegtmeier and Duffy, 2004.

[22]Dimitri and Oberholtzer, 2005.

[23]For example, Grau et al., 2003; Klooster, 2003; Aid & Grau, 2004; López et al., 2006; Hecht and Saatchi, 2007.

[24]Vandermeer et al., 2008b; Vandermeer and Perfecto, 2005b.

[25]Garcia-Barrios et al., 2009.

[26]Altieri, 1990.

[27]Runnels, 1995.

[28]*Agriculture, Ecosystems and Environment*, Vol, 1997; *Applied Soil Ecology*, Vol 6, 1997; see also Vandermeer et al., 1998.

[29]In personal conversation as well as many public presentations, Richard Levins has repeatedly made this point about agriculture.

[30]Badgley et al., 2007.

[31]See, for example, Landell Mills, 1992.

[32]Pollan, 2006.

[33]Lampkin, 2002.

[34]Howard, 2009.

[35]For example, Ewel, 1999; Altieri, 1990; Aarts et al., 2000; Pretty, 1995.

[36]Jackson, 1985.

[37]Piper, 1998.

[38]For example, Gliessman, 2000; Altieri, 1990; Ewel, 1999.

[39]Rosset, 2006.

[40]See, for example, Cassman's rather incoherent attack on the Badgeley et al. article, Cassman, 2007.

[41]Organic Trade Association, 2006 annual survey (http://www.organicnewsroom.com/2006_press_releases/)

[42]Paul, 2007.

[43]Rosset and Altieri, 1997.

[44]In my 1995 review, I effectively used this framework, although I labeled the categories as (1) soil management, (2) pest management, and (3) integration. It will be clear from a reading of Vandermeer (1995) that the intent was really as presented herein.

[45]A chemostat is a laboratory apparatus in which a medium (usually a liquid) flows into a chamber. Organisms live in the chamber and use whatever nutrients are in the inflow, and then the medium flows out of the chamber. It is a continuous-flow device in which the rate of inflow and outflow of medium is balanced, and the difference between nutrient content between inflow and outflow is due to the action of whatever system is in the chamber.

[46]Vandermeer et al., 2008a; Perfecto and Vandermeer, 2008.

[47]van den Bosch, 1989; Vandermeer and Andow, 1986.

[48]Vandermeer, 1990.

[49]Peters, 2002.

[50]Lang and Gabriel, 1996.

[51]Costanza et al., 1992; Callow, 1995.

[52]Leopold, 1941.

[53]Suter, 1993; Kelly and Harwell, 1989.

[54]Rapport, 1995.

[55]Rapport, 1989.

[56]Rapport, 1995.

[57]Rapport, 1995; Calow, 1995; Jardine et al., 2007; Macdonald and Laurenson, 2006.

[58]Vandermeer et al., 2002b.

[59]Levins and Vandermeer, 1990.

[60]Rapport, 1995.

[61]Andow and Hidaka, 1989. For a mathematical formalism of Andow and Hidaka, see Vandermeer, 1997.

[62]This anecdote is due to Richard Levins.

[63]This was a point made in the preface to our earlier text (Carroll et al., 1990) and remains pertinent today.

REFERENCES

Aarts, N., Röling, N. G., Wagemakers, A. E., & Blum, A. (2000). *Facilitating sustainable agriculture.* Cambridge, UK: Cambridge University Press.

Abrams, P. A., & Wilson, W. G. (2004). Coexistence of competitors in metacommunities due to spatial variation in resource growth rates; does R* predict the outcome of competition? *Ecology Letters, 7,* 929–940.

AFRC. (1991). *Rothamsted Experimental Station: Guide to the classical field experiments.* Harpenden, Herts, UK: AFRC Institute of Arable Crops Research.

Agamuthu, P., & Broughton, W. J. (1985). Nutrient cycling within the developing oil palm-legume ecosystem. *Agriculture, Ecosystems and Environment, 13,* 111–123.

Aid, T. M., & Grau, H. R. (2004). Globalization, migration, and Latin American ecosystems. *Science, 305,* 1915–1916.

Alexander, M. (1977). *Introduction to soil microbiology,* 2nd ed. New York: John Wiley & Sons.

Altieri, M. A. (1990). Why study traditional agriculture? In C. R. Carroll, J. H. Vandermeer, & P. Rosset (eds.). *Agroecology.* New York: McGraw-Hill.

Andow, D. A. (1991). Vegetational diversity and arthropod population response. *Annual Review of Entomology, 36,* 561–586.

Andow, D., & Hidaka, K. (1989). Experimental natural history of sustainable agriculture: syndromes of production. *Agriculture, Ecosystems and Environment, 27,* 447–462.

Andrews, R. W., Peters, S. E., Janke, R. R., & Sahs, W. W. (1990). Converting to sustainable farming systems. In C. A. Francis, C. B. Flora, & L. D. King (eds.). *Sustainable agriculture in temperate zones* (pp. 281–314). New York: John Wiley & Sons.

Angelsen, A., & Kaimowitz, D. (2001a). When does technological change in agriculture promote deforestation? In D. R. Lee & C. B. Barrett (eds.). *Tradeoffs or synergies? Agricultural intensification, economic development and the environment* (pp. 89–114). Willingford, UK: CABI Publishing.

Angelsen, A., & Kaimowitz, D. (eds.). (2001b). *Agricultural technologies and deforestation.* Wallingford, Oxfordshire, UK: CABI Publishing and Center for International Forestry Research.

Anonymous. (1944). British Ecological Society: Easter meeting 1944: Symposium on "The Ecology of Closely Allied Species." *Journal of Animal Ecology, 13,* 176–177.

Arditi, R., & Berryman, A. A. (1991). The biological control paradox. *Trends in Ecology and Evolution, 6,* 32.

Armstrong, R. A., & McGehee, R. (1980). Competitive exclusion. *The American Naturalist, 115,* 151–170.

Arrhenius, O. (1921). Species and area. *Journal of Ecology, 9,* 95–99.

Ashton, M. S., & Montagnini, F. (eds.). (1999). *The silvicultural basis for agroforestry systems.* Boca Raton, FL: CRC Press.

Ayala, F. J. (1969). Experimental invalidation of the principle of competition exclusion. *Nature, 224,* 1076–1079.

Badgley, C., Moghtader, J., Quintero, E., Zakem, E., Chappell, M. J., Aviles-Vazquez, K., Samulon, A., & Perfecto, I. (2007). Organic agriculture and the global food supply. *Renewable Agriculture and Food Systems, 22,* 86–108.

Balfour, E. (1943). *The living soil.* London: Faber and Faber.

Beattie, A. J. (1985). *The evolutionary ecology of ant-plant mutualisms.* Cambridge, UK: Cambridge University Press.

Bennet, M., Schatz, M. F., Rockwood, H., & Wiesenfeld, K. (2002). Huygens's clocks. *Proceedings of the Royal Society London A, 458,* 563–579.

Berkenbusch, K., & Rowden, A. A. (2003). Ecosystem engineering: moving away from "just-so" stories. *New Zealand Journal of Ecology, 27,* 67–73.

Beus, C. E., & Dunlap, R. E. (1992). The alternative-conventional agriculture debate: where do agricultural faculty stand? *Rural Sociology, 57,* 363–380.

Billman, B. (2002). Irrigation and the origins of the southern Moche State on the north coast of Peru. *Latin American Antiquity, 13,* 371–400.

Black, H. I. J., & Okwakol, M. J. N. (1997). Agricultural intensification, soil biodiversity and agroecosystem function in the tropics: the role of termites. *Applied Soil Ecology, 6,* 37–53.

Boddey, R. M., & Dobereiner, J. (1988). Nitrogen fixation associated with grasses and cereals: recent results and perspectives for future research. *Plant and Soil, 108,* 53–65.

Boserup, E. (1965). *The conditions of agricultural growth.* New Brunswick, NJ: Transaction Publishers.

Brady, N. C. (1990). *The nature and property of soils,* 10th ed. New York: Macmillan.

Brady, N. D. C., & Weil, R. R. (2007). *The nature and properties of soils,* 14th ed. New York: Prentice Hall.

Brown, B. J., Hanson, M. E., Liverman, D. M., & Merideth, R. W., Jr. (1987). Global sustainability: toward definition. *Environmental Management, 11,* 713–719.

Brundrett, M. (1991). Mycorrhizas in natural ecosystems. *Advances in Ecological Research, 21,* 171–313.

Brussard, L. (1994). An appraisal of the Dutch programme on soil ecology of arable farming systems (1985–1992). *Agriculture, Ecosystems and Environment, 51,* 1–6.

Buttel, F. H. (1990). Social relations and the growth of modern agriculture. In R. Carroll, J. Vandermeer, & P. Rosset. *Agroecology.* New York: McGraw-Hill.

Cadish, G., & Giller, K. E. (1997). *Driven by Nature: Plant litter quality and decomposition.* Wallingford, UK: CABI Publishing.

Callow, P. (1995). Ecosystem health: A critical analysis of concepts. In: D. J. Rapport, C. Gaudet, & P. Calow (eds.). *Evaluating and monitoring the health of large-scale ecosystems* (Vol. 28, pp. xi + 454). NATO ASI Series 1: Global Environmental Change.

Carpenter, S. R., & Kitchell, J. F. (1996). *The trophic cascade in lakes.* Cambridge, UK: Cambridge University Press.

Carroll, R., Vandermeer, J., & Rosset, P. (eds.). (1990). *Agroecology.* New York: MacMillan.

Carter, H. W. (1989). Agricultural sustainability: an overview and research assessment. *California Agriculture, 43,* 16–18, 37.

Casanova, E. F. (1995). Agronomic evaluation of fertilizers with special reference to natural and modified phosphate rock. *Fertilizer Research, 41,* 211–218.

Cassman, K. (2007). Can organic agriculture feed the world: science to the rescue? *Renewable Agriculture and Food Systems, 22,* 109.

Chacon, J. C., & Gliessman, S. R. (1982). Use of the "non-weed" concept in traditional tropical agroecosystems of southeastern Mexico. *Agroecosystems, 8,* 1–11.

Chan, K. Y., & Heenan, D. P. (1993). Surface hydraulic properties of a red earth under continuous cropping with different management practices. *Australian Journal of Soil Research, 31,* 13–24.

Colborn, T., Dumanoski, D., & Meyers, J. P. (1997). *Our stolen future: are we threatening our fertility, intelligence, and survival?—a scientific detective story.* New York: Plume, The Penguin Group.

Coley, P. D., Bryant, J. P., & Chapin, F. S. (1985). Resource availability and plant antiherbivore defence. *Science, 230,* 895–899.

Collins, W. J. (1997). When the tide turned: immigration and the delay of the great black migration. *The Journal of Economic History, 57,* 607–632.

Conford, P. (2001). *The origins of the organic movement.* Edinburgh, UK: Floris Books.

Connell, J. H. (1971). On the role of natural enemies in preventing competitive exclusion in some marine animals and in rain forest trees. In P. J. den Boer & G .R. Gradwell (eds.). *Dynamics of numbers in populations. Proceedings of the Advanced Study Institute on dynamics of numbers in populations, Oosterbeek, 1970* (pp. 298–312). Wageningen, The Netherlands: Centre for Agricultural Publishing and Documentation.

Cornia, G. A. (1985). Farm size, land yields and the agricultural production function: An analysis for fifteen developing countries. *World Development, 13,* 513–534.

Cornia, G. A., & Court, J. (2001). *Inequality, growth and poverty in the era of liberalization and globalization.* Policy Brief No. 4, UNU World Institute for Development Economics Research.

Costanza, R., Norton, B. G., & Haskell, B. D. (eds.). (1992). *Ecosystem health: new goals for environmental management* (pp. x + 269). Washington, DC: Island Press.

Cui, M., & Caldwell, M. M. (1996). Facilitation of plant phosphate acquisition by arbuscular mycorrhizas from enriched soil patches, I. Roots and hyphawe exploiting the same soil volume. *New Phytol, 133,* 453–460.

Cullen, L., Jr., Alger, K., & Rambaldi, D. M. (2005). Land reform and biodiversity conservation in Brazil in the 1990s: conflict and the articulation of mutual interests. *Conservation Biology, 19,* 747–755.

Cushing, J. M., Costantino, R. F., Dennis, B., Desharnais, R. A., & Henson, S. M. (2002). *Chaos in ecology: experimental nonlinear dynamics.* Theoretical Ecology Series. San Diego: Academic Press.

Dann, P. R., Derrick, J. W., Dumaresq, D. C., Ryan M. H. (1996). The response of organic and conventionally grown wheat to superphosphate and reactive phosphate rock. *Australian Journal of Experimental Agriculture, 36,* 35, 71–78.

Darwin, C. 1881. *The formation of vegetable mould through the action of worms, with observations of their habits.* London: Murray.

Dash, M. (2001). *Tulipomania: the story of the world's most coveted flower and the extraordinary passions it aroused.* New York: Three Rivers Press.

Davis, M. (2001). *Late Victorian holocausts: El Niño famines and the making of the Third World.* London: Verso.

De Wolf, E. D., & Isard, S. A. (2007). Disease cycle approach to plant disease prediction. *Annual Review of Phytopathology, 45,* 203–220.

Desmarais, A. A. (2007). *La Via Campesina: globalization and the power of peasants.* London: Pluto Press.

Dimitri, C., & Oberholtzer, L. (2005). *Market-led versus government-facilitated growth: Development of the U.S. and EU organic agricultural sectors.* Washington, DC: WRS-05-05, USDA, Economic Research Service.

Donahue, R. L., Miller, R. W., & Shickluna, U. C. (1977). *Soils: An introduction to soils and plant growth.* Englewood Cliffs, NJ: Prentice-Hall.

Edwards, C. A., & Bohlen, P. J. (1996). *Biology and ecology of earthworms,* 3rd ed. London: Chapman and Hall.

Ereshefsky, M. (1992). *The units of evolution: Essays on the nature of species.* Cambridge, MA: MIT Press.

Ewel, J. (1999). Natural systems as models for the design of sustainable systems of land use. *Agroforestry Systems, 45,* 1–21.

Faeth, P., Repetto, R., Kroll, K., Dai, Q., & Helmers, G. (1991). *Paying the farm bill: U.S. agricultural policy and the transition to sustainable agriculture.* Washington, DC: World Resources Institute.

Farshad, A., & Zinck, J. A. (1993). Seeking agricultural sustainability. *Agriculture, Ecosystems and Environment, 47,* 1–12.

Fite, G. C. (1984). *Cotton fields no more: Southern agriculture, 1865–1980.* Lexington, KY: University Press of Kentucky.

Flannery, K. V. (1973). The origins of agriculture. *Annual Review of Anthropology, 2,* 271–310.

Flecker, A. S. (1996). Ecosystem engineering by a dominant detritivore in a diverse tropical stream. *Ecology, 77,* 1845–1854.

Forlani, G., Pastorelli, R., Branzoni, M., & Favilli, F. (1995). Root colonization efficiency, plant-growth-promoting activity and potentially related properties in plant-associated bacteria. *Journal of Genetics and Breeding, 49,* 4, 343–351.

Fragoso, C., Brown, G. G., Patrón, J. C., Anchart, E. B., Lavelle, P., Pashanasi, B., Senapati, B., & Kumar, T. (1997). Agricultural intensification, soil biodiversity and agroecosystem function in the tropics: the role of earthworms. *Applied Soil Ecology, 6,* 17–35.

Francis, C. A. (1986). *Multiple cropping: Practices and potentials.* New York: MacMillan.

Francis, C. A., & Youngberg, G. (1990). Sustainable agriculture: an overview. In C. A. Francis, C. B. Flora, & L. D. King (eds.). *Sustainable agriculture in temperate zones* (pp. 1–23). New York: John Wiley and Sons.

Friedland, W., & Barton, A. (1981). *Destalking the Wily Tomato: A Case Study in Social Consequences in California.* Agricultural Research, University of California at Davis, Department of Applied Behavioral Sciences: Research Monograph No. 15.

Funes, F., Garcia, L., Bourque, M., Perez, N., & Rosset, P. (2002). *Sustainable agriculture and resistance.* Oakland, CA: Food First Books, Institute for Food and Development Policy.

Galbraith, J. K. (1994). *A short history of financial euphoria.* New York: Penguin Books.

Garcia-Barrios, M., Galvan-Miyoshi, L., Y. M., Valdivieso-Pérez, I. A., Masera, O. R., Bocco, G., & Vandermeer, J. (2009). Agricultural intensification, and rural out-migration: The Mexican experience. *BioScience, 59,* 863–873.

Garcia-Barrios, L., Mayer-Foulkes, D., Franco, M., Urquijo-Vásquez, G., & Franco-Pérez, J. (2001). Development and validation of a spatially explicit individual-based mixed crop growth model. *Bulletin of Mathematical Biology, 63,* 507–526.

Garcia-Barrios, L., & Ong, C. K. (2004). Ecological interactions, management lessons and design tools in tropical agroforestry systems. *Agroforestry Systems, 61–62,* 221–236.

Gilpin, M. E., & Justice, K. E. (1973). Reinterpretation of the invalidation of the principle of competitive exclusion. *Nature, 236,* 273–301.

Gliessman, S. (2000). *Agroecology: Ecological processes in sustainable agriculture.* Boca Raton, FL: CRC Press.

Goldberg, D. E. (1990). Components of resource competition in plant communities. In J. Grace, D. Tilman (eds.). *Perspectives in plant competition* (pp. 27–49). New York: Academic Press.

Gomiero, T., Paoletti, M. G., & Pimentel, D. (2008). Energy and environmental issues in organic and conventional agriculture. *Critical Reviews in Plant Sciences, 27,* 239–254.

Goodman, D., Sorj, B., & Wilkinson, J. (1987). *From farming to biotechnology: A theory of agro-industrial development.* Oxford: Blackwell.

Goulart, F. F., Vandermeer, J., Perfecto, I., & de Matta-Machado, R. P. (2007) Aves em quintais agroflorestais do Pontal do Paranapanema, São Paulo: epistemologia, estructura de comunidade e frugivoria. Dissertacion, Federal Universidade de Minas Gerais, Belo Horizonte, Brazil. Available at: http://scholar.google.com/scholar?hl=en&q=Goulart%2C+F.+F.%2C+Vandermeer%2C+J.%2C+Perfecto&btnG=Search&as_ylo=&as_vis=0. Accessed November 27, 2009.

Grantham, G. W. (1978). The diffusion of the new husbandry in Northern France, 1815–1840. *Journal of Economic History, 38,* 311–337.

Grau, H. R., Aide, T. M., Zimmerman, K. J., Thomlinson, R. J., Helmer, E., & Zou, X. (2003). The ecological consequences of socioeconomic and land-use changes in postagriculture Puerto Rico. *Bioscience, 53,* 1159–1168.

Green, R. E., Cornell, S. J., Scharlemann, J. P. W., & Balmford, A. (2005). Farming and the fate of wild nature. *Science, 307*, 550–555.

Greenberg, R., Bichier, P., & Sterling, J. (1997). Bird populations in planted and rustic shade coffee plantations in Chiapas, Mexico. *Biotropica, 29*, 501–514.

Grime, J. P. (1977). Evidence for the existence of three primary strategies in plants and its relevance to ecological and evolutionary theory. *American Naturalist, 111*, 1169–1194.

Haber, L. F. (1971). *The chemical industry, 1900–1930: International growth and technological change.* Oxford: Clarendon Press.

Harris, T. W. (1841). *Report on the insects of Massachusetts, injurious to vegetation.* Cambridge: Folsom, Wells, and Thurston.

Harvey, D. (2007). *A brief history of neoliberalism.* Oxford, UK: Oxford University Press.

Hassebrook, C. (1990). Developing a socially sustainable agriculture. *American Journal of Alternative Agriculture, 5*, 50, 96.

Hassell, M. (1978). *The dynamics of arthropod predator–prey systems.* Princeton, NJ: Princeton University Press.

Havelka, U. D., Boyle, M. G., & Hardy, R. W. F. (1982). Biological nitrogen fixation. In F. J. Stevenson (ed.). *Nitrogen in agricultural soils.* Agronomy Series No. 22. Madison, WI: American Society of Agronomy.

Hawkins, B. A., & Cornell, H. V. (2001). *Theoretical approaches to biological control.* Cambridge, UK: Cambridge University Press.

Hayes, T., Haston, K., Tsui, M., Hoang, A., Haeffele, C., & Vonk, A. (2003). Atrazine-induced hermaphroditism at 0.1 PPB in American Leopard Frogs (*Rana pipiens*): laboratory and field evidence. *Environmental Health Perspectives, 111*, 568–575.

Haynes, R. J. (1980). Competitive aspects of the grass-legume association. *Advances in Agronomy, 33*, 227–261.

Heal, O. W., & MacLean, S. F. (1975). Comparative productivity in ecosystems: secondary productivity. In W. H. van Dobbn & R. H. Lowe-McConnell (eds.). *Unifying concepts in ecology* (pp. 136–148). Wageningen, The Netherlands: Centre for Agricultural Publishing and Documentation.

Hecht, S. B., & Saatchi, S. S. (2007). Globalization and forest resurgence: changes in forest cover in El Salvador. *BioScience, 57*, 663–672.

Hepperly, P., & Wilson, D. (2006). The world has changed: a look at nitrogen and corn economics. Available at http://www.newfarm.org/columns/research_paul/2006/0606/nitrocorn.shtml. Accessed November 25, 2009.

Hirsch, A. (2003). *Avaliação da fragmentação do habitat e seleção de areas prioritárias para a conservação dos primates na bacia do Rio Doce, Minas Gerais, através de aplicação de um sistema de informações Geográficas.* PhD Thesis, Universidade Federal de Minas Gerais, Brazil.

Hoeksema, K. J., & Jongerius, A. (1959). On the influence of earthworms on the soil structure in mulched orchards (pp. 188–194). Proceedings of the International Symposium of Soil Structure Ghent, 1958.

Holling, C.S. (1959). Some characteristics of simple types of predation and parasitism. *Canadian Entomologist, 91*, 385–398

Holyoak, M., Leibold, M. A., & Holt, R. D. (2005). *Metacommunities: Spatial dynamics and ecological communities.* Chicago: University of Chicago Press.

Homewood, K. M., & Rodgers, W. A. (1991). *Maasailand ecology: Pastoralist development and wildlife conservation in Ngorongoro, Tanzania (Cambridge Studies in applied ecology and resource management).* Cambridge, UK: Cambridge University Press.

Howard, P. (2009). Consolidation in the North American organic food processing sector, 1997 to 2007. *International Journal of Sociology of Agriculture and Food, 16*, 13–30.

Howard, S. (1940). *An agricultural testament.* Oxford, UK: Oxford University Press.

Hubbell, S. P. (2001). *The unified neutral theory of biodiversity and biogeography.* Princeton, NJ: Princeton University Press.

Huisman, J., & Weissing, F. J. (1999). Biodiversity of plankton by species oscillations and chaos. *Nature, 402*, 407–410.

Huisman, J., & Weissing, F. J. (2001). Biological conditions for oscillations and chaos generated by multispecies competition. *Ecology, 82*, 2682–2695.

Huston, M. A., Aarssen, L. W., Austin, M. P., Cade, B. S., Fridley, J. D., Garnier, E., Grime, J. P., Hodgson, J., Lauenroth, W. K., Thompson, K., Vandermeer, J. H., & Wardle, D. A. (2000). No consistent effect of plant diversity on productivity. *Science, 289*, 1255a.

IFOAM. (2009). The IFOAM Norms. Available at: http://www.ifoam.org/about_ifoam/standards/norms.html. Accessed November 25, 2009.

Irwin, D. A. (2001). Tariffs and growth in late nineteenth century America. *The World Economy, 24*, 15–30.

Jackson, W. (1985). *New roots for agriculture.* Lincoln, NE: University of Nebraska Press.

Janzen, D. H. (1970). Herbivores and the number of tree species in tropical forests. *The American Naturalist, 104*, 501–528.

Jardine, A., Speldewinde, P., Carver, S., & Weinstein, P. (2007). Dryland salinity and ecosystem distress syndrome: Human health implications. *EcoHealth, 4*, 1612–9210.

Jenny, H. (1941). *Factors of soil formation.* New York: McGraw-Hill.

Johnson, C. (2004). *Blowback: The costs and consequences of American empire.* New York: Macmillian.

Johnson, C. (2006). *Nemesis: The last days of the American republic.* New York: Macmillan.

Johnston, K. J. (2003). The intensification of pre-industrial cereal agriculture in the tropics: Boserup, cultivation lengthening, and the Classic Maya. *Anthropological Archaeology, 22*, 126–161.

Jones, C. G., Lawton, J. H., & Shachak, M. (1997). Positive and negative effects of organisms as physical ecosystem engineers. *Ecology, 78*, 1946–1957.

Jordahl, J. L., & Karlen, D. L. (1993). Comparison of alternative farming systems. III. Soil aggregate stability. *American Journal of Alternative Agriculture, 8*, 27–33.

Joschko, M., Söchtig, W., & Larink, O. (1992). Functional relationships between earthworm burrows and soil water movement in column experiments. *Soil Biology & Biochemistry, 24*, 1545–1547.

Juo, A. S. R., & Manu, A. (1996). Chemical dynamics in slash-and-burn agriculture. *Agriculture, Ecosystems and Environment, 58*, 49–60.

Kareiva, P. (1986). Trivial movement and foraging by crop colonizers. In M. Kogan (ed.). *Ecological theory and integrated pest management practice* (pp. 59–82). New York: Wiley Interscience.

Keeney, D. R. (1989). Toward a sustainable agriculture: Need for clarification of concepts and terminology. *American Journal of Alternative Agriculture, 4*, 101–105.

Kelly, J. R., & Harwell, M. A. (1989). Indicators of ecosystem response and recovery. In S. A. Levin, J. R. Kelly Jr., M. A. Harwell, K. P. Kindell (eds.). *Ecotoxicology: problems and approaches* (pp. 9–39). New York: Springer Verlag.

Kleinman, P. J. A., Pimentel, D., & Bryant, R. B. (1995). The ecological sustainability of slash-and-burn agriculture. *Agriculture, Ecosystems and Environment, 52*, 235–249.

Klooster, D. (2003). Forest transitions in Mexico: institutions and forests in a globalized countryside. *The Professional Geographer, 55*, 227–237.

Lacy, W. B. (1996). Research, extension and user partnerships: models for collaboration and strategies for change. *Agriculture and Human Values, 13*, 33–41.

Lampkin, N. (2002). *Organic Farming.* Ipswich, UK: Old Pond Publishing.

Landell Mills, J. (1992). *Organic farming in seven European countries.* Study prepared for the European Crop Protection Association, Brussels.

Lang, T., & Gabriel, Y. (1996). *Mad Consumers?* London, UK: Soundings Ltd.

Lansing, S. (2007). *Priests and programmers: technologies of power in the engineered landscape of Bali.* Princeton, NJ: Princeton University Press.

Larcher, W. (1995). *Physiological plant ecology: ecophysiology and stress physiology of functional groups* (p. 506), 3rd ed. Berlin: Springer.

Lavelle, P., Dangerfield, M., Fragoso, C., Eschenbrenner, V., Lopez-Hernandez, D., Pashanasi, B., & Brussard, L. (1994). The relationship between soil macrofauna and tropical soil fertility. In P. L. Woomer & M. J. Swift (eds.). *The biological management of tropical soil fertility* (pp. 137–170). New York: John Wiley and Sons.

Lawton, J. H., & May, R. M. (1995). *Extinction rates: why so little extinction?* Oxford: Oxford University Press.

Lee, R. B. (1972a). Population growth and the beginnings of sedentary life among the !Kung Bushmen. In B. Spooner (ed.). *Population growth: anthropological implications* (pp. 329–342). Cambridge, MA: MIT Press.

Lee, R. B. (1972b). The intensification of life among the !Kung Bushmen. In B. Spooner (ed.). *Population growth: anthropological implications* (pp. 343–350). Cambridge, MA: MIT Press.

Lee, T., Burch, J., Jung, Y., Coote, T., Pearce-Kelly, P., Foighil, D. Ó. (2006). Tahitian tree snail mitochondrial clades survived recent mass extirpation. *Current Biology, 17,* R502–R503.

Leibold, M. A., Holyoak, M., Mouquet, N., Amarasekare, P., Chase, J. M., Hoopes, M. F., Holt, R. D., Shurin, J. B., Law, R., Tilman, D., Loreau, M., & Gonzales, A. (2004). The metacommunity concept: a framework for multi-scale community ecology. *Ecology Letters, 7,* 601–613.

Leopold, A. (1941). Wilderness as a land laboratory. *Living Wilderness, 6,* 3.

Levins, R. (1979). Coexistence in a variable environment. *The American Naturalist, 114,* 765–783.

Levins, R., & Vandermeer, J. H. (1990). The agroecosystem embedded in a complex system. In C. R. Carroll, J. H. Vandermeer, & P. Rosset (eds.). *Agroecology.* New York: Macmillan.

Lewis, J. G. E. (1973). Concepts in fungal nutrition and the origin of biotrophy. *Biological Reviews, 48,* 261–278.

Lewontin, R. (1982). Agricultural research and the penetration of capital. *Science for the People, January–February,* 12–17.

Liere, H., & Perfecto, I. (2008). Cheating on a mutualism: Indirect benefits of ant attendance to a coccidophagous coccinellid. *Ecological Entomology, 37,* 143–149.

Lin, B. (2007). Agroforestry management as an adaptive strategy against potential microclimate extremes in coffee agriculture. *Agricultural and Forest Meteorology, 144,* 85–94.

Lockeretz, W. (1988). Open questions in sustainable agriculture. *American Journal of Alternative Agriculture, 3,* 174–181.

Lockeretz, W., Shearer, G., & Kohl, D. H. (1981). Organic farming in the corn belt. *Science, 211,* 540–547.

Logsdon, S. D., Radke, J. K., & Karlen, D. L. (1993). Comparison of alternative farming systems. I. Infiltration techniques. *American Journal of Alternative Agriculture, 8,* 15–20.

Loomis, R. S., & Connor, D. J. (1992). *Crop ecology: Productivity and management in agricultural systems.* Cambridge: Cambridge University Press.

López, E., Bocco, G., Mendoza, M., Velázquez, A., & Aguirre, R. (2006). Peasant emigration and land-use change at the watershed level: a GIS-based approach in Central Mexico. *Agricultural Systems, 90,* 62–78.

Loreau, M., Naeem, S., & Inchausti, P. (2002). *Biodiversity and ecosystem functioning: Synthesis and perspectives.* Oxford: Oxford University Press.

Loveless, A. R. (1961). A nutritional interpretation of sclerophylly based on differences in chemical composition of sclerophyllous and mesophytic leaves. *Annals of Botany of Nova Scotia, 25,* 168–176.

Lundgren, B. (1987). Agroforestry in third world countries. Proceedings of the IUFRO workshop on agroforestry for rural needs, February 22–26, 1987, New Delhi, India.

MacArthur, R., & Levins, R. (1967). The limiting similarity, convergence, and divergence of coexisting species. *The American Naturalist, 101,* 377.

Macdonald, D. W., & Laurenson, M. K. (2006). Infectious disease: inextricable linkages between human and ecosystem health. *Biological Conservation, 131,* 143–150.

Madden, J. P. (1990). The economics of sustainable low-input farming systems. In C. A. Francis, C. B. Flora, & L. D. King (eds.). *Sustainable agriculture in temperate zones* (pp. 315–341). New York: John Wiley and Sons.

Marschner, H. (1991). Plant-soil relationships: acquisition of mineral nutrients by roots from soils. In J. R. Porter and D. W. Lawlor (eds.). *Plant growth: Interactions with nutrition and environment* (pp. 125–155). Cambridge; Cambridge University Press.

May, R. M. (1988). How many species are there on Earth? *Science, 241,* 1441–1449.

May, R. M., & Nowak, M. A. (1994). Superinfection, metapopulation dynamics, and the evolution of diversity. *Journal of Theoretical Biology, 1700,* 95–114.

Meyers, R. J. K., van Noordwijk, M., & Vityakon, P. (1997). Synchrony of nutrient release and plant demand: plant litter quality, soil environment and farmer management options. In G. Cadish & K. E. Giller (eds.). *Driven by nature: Plant litter quality and decomposition* (pp. 215–229). Wallingford, UK: CABI Publishing.

Mikanova, O., & Kubat, J. (1994). *Phosphorus solubilization from hardly soluble phosphates by soil microflora. Rostl, 40,* 833–844.

Minc, L. D., & Vandermeer, J. H. (1989). The origin and spread of agriculture. In C. R. Carroll, J. H. Vandermeer, & P. Rosset (eds.). *Agroecology.* New York: McGraw-Hill.

Mishustin, E. N. (1970). The importance of non-symbiotic nitrogen-fixing microorganisms in agriculture. *Plant and Soil, 32,* 545–554.

Moguel, P., & Toledo, V. M. (1999). Biodiversity conservation in traditional coffee systems of Mexico. *Conservation Biology, 13,* 11–21.

Morales, H. (2002). Pest Management in traditional tropical agroecosystems: lessons for pest prevention research and extension. *Integrated Pest Management Reviews, 7,* 145–163.

Morales, H., & Perfecto, I. (2000). Traditional knowledge and pest control in the Guatemalan highlands. *Agriculture and Human Values, 17,* 49–63.

Morales, H., Perfecto, I., & Ferguson, B. (2001). Traditional fertilization and its effect on corn insect populations in the Guatemalan highlands. *Agriculture, Ecosystems and Environment, 84,* 145–155.

Mundt, C. (1990). Disease dynamics in agroecosystems. In C. R. Carroll, J. H. Vandermeer, & P. M. Rosset. *Agroecology.* New York: McGraw-Hill Publishing Co.

Mundt, C. C. (2002). Use of multiline cultivars and cultivar mixtures for disease management. *Annual Review of Phytopathology, 40,* 381–410.

Munyanziza, E., Kehri, H. K., & Bagyaraj, D. J. (1997). Agricultural intensification, soil biodiversity and agroecosystem function in the tropics: the role of decomposer biota. *Applied Soil Ecology, 6,* 77–85.

Murdoch, W. W., & Briggs, C. J. (1996). Theory for biological control: recent developments. *Ecology, 77,* 2001–2013.

Nair, K. P. P., Patel, U. K., Singh, R. P., & Kaushik, M. K. (1979). Evaluation of legume intercropping in conservation of fertilizer nitrogen in maize culture. *Journal of Agricultural Science, Cambridge, 93,* 189–194

Nash, L. (2006). *Inescapable ecologies: A history of environment, disease, and knowledge.* Berkley, CA: University of California Press.

Nassaur, J. I., Santelmann, M. V., & Scavia, D. (2007). *From the corn belt to the gulf: Societal and environmental implications of alternative agricultural futures.* Washington, DC: Resources for the Future Press.

National Research Council. (1989). *Alternative agriculture.* Washington, DC: National Academy of Sciences, Board on Agriculture, National Academy Press.

Nelson, K. (1994). Participation, empowerment, and farmer evaluations: a comparative analysis of IPM technology generation in Nicaragua. *Agriculture and Human Values, 11,* 109–125.

Niggli, U., & Lockeretz, W. (1996). *Fundamentals of organic agriculture.* IFOAM, Tholey-Theley, Germany.

Ohtonen, R., Aikio, S., & Vare, H. (1997). Ecological theories in soil biology. *Soil Biology & Biochemistry, 29,* 1613–1619.

Olding-Smee, G. J., Laland, K. N., & Feldman, M. W. (2004). *Niche Construction: The neglected process in evolution.* Princeton, NJ: Princeton University Press.

Ong, C. K. (1994). Alley cropping, ecological pie in the sky? *Agroforestry Today, 6,* 8–10.

Ortloff, C. R. (1988). The ancient canal builders of pre-Inca Peru. *Scientific American, 259,* 100–107.

Paine, R. T. (1966). Food web complexity and species diversity. *The American Naturalist, 100,* 65.

Palmer, M. W. (1994). Variation in species richness: towards a unification of hypotheses. *Folia Geobotanica, 29,* 1211–9520.

Patel, R. (2008). *Stuffed and starved: The hidden battle for the world food system.* Hoboken, NJ: Melville House.

Paul, J. (2007). China's organic revolution. *Journal of Organic Systems, 2,* 1–11.

Paul, E. A., & Clark, F. E. (1980). *Soil microbiology and biochemistry.* San Diego: Academic Press.

Peet, J. R. (1969). The spatial expansion of commercial agriculture in the nineteenth century: a VonThunen interpretation. *Economic Geography, 45,* 283–301.

Pendergast, M. (2000). *Uncommon grounds: The history of coffee and how it transformed our world.* New York: Basic Books.

Perfecto, I., & Armbrecht, I. (2003). The coffee agroecosystem in the Neotropics: Combining ecological and economic goals. In J. Vandermeer (ed.). *Tropical Agroecosystems* (pp. 159–194). Boca Raton, FL: CRC Press.

Perfecto, I., & Vandermeer, J. H. (1993). Distribution and turnover rate of a population of *Atta cephalotes* in a tropical rain forest in Costa Rica. *Biotropica, 25,* 316–321.

Perfecto, I., & Vandermeer, J. H. (2008). Spatial pattern and ecological process in the coffee agroecosystem. *Ecology, 89,* 915–920.

Perfecto, I., Rice, R., Greenberg, R., & Van der Voolt, M. (1996). Shade coffee as refuge of biodiversity. *BioScience, 46,* 589–608.

Perfecto, I., Vandermeer, J., Mas, A., & Soto Pinto, L. (2005). Biodiversity, yield and shade coffee certification. *Ecological Economics, 54,* 435–446.

Perfecto, I., Vandermeer, J., & Wright, A. (2009). *Nature's matrix.* London: Earthscan.

Peters, D. J. (2002). Revisiting the Goldschmidt Hypothesis: The effect of economic structure on socioeconomic conditions in the rural Midwest. Technical Paper P-0702-1. Missouri Department of Economic Development.

Peterson, G. E. (1967). The Discovery and Development of 2,4-D. *Agricultural History, 4,* 243–254.

Pike, D. G. (1975). *Anasazi: Ancient people of the rock.* Palo Alto, CA: American West Publishing Company.

Pikovsky, A., Rosenblum, M., & Kurths, J. (2003). *Synchronization: A universal concept in nonlinear sciences.* Cambridge, UK: Cambridge University Press.

Piper, J. K. (1998). Natural systems agriculture. In W. W. Collins & C. O. Qualset (eds.). *Biodiversity in agroecosystems.* Boca Raton, FL: CRC Press.

Pirozynski, K. A., & Dalpé, Y. (1989). Geological history of the Glomaceae with particular reference to mycorrhizal symbiosis. *Symbiosis, 7,* 1–36.

Pollan, M. (2006). *The omnivore's dilemma: A natural history of four meals.* Waterville, ME: Thorndike Press.

Prathibha, C. K., Alagawadi, A. R., & Sreenivasa, M. N. (1995). Establishment of inoculated organisms in rhizosphere and their influence on nutrient uptake and yield of cotton. *Karnataka Journal of Agricultural Science, 8,* 1, 22–27.

Pretty, J. N. (1995). Participatory learning for sustainable agriculture. *World Development, 23,* 1247–1263.

Rapport, D. J. (1989). Symptoms of pathology in the Gulf of Bothnia (Baltic Sea): ecosystem response to stress from human activity. *Biological Journal of the Linnean Society, 37,* 33–49.

Rapport, D. J. (1995). Ecosystem health: exploring the territory. *Ecosystem Health, 1,* 5–13.

Rasmussen, P. E., & Collins, H. P. (1991). Long term impacts of tillage, fertilizer and crop residue on soil organic matter in temperate semiarid regions. *Advances in Agronomy, 45,* 93–134.

Raven, J. A. (1985). Regulation of pH and generation of osmolarity in fascular land plants: costs and benefits in relation to efficiency of use of water, energy and nitrogen. *New Phytologist, 101,* 25–77.

Reganold, J. P., Elliot, L. F., & Unger, Y. L. (1987). Long-term effects of organic and conventional farming on soil erosion. *Nature, 330,* 370–372.

Relyea, R. A. (2005). The impact of insecticides and herbicides on the biodiversity and productivity of aquatic communities. *Ecological Applications, 15,* 618–627.

Relyea, R. A., & Mills, N. (2001). Predator—induced stress makes the pesticide carbaryl more deadly to gray treefrog tacpolers (Hyla versicolor). *Proceedings of the National Academy of Sciences, 98,* 2491–2496.

Risch, S. J., Andow, D., & Altierri, M. A. (1983). Agroecosystem diversity and pest control: Data, tentative conclusions and new research directions. *Environmental Entomology, 12,* 625–629.

Rival, L. (2002). *Trekking through history: The Huaorani of Amazonian Ecuador.* New York: Columbia University Press.

Robertson, G. P. (1982). Nitrification in forested ecosystems. *Philosophical Trans Royal Society of London, B296,* 445–457.

Root, R. (1973). Organization of a plant-arthropod association in simple and diverse habitats: The fauna of collards (*Brassica oleracea*). *Ecological Monographs, 43,* 95–124.

Rosset, P. (1999). The multiple functions and benefits of small farm agriculture. *Food First Policy Brief, 4.*

Rosset, P. M. (2006). *Food is different: Why we must get the WTO out of agriculture.* Plymouth, UK: Zed Books.

Rosset, P. M., & Altieri, M. A. (1997). Agroecology versus input substitution: a fundamental contradiction of sustainable agriculture. *Society and Natural Resources, 10,* 283–295.

Runnels, C. N. (1995). Environmental Degredatio in Ancient Greece. *Scientific American, March,* 72–75.

Russell, E. (1993). *War and nature: Fighting humans and insects with chemicals from World War I to Silent Spring (Studies in Environment and History).* Cambridge, UK: Cambridge University Press.

Sanchez, P. A. (1995). Science in agroforestry. *Agroforestry Systems, 30,* 5–55.

Sattelmacher, B., & Thoms, K. (1989). Root growth and 14C translocation into the roots of maize (*Zea mays L.*) as influenced by local nitrate supply. *Zeitschrift für Pflanzenernährung und Bodenkunde, 152,* 7–10.

Schjønning, P., Christensen, B. T., & Carstensen, B. (1994). Physical and chemical properties of a sandy loam receiving animal manure, mineral fertilizer or no fertilizer for 90 years. *European Journal of Soil Science, 45,* 257–268.

Schoonderbeek, D., & Schoute, J. F. T. (1994). Root and root-soil contact of winter wheat in relation to soil macroporosity. *Agriculture, Ecosystems, and Environment, 51,* 89–98.

Schroth, G., da Fonseca, G. A. B., Harvey, C. A., & Gascon, C. (2004). *Agroforestry and biodiversity conservation in tropical landscapes.* Washington, DC: Island Press.

Shabaev V.P., Smolin V.Y., Strekozova V.I. (1991). The effect of *Azospirillum brasilense Sp. 7* and *Azotobacter chroococcum* on nitrogen—balance in soil under cropping with oats (*Avena sativa L.*). *Biology and Fertility of Soils, 10,* 1423–1789.

Shiva, V. (1997). *Biopiracy: The plunder of nature and knowledge.* Boston: South End Press.

Silvertown, J. (2004). Plant coexistence and the niche. *TREE, 19,* 605–611.

Silvertown, J., & Law, R. (1987). Do plants need niches? Some recent developments in plant community ecology. *TREE, 2,* 24–26.

Simard, R. R., Angers, D. A., & Lapierre, C. (1994). Soil organic matter quality as influenced by tillage, lime, and phosphorus. *Biology and Fertility of Soils, 18,* 13–18.

Singh, B., & G. S. Sakhon. (1978/1979). Nitrate pollution of groundwater from farm use of nitrogen fertilizers—a review. *Agriculture and Environment, 4,* 207–225.

Smith, C. W., & Cothren, J. T. (1999). *Cotton: Origin, history, technology, and production.* Wiley series in Crop Science. New York: John Wiley and Sons.

Smith, S. E. (1980). Mycorrhizas of autotrophic higher plants. *Biological Reviews, 55,* 474–510.

Snaydon, R. W., & Harris, P. M. (1979). Interactions below ground: The use of nutrients and water. In R. W. Willey (ed.). *Proceedings of the International Workshop on Intercropping* (pp. 188–201). Hyderabasd, India: ICRISAT.

Soule, J., Carré, D., & Jackson, W. (1990). Ecological impact of modern agriculture. In C. R. Carroll, J. H. Vandermeer, & P. Rosset (eds.). *Agroecology.* New York: McGraw-Hill.

Steingraber, S. (1997). *Living downstream: An ecologist looks at cancer and the environment.* Reading, MA: Addison-Wesley.

Stinner, B. R., & House, G. J. (1989). The search for sustainable agroecosystems. *Journal of Soil and Water Conservation, 44,* 111–116.

Stout, J., & Vandermeer, J. (1975). Comparison of species richness for stream-inhabiting insects in tropical and mid-latitude streams. *The American Naturalist, 100,* 263–280.

Suter II, G. W. (1993). A critique of ecosystem health concepts and indexes. *Environmental Toxicology and Chemistry, 112,* 1533–1539.

Swift, M. J., Heal, O. W., & Anderson, J. M. (1979). *Decomposition in terrestrial ecosystems. Studies in ecology 5.* Oxford: Blackwell Scientific.

Tegtmeier, E. M., & Duffy, M. D. (2004). External costs of agricultural production in the United States. *International Journal of Agricultural Sustainability, 2,* 55–175.

The Debt Threat: How debt is destroying the developing world. (2005). Available at: http://www.democracynow.org/2005/1/13/the_debt_threat_how_debt_is. Accessed November 25, 2009.

Thornton, J. (2001). *Pandora's poison: Chlorine, health and a new environmental strategy.* Cambridge, MA: MIT Press.

Tiessen, H., Cuievas, E., & Chacon, P. (1994). The role of soil organic matter in sustaining soil fertility. *Nature, 371,* 783–785.

Timmer, C. P. (1969). The turnip, the new husbandry and the English Agricultural Revolution. *Quarterly Journal of Economics, 83,* 375–395.

Tinker, P. B., Ingram, J. S. I., & Struwe, S. (1996). Effects of slash-and-burn agriculture and deforestation on climate change. *Agriculture, Ecosystems and Environment, 58,* 13–22.

Tjørve, E. (2003). Shapes and functions of species-area curves: a review of possible models. *Journal of Biogeography, 30,* 827–835.

Tjørve, E., & Calf Tjørve, K. M. (2008). The species-are relationship, self-similarity and the true meaning of the z-value. *Ecology, 89*, 3528–3533.

Trenbath, B. R. (1976). Plant interactions in mixed crop communities. In R. I. Pappendick, P. A. Sanchez, & G. B. Triplett (eds.). *Multiple cropping* (pp. 129–170, Special Publication 27). Madison, WI: American Society of Agronomy.

Ugland, K. I., Gray, J. S., & Lambshead, P. J. D. (2004). Species accumulation curves analysed by a class of null models discovered by Arrhenius. *Oikos, 108*, 263–274.

Uhl, C. (1987). Factors controlling succession following slash-and-burn agriculture in Amazonia. *The Journal of Ecology, 75*, 377–407.

U.S. Department of Agriculture. (1999). *Soil taxonomy: A basic system of soil classification for making and interpreting soil surveys*. Agriculture Handbook Number 436. Washington, DC: U.S. Government Printing Office.

U.S. Department of Veteran Affairs. (2009). Agent Orange: Information about Agent Orange, possible health problems, and related VA benefits. Available at: http://www1.va.gov/AgentOrange/. Accessed November 25, 2009.

Van den Bosch, R. (1989). *The pesticide conspiracy*. Berkeley, CA: University of California Press.

van der Werf, E. (1993). Agronomic and economic potential of sustainable agriculture in South India. *American Journal of Alternative Agriculture, 4*, 185–191.

van Noordwijk, M., & de Willigen, P. (1986). Qualitative root ecology as element of soil fertility theory. *Netherlands Journal of Agricultural Science, 34*, 273–282.

van Noordwijk, M., Kooistra, M. J., Boone, F. R., Veen, B. W., & Schoonderbeek, D. (1993). Root-soil contact of maize, as measured by thin-section technique: I. Validity of the method. *Plant Soil, 139*, 109–118.

Vandermeer, J. H. (1981). The interference production principle: an ecological theory for agriculture. *Bioscience, 31*, 361–364.

Vandermeer, J. H. (1984). The interpretation and design of intercrop systems involving environmental modification by one of the components: a theoretical framework. *Journal of Biological Agriculture and Horticulture, 2*, 135–156.

Vandermeer, J. H. (1986). Mechanized agriculture and social welfare: the tomato harvester in Ohio. *Agriculture and Human Values, 3*, 21–25.

Vandermeer, J. H. (1989). *The ecology of intercropping systems*. Cambridge, UK: Cambridge University Press.

Vandermeer, J. H. (1990). Notes on agroecosystem complexity: chaotic price and production trajectories deducible from simple one-dimensional maps. *Journal of Biological Agriculture and Horticulture, 6*, 293–304.

Vandermeer, J. (1995). The ecological basis of alternative agriculture. *Annual Review of Ecology and Systematics, 26*, 201–224.

Vandermeer, J. H. (1997). Syndromes of production: an emergent property of simple agroecosystem dynamics. *Journal of Environmental Management, 51*, 59–72.

Vandermeer, J. H. (1998). Maximizing crop yield in alley crops. *Agroforestry Systems, 40*, 199–206.

Vandermeer, J. H. (2008). The niche construction paradigm in ecological time. *Ecological Modelling, 214*, 385–390.

Vandermeer, J., & Andow, D. (1986). Prophylactic and responsive components of an Integrated Pest Management Program. *Journal of Economic Entomology, 79*, 299–302.

Vandermeer, J., & Perfecto, I. (2005a). *Breakfast of biodiversity*. Oakland, CA: Food First Books.

Vandermeer, J., & Perfecto, I. (2005b). The future of farming and conservation. *Science, 308*, 1257.

Vandermeer, J., & Perfecto, I. (2007a). The agricultural matrix and the future paradigm for conservation. *Conservation Biology, 21*, 274–277.

Vandermeer, J., & Perfecto, I. (2007b). Tropical conservation and grass roots social movements: ecological theory and social justice. *Bulletin of the Ecological Society of America, 88*, 171–175.

Vandermeer, J. H., van Noordwijk, M., Anderson, J., Perfecto, I., & Ong, C. (1998). Global change and multi-species agroecosystems: concepts and issues. *Agriculture, Ecosystems and Environment, 67*, 1–22.

Vandermeer, J., Evans, M. A., Foster, P., Höök, T., Reiskind, M., & Wund, M. (2002a). Increased competition may promote species coexistence. *Proceedings of the National Academy of Sciences, 99*, 8731–8736.

Vandermeer, J., Lawrence, D., Symstad, A., & Hobbie, S. (2002b). Effect of biodiversity on ecosystem function in managed ecosystems. In M. Laureau, S. Naeem, & P. Inchausti (eds.). *Biodiversity and ecosystem functioning*. Oxford, UK: Oxford University Press.

Vandermeer, J., Perfecto, I., & Philpott, S. M. (2008a). Clusters of ant colonies and robust criticality in a tropical agroecosystem. *Nature, 451*, 457–459.

Vandermeer, J., Perfecto, I., Philpott, S., & Chappell, M. J. (2008b). Reenfocando la conservación en el paisaje: La importancia de la matriz. In C. A. Harvey & J. C. Sáenz (eds.). *Evaluación y conservacion de biodiversidad en paisajes fragmentados de Mesoamérica*. Costa Rica: Editorial INBio.

Vandermeer, J., Perfecto, I., Liere, H. (2009). Evidence for hyperparasitism of coffee rust (*Hemileia vastatrix*) by the entomogenous fungus, *Lecanicillium lecanii* through a complex ecological web. *Plant Pathology, 58*, 636–641.

Vanderplank, J. E. (1963). *Plant Diseases: Epidemics and Control*. New York: Academic Press.

Wallerstein, I. (1974, 1980, 1989). *The modern world system*. New York: Academic Press.

Wallerstein, I. (2004). *World systems analysis: An introduction*. Durham, NC: Duke University Press.

Wardel, D. A., & Giller, K. E. (1997). The quest for a contemporary ecological dimension to soil biology. *Soil biology and Biochemistry, 8*, 1579–1554.

Wilson, E. O. (1986). *Biodiversity: papers from the National Forum on Biodiversity held September 21–25, 1986 in Washington*. Washington, DC: National Academies Press.

Wilson, E. O. (1987). The little things that run the world (the importance and conservation of invertebrates). *Conservation Biology, 1*, 344–346.

Wilson, E. O. (1999). *The diversity of life*. New York: W.W. Norton.

Woodham-Smith, C. B. (1962). *The Great Hunger: Ireland: 1845–1849*. Hamish Hamilton, 1962 (Paperback by Penguin Books, 1991). Middlesex, UK: Harmondsworth.

Wright, A. (2005). *The death of Ramón González: The modern agricultural dilemma*. Austin, TX: University of Texas Press.

Wright, J. (2009). *Sustainable agriculture and food security in an era of oil scarcity: Lessons from Cuba*. London: Earthscan.

Yachi, S., & Loreau, M. (1999). Biodiversity and ecosystem productivity in a fluctuating environment: the insurance hypothesis. *Proceedings of the National Academy of Sciences, 96*, 1463–1468.

Young, A. (1989). *Agroforestry for soil conservation*. Wallingford, UK: CABI Publishing.

INDEX

Photo Credits

Chapter 1
Chapter Opener © Julija Sergeeva/Dreamstime.com; **1-1, 1-2, 1-4** Courtesy of John H. Vandermeer, University of Michigan

Chapter 2
Chapter Opener © Shi Yali/ShutterStock, Inc.; **2-3** Courtesy of Richard Lee/Anthro-Photo; **2-8** © Deborah Wolfe/ShutterStock, Inc.; **2-10** Courtesy of Peter Zwitser; **2-14, 2-15** Courtesy of John H. Vandermeer, University of Michigan; **2-16** Reproduced from S. Copland. Agriculture, ancient and modern: a historical account of its principles and practice, exemplified in their rise, progress, and development. London: Virtue and Company, 1866. Courtesy of the Boston Public Library [call #3140778]

Chapter 3
Chapter Opener Courtesy of Library of Congress, Prints & Photographs Division [reproduction number cph 3b03780]; **3-6, 3-11A** Courtesy of John H. Vandermeer, University of Michigan; **3-11B** Courtesy of Manfred Mielke, USDA Forest Service, Bugwood.org; **3-11C, 3-13A, 3-13B** Courtesy of John H. Vandermeer, University of Michigan; **3-14** © F. Hal Higgins Collection, D-056, Special Collections, University of California Library, Davis

Chapter 4
Chapter Opener © Alexan24/Dreamstime.com; **4-2** Courtesy of Library of Congress, Prints & Photographs Division [reproduction number cph 3c23763]; **4-3** © Pancaketom/Dreamstime.com; **4-4** Courtesy of Library of Congress, Prints & Photographs Division [reproduction number ppmsca 08372]

Chapter 5
Chapter Opener © mashe/ShutterStock, Inc.; **5-16A, 5-16B** Courtesy of John H. Vandermeer, University of Michigan

Chapter 6
Chapter Opener © Stefan Fierros/ShutterStock, Inc.; **6-4, 6-5** Courtesy of John H. Vandermeer, University of Michigan

Chapter 7
Chapter Opener © szefel/ShutterStock, Inc.; **7-6** Courtesy of John H. Vandermeer, University of Michigan

Chapter 8
Chapter Opener © ryasick photography/ShutterStock, Inc.; **8-1A, 8-1B, 8-1C, 8-1D** Courtesy of John H. Vandermeer, University of Michigan

Unless otherwise indicated, all photographs and illustrations are under copyright of Jones and Bartlett Publishers, LLC.